AF360835

Oral Cancer: From Pathophysiology to Novel Therapeutic Approaches

Oral Cancer: From Pathophysiology to Novel Therapeutic Approaches

Editor

Vui King Vincent-Chong

Basel • Beijing • Wuhan • Barcelona • Belgrade • Novi Sad • Cluj • Manchester

Editor
Vui King Vincent-Chong
Roswell Park Comprehensive
Cancer Center
Buffalo, NY
USA

Editorial Office
MDPI
St. Alban-Anlage 66
4052 Basel, Switzerland

This is a reprint of articles from the Special Issue published online in the open access journal *Biomedicines* (ISSN 2227-9059) (available at: https://www.mdpi.com/journal/biomedicines/special_issues/oral_cancer_therapy).

For citation purposes, cite each article independently as indicated on the article page online and as indicated below:

Lastname, A.A.; Lastname, B.B. Article Title. *Journal Name* **Year**, *Volume Number*, Page Range.

ISBN 978-3-0365-9790-4 (Hbk)
ISBN 978-3-0365-9791-1 (PDF)
doi.org/10.3390/books978-3-0365-9791-1

Cover image courtesy of Vui King Vincent-Chong

Contents

About the Editor

Vui King Vincent-Chong

Vui King Vincent-Chong is a Research Assistant Professor at the Department of Oral Oncology at Roswell Park Comprehensive Cancer Center, Buffalo, New York. He has trained and mentored research trainees (graduate students) in the field of head and neck cancer. He also serves as a reviewer on issues related to head and neck cancer, including translational cancer research, onco-immunology, and therapeutic and chemoprevention agents using preclinical models of head and neck cancer. His primary research interests include using clinically relevant models to address several key questions of critical importance for the successful clinical translation of treatment response and personalized medicine in the field of head and neck cancer.

Editorial

Editorial of Special Issue "Oral Cancer: From Pathophysiology to Novel Therapeutic Approaches"

Vui King Vincent-Chong

Department of Oral Oncology, Roswell Park Comprehensive Cancer Center, Buffalo, NY 14263, USA; vincentvuiking.chong@roswellpark.org; Tel.: +1-7164803829

Citation: Vincent-Chong, V.K. Editorial of Special Issue "Oral Cancer: From Pathophysiology to Novel Therapeutic Approaches". *Biomedicines* **2023**, *11*, 2748. https://doi.org/10.3390/biomedicines11102748

Received: 25 September 2023
Accepted: 27 September 2023
Published: 11 October 2023

1. Introduction

Oral squamous cell carcinoma (OSCC) is a heterogeneous type of malignancy that develops within the oral cavity comprising the lips, tongue, mouth floor, gums, and buccal mucosa, with more than 90% arising from the oral lining epithelium [1]. Tobacco smoking and excessive alcohol drinking are the major risk factors contributing to OSCC [2]. More recently, HPV virus infections have been documented as another of the risk factors [3]. In 2020, approximately 377,000 new cases and 177,700 OSCC deaths were reported according to Global Cancer Observatory (GLOBOCAN) [4]. According to National Comprehensive Cancer Network (NCCN) guidelines, the major treatment for OSCC is surgery, followed by radiation or chemotherapy or both, which is known as chemoradiation (CRT), depending on the clinical stage [5]. Despite advancements in multimodality treatments for OSCC, the clinical outcomes have remained poor over the past decades [6–9], with the main reason being the acquirement of resistance to chemo or radiation therapy [10,11]. Accumulating evidence indicates that the failure of cancer treatment or its recurrence may be caused by the cancer cells involved in the mechanisms related to DNA damage repair, cell cycle dysregulation, cancer stem cells that promote self-renewal ability, and epithelial–mesenchymal transition (EMT) [11,12]. This has prompted us to seek new technology and efforts to decipher the pathophysiology of oral cancer and improve the often-ineffective current treatment options. To overcome these challenges, the investigation of novel therapeutic approaches related to immunotherapy, targeted therapy, and precision medicine show much promise in enhancing tumor destruction in the OSCC management process and provide new insights into strategies for improving cancer treatment.

2. Pathophysiology of Oral Cancer

OSCC is a malignancy that arises from a multistep molecular mechanism derived from normal epithelium to dysplasia that eventually transforms into invasive SCC via multiple genetic and molecular alterations post exposure to chronic stimulation from environmental factors such as tobacco smoking, excessive alcohol drinking, or exposure to HPV [13]. These major risk factors induce mutation and genetic alterations that activate the cancer hallmarks as described in Hanahan and Weinberg, leading to uncontrolled cell proliferation, the evasion of growth suppressors, replicative immortality, angiogenesis, inflammation, resistance to cell death and invasion, and metastasis [14]. Following the completion of The Cancer Genome Atlas (TCGA) on head and neck SCC (HNSCC) which comprises OSCC, integrative exome sequencing data have revealed hotspot mutations which include *TP53*, *FAT1*, *EPHA2*, *NOTCH1*, *CASP8*, and *PIK3CA* [15,16]. In addition, studies have shown that OSCC is driven by essential cancer pathways related to the EGFR/PI3K/AKT/mTOR cascade which play a significant role in the pathogenesis of OSCC [17]. This has led to studies investigating potential molecular targets from these OSCC-related cancer pathways. The EGFR/PI3K/AKT/mTOR cascade is activated by EGF ligands binding to EGFR, leading to PI3K activation. PI3K activation induces Akt phosphorylation, which activates mTORC1 with phosphorylating key components which trigger downstream activities involved in apoptosis, metabolism, cell proliferation, and cell growth. Numerous efforts have been initiated to target EGFR pathways to halt the PI3K/AKT/mTOR cascade through monoclonal antibodies targeting the ligand binding domain of the receptor and the small-molecule tyrosine kinase inhibitors (TKIs) in HNSCC clinical trials [18,19]. However, overall survival rates from the random clinical trial (RCT) studies focusing on recurrent/metastasis (R/M) HNSCC patients remain below 50%. This highlights the importance of identifying biomarkers for predicting the clinical responses of these targeted therapies in HNSCC patients who could benefit from these regimens.

3. Treatment Options for Oral Cancer

According to NCCN guidelines, the standard of care for OSCC patients depends on the clinical staging comprising surgery followed by radiation therapy and/or chemotherapy, known as CRT [5]. Patients diagnosed with recurrent/metastasis are treated with immunotherapy as an alternative standard of care. Generally, cisplatin remains the main chemotherapeutic agent for OSCC patients [5]. Despite the potent anti-cancer activities of cisplatin shown in preclinical OSCC models, the development of cisplatin resistance in cancer cells impairs the cytotoxicity effect of this regimen and leads to ineffective treatment as evidenced by a meta-analysis report showing that more than 30% of HNSCC patients acquired cisplatin resistance [10,11]. Over the past decades, there has been growing interest in developing novel therapeutic interventions for OSCC aimed at selectively targeting cancer cells while sparing normal cells to reduce the adverse effects associated with conventional standards of care [20–22]. The RCT CHECKMATE 141 studies in R/M HNSCC patients who failed cisplatin treatment report that nivolumab treatment prolongs overall survival rates among such patients compared to conventional standards of care. This has made immunotherapy another promising regimen for such patients despite the response rate being < 20% [9]. Nivolumab, also known as an immune checkpoint blockade (ICB), plays a role in preventing immune evasion by blocking the engagement of PD-L1 in tumors with the PD-1 receptor in CD8 T cells, which could lead to exhaustion in promoting immune surveillance [23]. The below 20% response rate for a single agent of Nivolumab has prompted the emergence of targeted therapies for druggable candidates exposed to therapeutic vulnerabilities in OSCC using personalized medicine based on genomic approaches, transcriptomic profiling, and single-cell sequencing [24]. This facilitates the identification of exclusive druggable hotspot mutations from OSCC patients' samples and allows them to receive targeted therapies aimed specifically at the genetic mutations or proteins that favor tumor cell growth and metastasis.

4. Artificial Intelligence (AI)

Over the past few years, artificial intelligence, an area that involves the capacity of machines to compete with human intellectual capacity, has been widely used in the field of OSCC research to provide more accurate predictions for early diagnosis and prognosis outcomes to improve OSCC management [25]. Generally, AI has two subfields, which are machine and deep learning. Machine learning involves the use of algorithms and computer processes to recognize patterns and provide diagnoses according to input information. Briefly, there are two types of approaches, namely supervised and unsupervised methods. The supervised approach employs a labeled set of training data to map input-to-output data, whereas the unsupervised approach involves analyses of unlabeled datasets using algorithms to discover hidden patterns without human supervision. A systematic review has reported machine learning as a promising emerging approach to accurately classify the differentiation of OSCC and predict prognosis and lymph node metastasis in OSCC patients [26,27]. Deep learning, on the other hand, uses a machine learning method called artificial neural networks to extract patterns and provide predictions from large datasets. Studies have used deep learning approaches in applications for predicting/detecting early diagnosis, metastases and prognosis based on genomic, methylation and transcriptomic data, and pathological and radiographic images of OSCC [28,29]. Overall, deep learning algorithms allow for the identification of biomarkers to predict patient prognosis and guide personalized treatment plans to reduce OSCC burdens and improve the quality of life.

5. Limitations, Challenges, and Conclusions

To date, cetuximab and ICB remain the only systemic targeted therapies approved for the treatment of OSCC. However, the clinical benefits of these therapies remain poor with a response rate of less than 20%. This has prompted the search for novel treatment strategies to improve the clinical outcomes for OSCC patients. The lack of significant biomarkers or prognosticators for OSCC remains a challenge in early and advanced stage detection, leading to poor clinical outcomes. Newly emerging interests in genomic, proteomic, transcriptomic, and metabolomic markers for prognostic and predictive markers may facilitate treatment selection. Another challenge for identifying biomarkers in OSCC is the heterogeneity found in tumor characteristics that provide variables of tumor mutational burden. Therefore, non-invasive biomarkers such as tumor oxygenation and signals from autofluorescence imaging, intensity signals from optical coherence tomography, and ultrasound wave signals from surface tumor tissues have been shown to provide accurate predictions for early detection and prognosis assessments in OSCC [30]. To overcome these challenges, this Special Issue has accepted a few publications that highlighted the potential biomarkers that can be used for diagnostic and therapeutic purposes [31–40]. The diagnostic section can be divided into biomarkers for OSCC and prognosis. In this case, studies have demonstrated that the protein expression of C44Mab 34 [32] can solely detect the OSCC cell as opposed to the stroma or immune cell, and the downregulation of IFIT2 has been identified as a poor prognostic marker for OSCC [33]. A review paper conducted by Yeh et al. [40] summarized evidence showing that TERT promoter mutations were associated with poor survival in OSCC. Apart from the mutations and protein expression of the markers, this Special Issue also provides evidence of using the clinicopathological parameter (lymph node ratio) as a prognosticator to access patient clinical outcome post surgery [36]. Similarly, Chen et al. provided a longitudinal study that demonstrated the significance of using psychological behaviors as a marker to access patients' requirements for follow-up for appropriate supportive interventions to strengthen patients' psychosocial adjustments in OSCC patients who had undergone surgery [35]. Another clinical study also demonstrated the potential of facilitating the platelet–lymphocyte ratio (PLR) in OSCC patients prior to surgery treatment as a prognosticator for disease-specific survival (DSS) and progression-free survival (PFS) of patients [38]. This Special Issue also published a comprehensive review that provides an update on recent studies using preclinical models of HNSCC to investigate the role of natural products as therapeutic or chemo preventive agents to improve patient clinical outcomes [39]. In addition to the potential of natural products as therapeutic agents for HNSCC, this Special Issue also published an original article highlighting the role of mometasone furoate (MF) as a potential therapeutic agent targeting PTPN11, which halts tumorigenesis in vitro and impairs tumor growth in vivo using a preclinical model of HNSCC [31]. For the diagnostic approach that focuses on the early detection in OSCC, Surendran et al. [37] demonstrated the gradual increments in the infiltration of Tregs (CD4, CD25, FOXP3), exhausted CD8 T cells (CD8 and PD-1) in the immune microenvironment, and the expression of PD-L1 in epithelial cells in the oral carcinogenesis mechanism, from normal epithelium to dysplasia and eventually to invasive SCC. With the advancement of technology, it is no surprise that our Special Issue also successfully published a paper that focuses on using a deep-learning-based model to automate the identification and visualization of OSCC in optical coherence tomography images from the extracted tumor tissues to classify the normal epithelium, dysplasia, and OSCC based on the features extracted from the OCT images [34]. Using OCT imaging in combination with a deep learning model can act as an adjunct tool for identifying OSCC and margin and sparing the normal region during surgical excision.

In conclusion, OSCC is a deadly malignancy and a major public health issue worldwide. While significant efforts have been made to gain better insights into the pathophysiology of OSCC, current standards of care remain limited and ineffective. Studies that focus on establishing novel therapeutic approaches that involve overcoming chemoresistance and eliminating tumor evasion from immune surveillance are needed to improve the clinical outcomes for OSCC patients.

Conflicts of Interest: The authors declare no conflict of interest.

References

1. Neville, B.W.; Day, T.A. Oral cancer and precancerous lesions. *CA Cancer J. Clin.* **2002**, *52*, 195–215. [CrossRef]
2. Johnson, D.E.; Burtness, B.; Leemans, C.R.; Lui, V.W.Y.; Bauman, J.E.; Grandis, J.R. Head and neck squamous cell carcinoma. *Nat. Rev. Dis. Primers* **2020**, *6*, 92. [CrossRef]
3. Sabatini, M.E.; Chiocca, S. Human papillomavirus as a driver of head and neck cancers. *Br. J. Cancer* **2020**, *122*, 306–314. [CrossRef]
4. Sung, H.; Ferlay, J.; Siegel, R.L.; Laversanne, M.; Soerjomataram, I.; Jemal, A.; Bray, F. Global Cancer Statistics 2020: GLOBOCAN Estimates of Incidence and Mortality Worldwide for 36 Cancers in 185 Countries. *CA Cancer J. Clin.* **2021**, *71*, 209–249. [CrossRef] [PubMed]
5. Caudell, J.J.; Gillison, M.L.; Maghami, E.; Spencer, S.; Pfister, D.G.; Adkins, D.; Birkeland, A.C.; Brizel, D.M.; Busse, P.M.; Cmelak, A.J.; et al. NCCN Guidelines(R) Insights: Head and Neck Cancers, Version 1.2022. *J. Natl. Compr. Canc. Netw.* **2022**, *20*, 224–234. [CrossRef] [PubMed]
6. Zandberg, D.P.; Cullen, K.; Bentzen, S.M.; Goloubeva, O.G. Definitive radiation with concurrent cetuximab vs. radiation with or without concurrent cytotoxic chemotherapy in older patients with squamous cell carcinoma of the head and neck: Analysis of the SEER-medicare linked database. *Oral Oncol.* **2018**, *86*, 132–140. [CrossRef]
7. Xiang, M.; Holsinger, F.C.; Colevas, A.D.; Chen, M.M.; Le, Q.T.; Beadle, B.M. Survival of patients with head and neck cancer treated with definitive radiotherapy and concurrent cisplatin or concurrent cetuximab: A Surveillance, Epidemiology, and End Results-Medicare analysis. *Cancer* **2018**, *124*, 4486–4494. [CrossRef]
8. Pignon, J.P.; Bourhis, J.; Domenge, C.; Designe, L. Chemotherapy added to locoregional treatment for head and neck squamous-cell carcinoma: Three meta-analyses of updated individual data. MACH-NC Collaborative Group. Meta-Analysis of Chemotherapy on Head and Neck Cancer. *Lancet* **2000**, *355*, 949–955. [CrossRef]
9. Ferris, R.L.; Blumenschein, G., Jr.; Fayette, J.; Guigay, J.; Colevas, A.D.; Licitra, L.; Harrington, K.; Kasper, S.; Vokes, E.E.; Even, C.; et al. Nivolumab for Recurrent Squamous-Cell Carcinoma of the Head and Neck. *N. Engl. J. Med.* **2016**, *375*, 1856–1867. [CrossRef] [PubMed]
10. Atashi, F.; Vahed, N.; Emamverdizadeh, P.; Fattahi, S.; Paya, L. Drug resistance against 5-fluorouracil and cisplatin in the treatment of head and neck squamous cell carcinoma: A systematic review. *J. Dent. Res. Dent. Clin. Dent. Prospect.* **2021**, *15*, 219–225. [CrossRef]
11. Alsahafi, E.; Begg, K.; Amelio, I.; Raulf, N.; Lucarelli, P.; Sauter, T.; Tavassoli, M. Clinical update on head and neck cancer: Molecular biology and ongoing challenges. *Cell Death Dis.* **2019**, *10*, 540. [CrossRef]
12. Liu, Y.P.; Zheng, C.C.; Huang, Y.N.; He, M.L.; Xu, W.W.; Li, B. Molecular mechanisms of chemo- and radiotherapy resistance and the potential implications for cancer treatment. *MedComm (2020)* **2021**, *2*, 315–340. [CrossRef] [PubMed]
13. Choi, S.; Myers, J.N. Molecular pathogenesis of oral squamous cell carcinoma: Implications for therapy. *J. Dent. Res.* **2008**, *87*, 14–32. [CrossRef] [PubMed]
14. Hanahan, D. Hallmarks of Cancer: New Dimensions. *Cancer Discov.* **2022**, *12*, 31–46. [CrossRef]
15. Lin, L.H.; Chou, C.H.; Cheng, H.W.; Chang, K.W.; Liu, C.J. Precise Identification of Recurrent Somatic Mutations in Oral Cancer Through Whole-Exome Sequencing Using Multiple Mutation Calling Pipelines. *Front. Oncol.* **2021**, *11*, 741626. [CrossRef]
16. Cancer Genome Atlas, N. Comprehensive genomic characterization of head and neck squamous cell carcinomas. *Nature* **2015**, *517*, 576–582. [CrossRef]
17. Freudlsperger, C.; Burnett, J.R.; Friedman, J.A.; Kannabiran, V.R.; Chen, Z.; Van Waes, C. EGFR-PI3K-AKT-mTOR signaling in head and neck squamous cell carcinomas: Attractive targets for molecular-oriented therapy. *Expert Opin. Ther. Targets* **2011**, *15*, 63–74. [CrossRef]
18. Seiwert, T.Y.; Fayette, J.; Cupissol, D.; Del Campo, J.M.; Clement, P.M.; Hitt, R.; Degardin, M.; Zhang, W.; Blackman, A.; Ehrnrooth, E.; et al. A randomized, phase II study of afatinib versus cetuximab in metastatic or recurrent squamous cell carcinoma of the head and neck. *Ann. Oncol.* **2014**, *25*, 1813–1820. [CrossRef]
19. Soulieres, D.; Senzer, N.N.; Vokes, E.E.; Hidalgo, M.; Agarwala, S.S.; Siu, L.L. Multicenter phase II study of erlotinib, an oral epidermal growth factor receptor tyrosine kinase inhibitor, in patients with recurrent or metastatic squamous cell cancer of the head and neck. *J. Clin. Oncol.* **2004**, *22*, 77–85. [CrossRef] [PubMed]
20. Longton, E.; Schmit, K.; Fransolet, M.; Clement, F.; Michiels, C. Appropriate Sequence for Afatinib and Cisplatin Combination Improves Anticancer Activity in Head and Neck Squamous Cell Carcinoma. *Front. Oncol.* **2018**, *8*, 432. [CrossRef] [PubMed]
21. Brands, R.C.; De Donno, F.; Knierim, M.L.; Steinacker, V.; Hartmann, S.; Seher, A.; Kubler, A.C.; Muller-Richter, U.D.A. Multi-kinase inhibitors and cisplatin for head and neck cancer treatment in vitro. *Oncol. Lett.* **2019**, *18*, 2220–2231. [CrossRef]

22. Robinson, A.M.; Rathore, R.; Redlich, N.J.; Adkins, D.R.; VanArsdale, T.; Van Tine, B.A.; Michel, L.S. Cisplatin exposure causes c-Myc-dependent resistance to CDK4/6 inhibition in HPV-negative head and neck squamous cell carcinoma. *Cell Death Dis.* **2019**, *10*, 867. [CrossRef]
23. Jiang, Y.; Chen, M.; Nie, H.; Yuan, Y. PD-1 and PD-L1 in cancer immunotherapy: Clinical implications and future considerations. *Hum. Vaccin. Immunother.* **2019**, *15*, 1111–1122. [CrossRef] [PubMed]
24. Chang, J.Y.F.; Tseng, C.H.; Lu, P.H.; Wang, Y.P. Contemporary Molecular Analyses of Malignant Tumors for Precision Treatment and the Implication in Oral Squamous Cell Carcinoma. *J. Pers. Med.* **2021**, *12*, 12. [CrossRef] [PubMed]
25. Sultan, A.S.; Elgharib, M.A.; Tavares, T.; Jessri, M.; Basile, J.R. The use of artificial intelligence, machine learning and deep learning in oncologic histopathology. *J. Oral Pathol. Med.* **2020**, *49*, 849–856. [CrossRef] [PubMed]
26. López-Cortés, X.A.; Matamala, F.; Venegas, B.; Rivera, C. Machine-Learning Applications in Oral Cancer: A Systematic Review. *Appl. Sci.* **2022**, *12*, 5715. [CrossRef]
27. Adeoye, J.; Tan, J.Y.; Choi, S.W.; Thomson, P. Prediction models applying machine learning to oral cavity cancer outcomes: A systematic review. *Int. J. Med. Inform.* **2021**, *154*, 104557. [CrossRef]
28. Wang, X.; Li, B.B. Deep Learning in Head and Neck Tumor Multiomics Diagnosis and Analysis: Review of the Literature. *Front. Genet.* **2021**, *12*, 624820. [CrossRef]
29. Chu, C.S.; Lee, N.P.; Ho, J.W.K.; Choi, S.W.; Thomson, P.J. Deep Learning for Clinical Image Analyses in Oral Squamous Cell Carcinoma: A Review. *JAMA Otolaryngol. Head Neck Surg.* **2021**, *147*, 893–900. [CrossRef]
30. Romano, A.; Di Stasio, D.; Petruzzi, M.; Fiori, F.; Lajolo, C.; Santarelli, A.; Lucchese, A.; Serpico, R.; Contaldo, M. Noninvasive Imaging Methods to Improve the Diagnosis of Oral Carcinoma and Its Precursors: State of the Art and Proposal of a Three-Step Diagnostic Process. *Cancers* **2021**, *13*, 2864. [CrossRef]
31. Qiu, L.; Gao, Q.; Tao, A.; Jiang, J.; Li, C. Mometasone furoate inhibits the progression of head and neck squamous cell carcinoma via regulating PTPN11. *Biomedicines* **2023**, *11*, 2597. [CrossRef]
32. Suzuki, H.; Ozawa, K.; Tanaka, T.; Kaneko, M.K.; Kato, Y. Development of a Novel Anti-CD44 Variant 7/8 Monoclonal Antibody, C(44)Mab-34, for Multiple Applications against Oral Carcinomas. *Biomedicines* **2023**, *11*, 1099. [CrossRef] [PubMed]
33. Lai, K.C.; Regmi, P.; Liu, C.J.; Lo, J.F.; Lee, T.C. IFIT2 Depletion Promotes Cancer Stem Cell-like Phenotypes in Oral Cancer. *Biomedicines* **2023**, *11*, 896. [CrossRef] [PubMed]
34. Yang, Z.; Pan, H.; Shang, J.; Zhang, J.; Liang, Y. Deep-Learning-Based Automated Identification and Visualization of Oral Cancer in Optical Coherence Tomography Images. *Biomedicines* **2023**, *11*, 802. [CrossRef] [PubMed]
35. Chen, Y.W.; Lin, T.R.; Kuo, P.L.; Lee, S.C.; Wu, K.F.; Duong, T.V.; Wang, T.J. Psychosocial Adjustment Changes and Related Factors in Postoperative Oral Cancer Patients: A Longitudinal Study. *Biomedicines* **2022**, *10*, 3231. [CrossRef]
36. Suzuki, H.; Beppu, S.; Nishikawa, D.; Terada, H.; Sawabe, M.; Hanai, N. Lymph Node Ratio in Head and Neck Cancer with Submental Flap Reconstruction. *Biomedicines* **2022**, *10*, 2923. [CrossRef]
37. Surendran, S.; Aboelkheir, U.; Tu, A.A.; Magner, W.J.; Sigurdson, S.L.; Merzianu, M.; Hicks, W.L., Jr.; Suresh, A.; Kirkwood, K.L.; Kuriakose, M.A. T-Cell Infiltration and Immune Checkpoint Expression Increase in Oral Cavity Premalignant and Malignant Disorders. *Biomedicines* **2022**, *10*, 1840. [CrossRef]
38. Cho, U.; Sung, Y.E.; Kim, M.S.; Lee, Y.S. Prognostic Role of Systemic Inflammatory Markers in Patients Undergoing Surgical Resection for Oral Squamous Cell Carcinoma. *Biomedicines* **2022**, *10*, 1268. [CrossRef]
39. Liew, Y.X.; Karen-Ng, L.P.; Vincent-Chong, V.K. A Comprehensive Review of Natural Products as Therapeutic or Chemopreventive Agents against Head and Neck Squamous Cell Carcinoma Cells Using Preclinical Models. *Biomedicines* **2023**, *11*, 2359. [CrossRef]
40. Yeh, T.J.; Luo, C.W.; Du, J.S.; Huang, C.T.; Wang, M.H.; Chuang, T.M.; Gau, Y.C.; Cho, S.F.; Liu, Y.C.; Hsiao, H.H.; et al. Deciphering the Functions of Telomerase Reverse Transcriptase in Head and Neck Cancer. *Biomedicines* **2023**, *11*, 691. [CrossRef]

Article

The Prognosis Performance of a Neutrophil- and Lymphocyte-Associated Gene Mutation Score in a Head and Neck Cancer Cohort

Tsung-Jang Yeh [1,2,3], Hui-Ching Wang [1,2,4], Shih-Feng Cho [1,3,4], Chun-Chieh Wu [5], Tzu-Yu Hsieh [1], Chien-Tzu Huang [1,2], Min-Hong Wang [1,2], Tzer-Ming Chuang [1], Yuh-Ching Gau [1,2], Jeng-Shiun Du [1,2], Yi-Chang Liu [1,3,4], Hui-Hua Hsiao [1,3,4], Mei-Ren Pan [2,6,7], Li-Tzong Chen [3,8,9] and Sin-Hua Moi [2,10,*]

[1] Division of Hematology & Oncology, Department of Internal Medicine, Kaohsiung Medical University Hospital, Kaohsiung Medical University, Kaohsiung 807, Taiwan; aw7719@gmail.com (T.-J.Y.); joellewang66@gmail.com (H.-C.W.); sifong96@gmail.com (S.-F.C.); cathyvioletevergarden@gmail.com (T.-Y.H.); gankay18@hotmail.com (C.-T.H.); dhlsy01128@gmail.com (M.-H.W.); benjer6@gmail.com (T.-M.C.); cheesecaketwin@gmail.com (Y.-C.G.); ashiun@gmail.com (J.-S.D.); ycliu@cc.kmu.edu.tw (Y.-C.L.); huhuhs@kmu.edu.tw (H.-H.H.)
[2] Graduate Institute of Clinical Medicine, College of Medicine, Kaohsiung Medical University, Kaohsiung 807, Taiwan; pan.meiren.0324@gmail.com
[3] Center for Cancer Research, Kaohsiung Medical University, Kaohsiung 807, Taiwan; leochen@nhri.org.tw
[4] Faculty of Medicine, College of Medicine, Kaohsiung Medical University, Kaohsiung 807, Taiwan
[5] Department of Pathology, Kaohsiung Medical University Hospital, Kaohsiung Medical University, Kaohsiung 807, Taiwan; lazzz.wu@gmail.com
[6] Drug Development and Value Creation Research Center, Kaohsiung Medical University, Kaohsiung 807, Taiwan
[7] Department of Medical Research, Kaohsiung Medical University Hospital, Kaohsiung Medical University, Kaohsiung 807, Taiwan
[8] Division of Gastroenterology, Department of Internal Medicine, Kaohsiung Medical University Hospital, Kaohsiung Medical University, Kaohsiung 807, Taiwan
[9] National Institute of Cancer Research, National Health Research Institutes, Tainan 704, Taiwan
[10] Research Center for Precision Environmental Medicine, Kaohsiung Medical University, Kaohsiung 807, Taiwan
* Correspondence: moi9009@gmail.com; Tel.: +886-7-3121101 (ext. 2512)

Citation: Yeh, T.-J.; Wang, H.-C.; Cho, S.-F.; Wu, C.-C.; Hsieh, T.-Y.; Huang, C.-T.; Wang, M.-H.; Chuang, T.-M.; Gau, Y.-C.; Du, J.-S.; et al. The Prognosis Performance of a Neutrophil- and Lymphocyte-Associated Gene Mutation Score in a Head and Neck Cancer Cohort. *Biomedicines* **2023**, *11*, 3113. https://doi.org/10.3390/biomedicines11123113

Academic Editor: Vui King Vincent-Chong

Received: 8 May 2023
Revised: 16 November 2023
Accepted: 21 November 2023
Published: 22 November 2023

Abstract: The treatment of head and neck squamous cell carcinomas (HNSCCs) is multimodal, and chemoradiotherapy (CRT) is a critical component. However, the availability of predictive or prognostic markers in patients with HNSCC is limited. Inflammation is a well-documented factor in cancer, and several parameters have been studied, with the neutrophil-to-lymphocyte ratio (NLR) being the most promising. The NLR is the most extensively researched clinical biomarker in various solid tumors, including HNSCC. In our study, we collected clinical and next-generation sequencing (NGS) data with targeted sequencing information from 107 patients with HNSCC who underwent CRT. The difference in the NLR between the good response group and the poor response group was significant, with more patients having a high NLR in the poor response group. We also examined the genetic alterations linked to the NLR and found a total of 41 associated genes across eight common pathways searched from the KEGG database. The overall mutation rate was low, and there was no significant mutation difference between the low- and high-NLR groups. Using a multivariate binomial generalized linear model, we identified three candidate genes (*MAP2K2*, *MAP2K4*, and *ABL1*) that showed significant results and were used to create a gene mutation score (GMS). Using the NLR-GMS category, we noticed that the high-NLR-GMS group had significantly shorter relapse-free survival compared to the intermediate- or low-NLR-GMS groups.

Keywords: genomic mutation signature; neutrophil to lymphocyte ratio; head and neck cancer; prognostic biomarker

1. Introduction

Head and neck squamous cell carcinoma (HNSCC) is mostly derived from the mucosal epithelium of the oral cavity, pharynx, and larynx [1]. It is the most common malignancy of the upper aerodigestive tract and was ranked fourth in a Taiwanese male cohort study [2,3]. HNSCC treatment is generally multimodal and differs according to the disease stage, anatomical location, and surgical accessibility to achieve the most curative approach while optimizing the preservation of function [1,4]. Some early-stage diseases are curable with surgery or definitive radiotherapy; however, more than 60% of patients present with locally advanced disease upon diagnosis [5].

Concurrent platinum-based chemotherapy and radiotherapy, also referred to as chemoradiotherapy (CRT) or concurrent chemoradiotherapy (CCRT), plays an important role in HNSCC treatment. When considering organ preservation, definite CRT is recommended as a nonsurgical treatment for most patients with advanced pharyngeal and laryngeal cancers [6]. In postoperative management, the efficacy of adjuvant CRT has been proven in two multicenter randomized trials (EORTC 22931 and RTOG 9501) for high-risk patients with HNSCC, especially those with extranodal extension or positive surgical margins [7,8]. Despite the development of risk-adapted curative treatment strategies and other progress in therapeutic modalities, the overall 5-year survival is only 50%, and 65% of patients with an advanced stage of the disease have significantly compromised survival [9]. Therefore, in addition to the development of novel treatment approaches, the search for predictive or prognostic markers in patients with HNSCC is necessary.

To date, approximately 70 markers have been evaluated and reported from either blood or tumor tissues [10–12]. In the conventional treatment era, epidermal growth factor receptor (EGFR), p16, human papillomavirus (HPV), cyclin D1 (CCND1), B cell lymphoma-extra large (Bcl-xL)/Bcl-2 and excision repair cross complementation group 1 (ERCC1) have been identified as possible prognostic markers in clinical trials [11,12]. In the current immunotherapy era, programmed death ligand-1 (PD-L1) expression, tumor mutational burden (TMB), microsatellite instability (MSI), HPV status, smoking status, circulating tumor cells (CTCs), circulating tumor DNA (ctDNA), gut or oral cavity microbiota, and tumor-microenvironment-related gene expression profiles have been suggested as potential immune biomarkers to predict the efficacy of immune checkpoint inhibitors [12].

In 1863, Virchow established a connection between inflammation and cancer based on his observations [13]. Since then, research on the association between inflammation and carcinogenesis has increased, supporting Virchow's theory [14]. Inflammation is now characterized as a critical component of tumor progression based on its contribution to the multiple hallmarks of tumorigenesis [15–18]. Several inflammatory parameters have been reported, such as C-reactive protein, the neutrophil-to-lymphocyte ratio (NLR), the platelet-to-lymphocyte ratio (PLR), and the lymphocyte-to-monocyte ratio (LMR) [19–24]. Among these, the NLR is the most studied and promising clinical biomarker and has been shown to be prognostic in many solid tumors, including HNSCC [10,25–33]. However, there are still many unknown areas to explore, and research on inflammatory biomarkers and cancer genomic mutations is lacking. This study aimed to decipher the predictive value of the NLR in patients with locally advanced HNSCC treated with CRT and to explore the associations between the NLR and the cancer genomic landscape.

2. Materials and Methods

2.1. Data Source

In this retrospective cohort study, all data were collected via the health information system of Kaohsiung Medical University Hospital under an approved protocol (KMUHIRB-E(I)-20210401). Eligible patients with histologically proven HNSCC (grades 1 to 3) originating in the oral cavity (OC), oropharynx (OPC), hypopharynx (HPC), or larynx (LC) were recruited between 2016 and 2022 at Kaohsiung Medical University Hospital, Taiwan. The Head and Neck Cancer Committee confirmed the tumor stage according to the 8th edition of the American Joint Committee on Cancer (AJCC)'s staging system.

All patients underwent CRT, including 75 who received postoperative adjuvant therapy and 32 who received initial definitive treatment. CRT treatment included a total radiotherapy dose of 60–70 Gy and cisplatin-based chemotherapy. Other clinical data included detailed information on patient age, sex, tumor location, histological grade, clinical staging, body weight before and after CRT, changes in body mass index (BMI), and laboratory findings.

2.2. Treatment Response

All patients were followed up with regularly at the Medical Oncology and Otorhinolaryngology outpatient departments. Disease status evaluation included tumor site inspection, laboratory examinations, and imaging studies. Treatment response was assessed and determined using computed tomography or magnetic resonance imaging at baseline and at three- to six-month intervals after treatment initiation. The treatment response of the patients was evaluated using Response Evaluation Criteria in Solid Tumors (RECIST) 1.1-measurable lesions and classified into four categories: complete response (CR), partial response (PR), stable disease (SD), and progressive disease (PD). CR and PR were classified as good responses, whereas SD and PD were classified as poor responses. The median follow-up duration in this cohort was 16.6 (range 2.2–80.9) months.

2.3. Neutrophil-to-Lymphocyte Ratio (NLR)

The NLR was calculated as the simple ratio between the neutrophil and lymphocyte counts measured in the peripheral blood. Absolute lymphocyte count (ALC) and absolute neutrophil count (ANC) data were retrieved from four weeks prior to commencing radiotherapy. If multiple values were available before treatment, those closest to the start date of radiotherapy were selected. The neutrophil-to-lymphocyte ratio (NLR) was computed using neutrophil and lymphocyte measurements and dichotomized into low- and high-NLR categories using a receiver operating characteristic (ROC) analysis.

2.4. Somatic Gene Mutation Profiles and Candidate Genes

The somatic gene mutation profiles of the study cohort were determined using next-generation sequencing (NGS) and FoundationOne CDx (F1CDx), according to the Illumina® HiSeq 4000 platform, using formalin-fixed paraffin-embedded (FFPE) HNSCC tissue specimens. The F1CDx-targeted NGS platform method was validated previously [34]. Neutrophil- and lymphocyte-associated pathways were identified, and the genes involved in these pathways were retrieved from the Kyoto Encyclopedia of Genes and Genomes (KEGG) database. The results of the somatic gene mutation profiles in our study cohort were subsequently mapped. A total of 41 genes associated with lymphocyte and neutrophil signaling pathways were mapped from the somatic gene mutation profiles of the study cohort and considered as candidate gene panels for later analyses. The somatic mutation rates of the candidate genes in the study cohort and the NLR categories were summarized, and the difference between the NLR categories was estimated using Fisher's exact or Pearson's chi-squared test.

2.5. Gene Mutation Score (GMS)

A multivariate binomial generalized linear model was used to evaluate the association between the NLR and somatic mutations in the candidate genes. Candidate genes with significant results derived from the multivariate model were further selected to generate the GMS, which was calculated by multiplying the estimated coefficient in the multivariate model by the gene mutation status (wild as 0, mutated as 1). Subsequently, the study cohort was dichotomized into low- and high-GMS categories using an ROC analysis. NLR-GMS categories were generated using both the NLR and GMS, and the study cohort was reassigned into low-, intermediate-, and high-NLR-GMS categories based on their NLR and GMS categories. Patients with both a low NLR and GMS were categorized as having a low NLR-GMS, those with both a high NLR and GMS were categorized as having a high NLR-GMS, and the remaining patients were considered to have an intermediate NLR-GMS.

2.6. Statistical Analysis

The baseline characteristics of the study cohort were summarized using frequencies and percentages, and laboratory measurements were summarized using medians and interquartile ranges. Differences in baseline characteristics and laboratory measurements between the good- and poor-response subgroups were estimated using Fisher's exact test, Pearson's chi-squared test, or the Wilcoxon rank-sum test. The predictive performances of both the NLR and NLR-GMS for treatment response and short-term relapse-free survival within 36 months were evaluated using an ROC analysis. The area under the ROC curve (AUC) was used to determine the predictive performance of both the NLR and NLR-GMS for progression-free survival (PFS). A higher AUC indicated a better predictive performance. The survival rate of each NLR and NLR-GMS category was estimated using the Kaplan–Meier estimator, and the survival difference between subcategories was estimated using the log-rank test. All p-values were two-sided, and $p < 0.05$ was considered statistically significant. All analyses were performed using R 4.1.2 software (R Core Team, 2021, Vienna, Austria).

2.7. Immunohistochemistry

The HNSCC specimens were fixed on paraffin-embedded biopsies and sectioned. The slices were deparaffinized using xylene and then dehydrated with ethanol. Endogenous peroxidase activity was quenched with 3% hydrogen peroxide containing methanol for 15 min. The sections were heated in 100 mmol/L citrate buffer for 10 min to revive the antigens. The tissues were incubated with 3 primary antibodies at room temperature for 30 min and then rinsed three times with phosphate-buffered saline (PBS) according to the manufacturer's protocol. Following color development, we applied cover slips to the sections and observed them under a microscope. Staining intensity in the cancer tissue was independently examined by pathologists who were blinded to the patients' clinical features and outcomes. The following primary antibodies were used: anti-ABL1 (1:100, Elabscience, Houston, TX, USA), anti-MAP2K2 (1:100, Elabscience, Houston, TX, USA), and anti-MAP2K4 (1:100, Elabscience, State of Texas, USA). In the assessment of IHC staining, staining intensity ranged from 0 (negative) to 3+ (high strength) with the percentage of positively labeled cells.

3. Results

3.1. Baseline Characteristics of Patients

Between 2016 and 2022, 107 patients were enrolled in this study. All patients were diagnosed with HNSCC and underwent CRT. Patient characteristics, including age, sex, tumor location, pathological grade, stage, pre-CRT BMI, post-CRT BMI, body weight loss, white blood cell count, ANC, ALC, and the NLR, are summarized in Table 1. The majority of the patients were middle-aged (65.4% between 45 and 64 years), male (93.5%), had oral cavity cancer (67.3%), had grade 2 disease (55.8%), and were stage IV (85.0%). Overall, 84 patients had a good response to CRT, and 23 patients had a poor response to CRT.

There were no statistically significant differences between the good- and poor-response groups with regards to age, sex, pathological grade, clinical cancer stage, pre-CRT BMI, post-CRT BMI, body weight loss, white blood cell count, ANC, and ALC. However, there were significantly more oral cavity cancers in the poor-response group ($p = 0.038$). The difference in the NLR between the good-response group and poor-response group was also prominent ($p = 0.037$), with more patients having a high NLR (≥ 2.7) in the poor-response group (87.0% vs. 64.3%).

Table 1. Baseline characteristics of HNSCC cohort.

Characteristics	Overall, $n = 107$	Good Response, $n = 84$	Poor Response, $n = 23$	p
Age				0.804
<45	8 (7.5%)	6 (7.1%)	2 (8.7%)	
>65	29 (27.1%)	22 (26.2%)	7 (30.4%)	
45–64	70 (65.4%)	56 (66.7%)	14 (60.9%)	
Sex				0.168
Female	7 (6.5%)	4 (4.8%)	3 (13.0%)	
Male	100 (93.5%)	80 (95.2%)	20 (87.0%)	
Location				**0.038**
Hypopharynx	12 (11.2%)	12 (14.3%)	0 (0.0%)	
Larynx	2 (1.9%)	2 (2.4%)	0 (0.0%)	
Oral cavity	72 (67.3%)	51 (60.7%)	21 (91.3%)	
Oropharyx	21 (19.6%)	19 (22.6%)	2 (8.7%)	
Grade				0.617
Grade 1	28 (26.9%)	20 (24.7%)	8 (34.8%)	
Grade 2	58 (55.8%)	46 (56.8%)	12 (52.2%)	
Grade 3	18 (17.3%)	15 (18.5%)	3 (13.0%)	
Unknown	3 (2.8%)	3 (3.6%)	0 (0.0%)	
Stage				0.670
Stage I	3 (2.8%)	3 (3.6%)	0 (0.0%)	
Stage II	4 (3.7%)	4 (4.8%)	0 (0.0%)	
Stage III	9 (8.4%)	8 (9.5%)	1 (4.3%)	
Stage IV	91 (85.0%)	69 (82.1%)	22 (95.7%)	
BMI (Pre-CRT)	23.2 (14.6–34.1)	23.2 (15.1–34.0)	23.2 (14.6–34.1)	0.601
BMI (Post-CRT)	22.1 (13.9–33.6)	22.1 (13.9–33.6)	21.9 (14.5–33.0)	0.900
Body weight loss	−2.3 (−18.5–7.2)	−2.0 (−18.5–7.2)	−2.8 (−13.7–2.5)	0.377
White blood cell (/µL)	6810 (3020–35,150)	6660 (3020–15,170)	6970 (3710–35,150)	0.585
Neutrophils (Neu) (/µL)	68.8 (33.2–96.1)	68.4 (33.2–88.4)	70.0 (50.4–96.1)	0.147
Lymphocytes (Lym) (/µL)	20.8 (1.0–53.6)	21.0 (2.0–53.6)	20.7 (1.0–29.9)	0.147
NLR (Neu/Lym)				**0.037**
Low (<2.7)	33 (30.8%)	30 (35.7%)	3 (13.0%)	
High (≥2.7)	74 (69.2%)	54 (64.3%)	20 (87.0%)	

Abbreviation: BMI, body mass index; CRT, chemoradiotherapy; NLR, neutrophil-to-lymphocyte ratio.

3.2. Signaling Pathways Associated with Lymphocytes and Neutrophils

After confirming that a high NLR was related to a poor treatment response, we further analyzed the genetic alterations associated with lymphocytes and neutrophils. Table 2 summarizes eight common pathways associated with lymphocytes and neutrophils which were identified using the KEGG database, including phosphoinositide 3-kinase (PI3K) and Fc gamma receptor IIb (FcγRIIb) signaling in B lymphocytes, protein kinase C (PKC) and 4-1BB signaling in T lymphocytes, the regulation of IL-2 expression in activated and anergic T lymphocytes, cytotoxic T-lymphocyte antigen 4 (CTLA4) signaling in cytotoxic T lymphocytes, CD27 signaling in lymphocytes, and N-formyl methionyl-leucyl-phenylalanine (fMLP) signaling in neutrophils.

Table 2 reports the pathways associated with the lymphocytes and neutrophils derived using the mutated genes detected in this cohort. The "involved genes" column indicates the mutated genes found in our study cohort, which are simultaneously involved in the corresponding pathways. The "genes in pathways" column reported the overall number of genes involved in the correspond pathways, and the "pathway percentage" was computed by dividing the "involved genes" column by the "genes in pathways" column, denoting the percentage of mutated genes found in this study cohort involving a correspondence pathway. The raw NGS data for the somatic mutation profiles of this study cohort are in Supplementary Table S1.

Table 2. Pathways associated with lymphocytes and neutrophils derived using mutated genes in the HNSCC cohort.

Involved Pathway	Genes in Pathway	Involved Genes	Pathway Percentage	Gene Symbol
PI3K signaling in B lymphocytes	122	25	20.5	*LYN;IKBKE;AKT2;PIK3CA;AKT1;CD79A;IRS2;MAP2K2; PRKCI;CD79B;PTEN;NFKBIA;CBL; MAPK1;ABL1;PIK3CB;RAC1;SYK;HRAS;JUN;AKT3; MAP2K1;RAF1;KRAS;PIK3R1*
FcγRIIb signaling in B lymphocytes	41	14	34.1	*LYN;PIK3C2G;PIK3CA;AKT1;CD79A;ATM; CD79B;MAP2K4;PIK3CB;SYK;HRAS;PIK3C2B; KRAS;PIK3R1*
Regulation of IL-2 expression in activated and anergic T lymphocytes	75	16	21.3	*IKBKE;TGFBR2;SMAD2;CARD11;MAP2K2; NFKBIA;MAP2K4;MAPK1;RAC1;HRAS; MAP3K1;JUN;MAP2K1;RAF1;KRAS;SMAD4*
PKC signaling in T lymphocytes	107	17	15.9	*MAP3K13;PIK3C2G;IKBKE;CARD11;PIK3CA; ATM;NFKBIA;MAP2K4;MAPK1;PIK3CB;RAC1; HRAS;MAP3K1;PIK3C2B;JUN;KRAS; PIK3R1*
fMLP signaling in neutrophils	106	16	15.1	*GNAS;PIK3C2G;PIK3CA;MAP2K2;ATM;PRKCI; NFKBIA;MAPK1;PIK3CB;RAC1;HRAS; PIK3C2B;MAP2K1;RAF1;KRAS;PIK3R1*
CTLA4 signaling in cytotoxic T lymphocytes	82	13	15.9	*JAK2;PIK3C2G;AKT2;PIK3CA;AKT1;ATM; PIK3CB;SYK;PIK3C2B;AKT3;PTPN11;PPP2R2A; PIK3R1*
CD27 signaling in lymphocytes	51	10	19.6	*CASP8;MAP3K13;IKBKE;MAP2K2;NFKBIA; MAP2K4;MAP3K1;JUN;MAP2K1;BCL2L1*
4-IBB signaling in T lymphocytes	31	7	22.6	*IKBKE;MAP2K2;NFKBIA;MAP2K4;MAPK1; JUN;MAP2K1*

Abbreviation: PI3K, phosphoinositide 3-kinase; FcγRIIb, Fc gamma receptor IIb; PKC, protein kinase C; fMLP, N-formyl methionyl-leucyl-phenylalanine; CTLA4, cytotoxic T-lymphocyte antigen 4.

3.3. Somatic Mutation Profiles of Candidate Genes

We matched the genes in these eight pathways with the somatic gene mutation profiles of the patients. A total of 41 genes were identified and were considered candidate gene panels for subsequent analyses, as shown in Table 3. Overall, gene mutation rates were low. Only nine genes had mutation rates of over 10%: *PIK3CA* (23.4%), *PRKCI* (18.7%), *CASP8* (18.7%), *PIK3C2G* (15.0%), *MAP3K13* (13.1%), *ATM* (11.2%), *JAK2* (11.2%), *PTEN* (10.3%), and *CARD11* (10.3%). Among these 41 genes, there were no mutation differences between the low-NLR group and the high-NLR group.

3.4. Predictive Performance of GMS and NLR-GMS

Due to the relatively low mutation rate, we used a multivariate binomial generalized linear model to evaluate the association between the NLR and somatic mutations of the candidate genes (Table 4). Three candidate genes (*MAP2K2*, *MAP2K4*, and *ABL1*) exhibiting significant results derived from the multivariate model were selected to generate the GMS. *MAP2K2* mutations were associated with a low NLR, whereas *ABL1* and *MAP2K4* mutations were associated with a high NLR.

Table 3. Gene mutation rate of 41 genes associated with lymphocyte and neutrophil signaling pathways.

Genes	Overall, $n = 107$	Low NLR, $n = 33$	High NLR, $n = 74$	p Value
PIK3CA	25 (23.4%)	10 (30.3%)	15 (20.3%)	0.257
PRKCI	20 (18.7%)	7 (21.2%)	13 (17.6%)	0.655
CASP8	20 (18.7%)	3 (9.1%)	17 (23.0%)	0.089
PIK3C2G	16 (15.0%)	4 (12.1%)	12 (16.2%)	0.771
MAP3K13	14 (13.1%)	5 (15.2%)	9 (12.2%)	0.758
ATM	12 (11.2%)	5 (15.2%)	7 (9.5%)	0.508
JAK2	12 (11.2%)	3 (9.1%)	9 (12.2%)	0.751
PTEN	11 (10.3%)	4 (12.1%)	7 (9.5%)	0.735
CARD11	11 (10.3%)	1 (3.0%)	10 (13.5%)	0.167
GNAS	9 (8.4%)	2 (6.1%)	7 (9.5%)	0.718
HRAS	8 (7.5%)	3 (9.1%)	5 (6.8%)	0.700
CBL	7 (6.5%)	1 (3.0%)	6 (8.1%)	0.433
IRS2	6 (5.6%)	4 (12.1%)	2 (2.7%)	0.071
MAP2K2	6 (5.6%)	4 (12.1%)	2 (2.7%)	0.071
MAP3K1	6 (5.6%)	0 (0.0%)	6 (8.1%)	0.174
NFKBIA	5 (4.7%)	2 (6.1%)	3 (4.1%)	0.643
PIK3CB	5 (4.7%)	2 (6.1%)	3 (4.1%)	0.643
KRAS	5 (4.7%)	1 (3.0%)	4 (5.4%)	1.000
TGFBR2	5 (4.7%)	2 (6.1%)	3 (4.1%)	0.643
LYN	4 (3.7%)	1 (3.0%)	3 (4.1%)	1.000
IKBKE	3 (2.8%)	1 (3.0%)	2 (2.7%)	1.000
AKT1	3 (2.8%)	2 (6.1%)	1 (1.4%)	0.224
CD79B	3 (2.8%)	1 (3.0%)	2 (2.7%)	1.000
MAPK1	3 (2.8%)	2 (6.1%)	1 (1.4%)	0.224
JUN	3 (2.8%)	0 (0.0%)	3 (4.1%)	0.551
MAP2K4	3 (2.8%)	0 (0.0%)	3 (4.1%)	0.551
PIK3C2B	3 (2.8%)	2 (6.1%)	1 (1.4%)	0.224
ABL1	2 (1.9%)	0 (0.0%)	2 (2.7%)	1.000
RAC1	2 (1.9%)	0 (0.0%)	2 (2.7%)	1.000
SYK	2 (1.9%)	1 (3.0%)	1 (1.4%)	0.524
AKT3	2 (1.9%)	0 (0.0%)	2 (2.7%)	1.000
RAF1	2 (1.9%)	1 (3.0%)	1 (1.4%)	0.524
PIK3R1	2 (1.9%)	0 (0.0%)	2 (2.7%)	1.000
SMAD4	2 (1.9%)	2 (6.1%)	0 (0.0%)	0.093
PTPN11	2 (1.9%)	0 (0.0%)	2 (2.7%)	1.000
BCL2L1	2 (1.9%)	0 (0.0%)	2 (2.7%)	1.000
AKT2	1 (0.9%)	1 (3.0%)	0 (0.0%)	0.308
CD79A	1 (0.9%)	0 (0.0%)	1 (1.4%)	1.000
MAP2K1	1 (0.9%)	0 (0.0%)	1 (1.4%)	1.000
SMAD2	1 (0.9%)	1 (3.0%)	0 (0.0%)	0.308
PPP2R2A	1 (0.9%)	0 (0.0%)	1 (1.4%)	1.000

p-value is estimated using Fisher's exact test or Pearson chi-squared test.

Table 4. Binomial generalized linear model result for the association between the NLR category and the mutation status of 41 associated genes.

Genes	Coefficients	SE	t	p
LYN	−0.386	0.336	−1.150	0.254
IKBKE	0.036	0.278	0.130	0.897
AKT2	−1.030	0.560	−1.842	0.070
PIK3CA	−0.486	0.251	−1.942	0.057
AKT1	−0.194	0.422	−0.461	0.646
CD79A	0.441	0.871	0.507	0.614
IRS2	−0.367	0.244	−1.506	0.137
MAP2K2	−0.749	0.315	−2.373	**0.021**
PRKCI	0.342	0.277	1.233	0.222
CD79B	−0.510	0.421	−1.212	0.230
PTEN	0.001	0.211	0.006	0.996
NFKBIA	0.384	0.359	1.069	0.289
CBL	0.219	0.277	0.791	0.432
MAPK1	0.088	0.447	0.196	0.845
ABL1	0.986	0.449	2.197	**0.032**
PIK3CB	0.149	0.298	0.501	0.618
RAC1	0.602	0.619	0.973	0.334
SYK	0.130	0.418	0.311	0.757
HRAS	−0.210	0.223	−0.942	0.350
JUN	0.698	0.466	1.500	0.138
AKT3	0.102	0.530	0.193	0.848
MAP2K1	−0.101	0.521	−0.193	0.848
RAF1	−0.011	0.384	−0.027	0.978
KRAS	0.198	0.305	0.649	0.519
PIK3R1	−0.009	0.492	−0.019	0.985
PIK3C2G	0.085	0.153	0.554	0.582
ATM	0.037	0.202	0.181	0.857
MAP2K4	0.730	0.339	2.153	**0.035**
PIK3C2B	−0.217	0.308	−0.704	0.484
TGFBR2	−0.360	0.342	−1.055	0.295
SMAD2	−0.768	0.656	−1.169	0.246
CARD11	0.009	0.179	0.052	0.959
MAP3K1	0.156	0.270	0.577	0.566
SMAD4	−0.523	0.429	−1.219	0.227
MAP3K13	−0.044	0.247	−0.177	0.860
GNAS	0.111	0.193	0.576	0.566
JAK2	0.236	0.158	1.498	0.139
PTPN11	−0.450	0.593	−0.759	0.450
PPP2R2A	−0.013	0.718	−0.017	0.986
CASP8	0.203	0.148	1.371	0.175
BCL2L1	0.522	0.532	0.982	0.330

SE, standard error. The significant genes were abstracted to generate a GMS (gene mutation score) = coefficient × gene mutation status (0: wild, 1: mut).

NLR-GMS categories were generated using both the NLR and GMS, and the study cohort was reassigned into low-, intermediate-, and high-NLR-GMS categories based on their NLR and GMS categories. Figure 1A,B display ROC plots for the NLR category and the NLR-GMS for treatment response and relapse events within 36 months, respectively.

Both exhibited improved predictive performance using the NLR-GMS categories compared to the NLR categories.

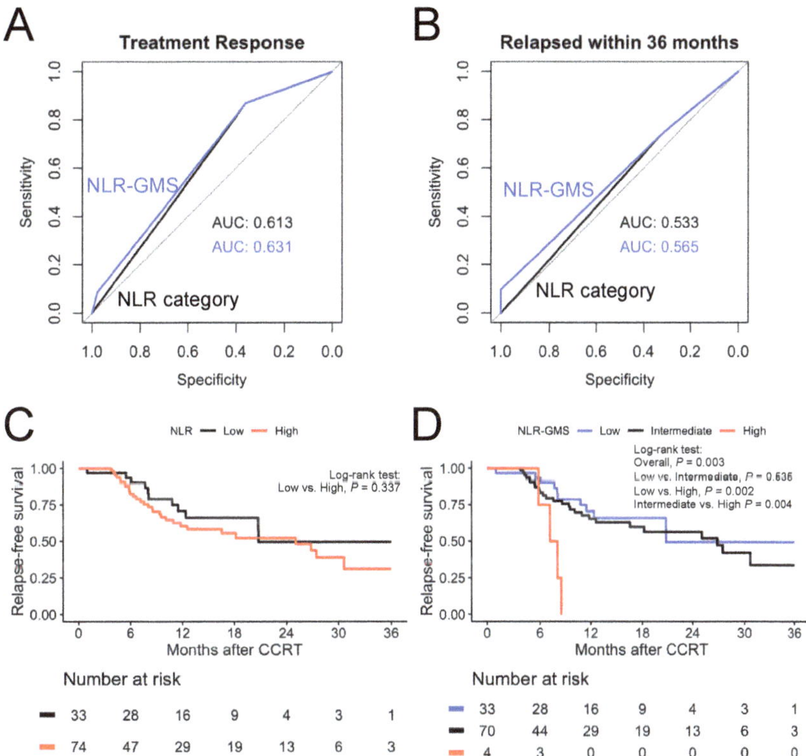

Figure 1. Predictive performance of NLR and NLR-GMS in treatment response and short-term relapse-free survival (within 36 months). ROC plots for NLR category and NLR-GMS for (**A**) treatment response and (**B**) relapse event within 36 months. Kaplan–Meier plot for short-term relapse-free survival comparison according to (**C**) NLR and (**D**) NLR-GMS category. ROC, receiver operating characteristics. AUC, area under ROC curve.

Although the high-NLR group exhibited poorer survival than the low-NLR group, no significant survival difference was found between the groups (Figure 1C). However, when we used the NLR-GMS categories, the difference in relapse-free survival between the NLR-GMS groups was significant (overall $p = 0.003$; Figure 1D). The high NLR-GMS group had a significantly shorter relapse-free survival period than the intermediate ($p = 0.004$) and low ($p = 0.002$) NLR-GMS groups.

3.5. Somatic Mutation Validation via Immunohistochemistry Staining

Based on the result of the multivariate binomial generalized linear model, three candidate genes (*MAP2K2*, *MAP2K4* and *ABL1*) exhibited significant results and were used to generate the GMS. We further validated the expression of three candidate genes by immunohistochemistry (IHC) staining cancer tissues from HNSCC patients. Figure 2 shows representative images of IHC staining for the mutated and wild type of each candidate gene. The yellow words in the upper-left corner of each picture indicate the staining intensity and the percentage of positively labeled cells. Cancer tissues with somatic mutations of the three candidate genes showed a higher staining intensity and percentage of positive cells compared to wild-type specimens. Overall, it showed an increasing IHC intensity in somatic mutated samples.

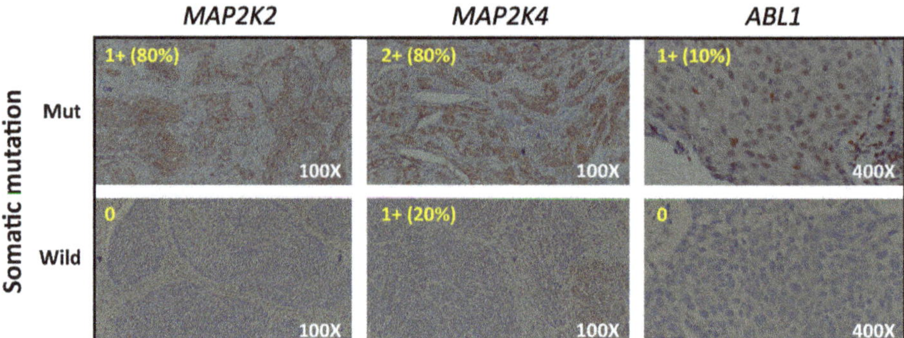

Figure 2. Representative immunohistochemistry (IHC) images of *MAP2K2*, *MAP2K4*, and *ABL1* expression from cancer tissues from HNSCC patients.

4. Discussion

The relationship between inflammation and cancer is highly associated. Inflammation predisposes patients to cancer development and promotes all stages of tumorigenesis [15–18]. Several acquired factors are already proven to be carcinogenic and are often associated with chronic inflammation, including chronic bacterial and viral infections, autoimmune diseases, environmental factors (asbestos exposure), lifestyle factors (obesity, tobacco smoking, and excessive alcohol consumption), and aging, which are thought to promote tumor-extrinsic inflammation [14,35]. In contrast, tumor-intrinsic inflammation, also referred to as cancer-elicited inflammation, is induced after tumor initiation and contributes to malignant progression by recruiting and activating inflammatory cells [14,17,36]. Moreover, anticancer therapies can also induce inflammation via the necrosis and necroptosis of cancer cells [14,17]. Overall, cancer cells and surrounding stromal and inflammatory cells engage in well-orchestrated reciprocal interactions to form an inflammatory tumor microenvironment [17].

Several parameters have been reported to represent the degree of systemic inflammation, such as C-reactive protein and the NLR, PLR, and LMR [19–24]. The NLR is the most promising clinical biomarker and has been shown to be prognostic in many solid tumors, including HNSCC [25–33]. The NLR reflects the dynamic relationship between neutrophils (innate immune response) and lymphocytes (adaptive immune response). During stress, trauma, surgery, systemic infection, inflammation, sepsis, or critical illness, the dysregulation of innate and adaptive immune responses results in neutrophilia and lymphocytopenia [37]. Initially, the NLR was used as an index of systemic inflammatory response syndrome (SIRS) and stress in critically ill patients [38]. However, more recently, the NLR has been applied in almost all medical scenarios as a reliable and easily available marker of immune response to various stimuli [37].

The NLR was first applied as a prognostic factor for colorectal cancer in 2005 by Walsh et al. [39]. Two meta-analyses [25,26] and various studies have demonstrated that the NLR is a prognostic factor in different solid tumors, including esophageal cancer [40], gastric cancer [41,42], pancreatic cancer [43], and biliary tract cancer [44], etc. The role of the NLR in HNSCC was also evaluated in many studies, and all of them concluded that the NLR is a reliable prognostic marker [28–30,33,45–54]. Similarly, our cohort demonstrated patients with a high NLR in the poor-response group (87.0% vs. 64.3%), although there was no difference in the 36-month relapse-free survival between the high- and low-NLR groups.

To further evaluate the genetic alterations of the immune response associated with the NLR, we analyzed the NGS data of HNSCC patients and matched them with the

KEGG database. Eight pathways and forty-one associated genes were identified. Different statistical methods were applied to determine the association between the NLR and somatic mutations in these candidate genes. Three candidate genes (*MAP2K2*, *MAP2K4*, and *ABL1*) were selected using a multivariate binomial generalized linear model and were further advanced to GMS generation.

MAP2K2 and *MAP2K4* both belong to the mitogen-activated protein kinase (MAPK) family. *MAP2K2* regulates the phosphorylation and activation of extracellular signal-regulated kinases (ERKs) [55]. *MAP2K4* can activate p38 via the phosphorylation of c-Jun-N-terminal kinases (JNK) [56]. MAPK cascades regulate a wide variety of cellular processes including proliferation, differentiation, transcriptional regulation, and stress responses [55,57]. MAPK activation also plays critical roles in the production of pro-inflammatory cytokines and the induction of the expression of multiple inflammatory-associated genes [58–62]. A variety of pharmacological inhibitors have been developed to specifically block MAPK kinase 1 and 2 (MEK1/2) [63,64]. Four MEK inhibitors, Trametinib, Cobimetinib, Binimetinib, and Selumetinib, have already been approved by the FDA to date [65]. An emerging technology, named proteolysis targeting chimera (PROTAC), could break the limitation of acquired resistance during long-term treatment by inducing MEK1/2 degradation [65]. Unlike *MAP2K2*, inhibitors specifically targeting *MAP2K4* are few. PLX8725, a novel *MAP2K4* inhibitor, demonstrates promising in vivo activity against patient-derived xenografts of uterine leiomyosarcomas harboring gain-of-function alterations in the *MAP2K4* gene [66]. In fact, *MAP2K4* mutations are sensitive to MEK inhibitors in multiple cancer models [67]. There are also a variety of potent inhibitors of the p38 MAP kinases have been developed, such as SB203580, SB202190, and BIRB-796 [64].

ABL1 is a proto-oncogene that encodes a protein tyrosine kinase involved in a variety of cellular processes, including cell division, adhesion, differentiation, and response to stress. *ABL1* is involved in the occurrence and development of several types of cancers including colon, kidney, and breast cancer [68,69]. Some inflammatory conditions are also associated with *ABL1* [70]. *ABL1* mediates inflammation by regulating the NF-κB and STAT3 signaling pathways. The blockade of *ABL1* suppresses inflammatory signaling and cytokines [71]. To date, there are several approved *ABL1* inhibitors, such as imatinib, nilotinib, dasatinib, bosutinib, ponatinib, and so on. Most of them are treatments for chronic myeloid leukemia, which is *BCR-ABL1*-positive.

In short, three candidate genes (*MAP2K2*, *MAP2K4*, and *ABL1*) are all associated with inflammation and carcinogenesis. In our study, by using the NLR-GMS categories to divide patients into three groups, we noticed that the high-NLR-GMS group had a significantly shorter relapse-free survival than the intermediate- and low-NLR-GMS groups. In conclusion, the GMS could further target extremely high-risk patients with worse short-term survival based on the NLR-GMS categories.

This article possesses some strengths and limitations. Initially, the previous publications only affirmed the association between the NLR and treatment response. However, in this study, we conducted a comprehensive genetic investigation along with an investigation of clinical biomarkers in patients with HNSCC who underwent CRT and devised NLR-GMS categories to predict survival. Nonetheless, this cohort has a relatively small sample size, and there were differences in the tumor locations between the two groups. Specifically, the poor-response group had more oral cavity cancers, while the good-response group had more oropharynx cancers. Since different anatomical sites represent distinct etiological factors and background genetic alterations, this may have affected the study results. The longitudinal pattern (or variation) in the NLR of the study population was not investigated due to the nature of the study's design. Additionally, further experiments are necessary to elucidate the role, influence, and mechanism of the candidate genes (*MAP2K2*, *MAP2K4*, and *ABL1*), which could serve as a valuable theme for future research.

5. Conclusions

In our cohort of HNSCC patients that underwent CRT, a high pre-treatment NLR is linked to a poor response. We conducted a further analysis on the genetic alterations associated with the NLR using somatic gene mutation profiles and identified 41 associated genes. However, there was no significant difference observed between the high- and low-NLR groups. Through the use of a multivariate binomial generalized linear model, we selected *MAP2K2*, *MAP2K4*, and *ABL1* to develop a GMS. By utilizing the NLR-GMS categories, we could effectively target high-risk patients with an extremely poor short-term survival outcome.

Supplementary Materials: The following supporting information can be downloaded at: https://www.mdpi.com/article/10.3390/biomedicines11123113/s1, Table S1: Raw NGS data of the somatic mutation profiles.

Author Contributions: Conceptualization, T.-J.Y., H.-C.W., L.-T.C., M.-R.P. and S.-H.M.; formal analysis, H.-C.W. and S.-H.M.; data curation, T.-J.Y., H.-C.W. and S.-H.M.; writing—original draft preparation, T.-J.Y., T.-Y.H., T.-M.C., J.-S.D. and Y.-C.G.; writing—review and editing, T.-J.Y., H.-C.W., S.-F.C., C.-T.H., M.-H.W., Y.-C.L. and S.-H.M.; validation, C.-C.W. and M.-R.P.; supervision, H.-H.H., L.-T.C., M.-R.P. and S.-H.M. All authors have read and agreed to the published version of the manuscript.

Funding: We acknowledge the support from the following grants: (1) 112-2321-B-037-004-, 111-2321-B-037-002, 111-2628-B-037-008, 111-2320-B-037-030, and 110-2327-B-037-001 from the Ministry of Science and Technology, Taiwan; and (2) KMUH111-1M14, KMUH111-1M15, and KMUH110-0T04 from the Kaohsiung Medical University Hospital; (3) KMU-KI110001, KMU-TC109A03, KMU-TC108A03-6, KMU-TC111A04, KMU-TC112A04, and KMU-TC109B05 from Kaohsiung Medical University Research Center Grant.

Institutional Review Board Statement: This study was approved by the Institutional Review Board and Ethics Committee of Kaohsiung Medical University Hospital (KMUHIRB-E(I)-20210401). The data were analyzed anonymously and therefore, no additional informed consent was required. All methods were performed in accordance with the approved guidelines and regulations.

Informed Consent Statement: Not applicable.

Data Availability Statement: The data that support the findings of this study are available upon reasonable request (e.g., for research purposes) from the authors.

Conflicts of Interest: The authors declare no conflict of interest.

References

1. Johnson, D.E.; Burtness, B.; Leemans, C.R.; Lui, V.W.Y.; Bauman, J.E.; Grandis, J.R. Head and neck squamous cell carcinoma. *Nat. Rev. Dis. Primers* **2020**, *6*, 92. [CrossRef] [PubMed]
2. Hsu, W.-L.; Yu, K.J.; Chiang, C.-J.; Chen, T.-C.; Wang, C.-P. Head and neck cancer incidence trends in Taiwan, 1980–2014. *Int. J. Head Neck Sci.* **2017**, *1*, 180–189.
3. Bray, F.; Ferlay, J.; Soerjomataram, I.; Siegel, R.L.; Torre, L.A.; Jemal, A. Global cancer statistics 2018: GLOBOCAN estimates of incidence and mortality worldwide for 36 cancers in 185 countries. *CA Cancer J. Clin.* **2018**, *68*, 394–424. [CrossRef] [PubMed]
4. Yeh, T.J.; Chan, L.P.; Tsai, H.T.; Hsu, C.M.; Cho, S.F.; Pan, M.R.; Liu, Y.-C.; Huang, C.-J.; Wu, C.-W.; Du, J.-S.; et al. The Overall Efficacy and Outcomes of Metronomic Tegafur-Uracil Chemotherapy on Locally Advanced Head and Neck Squamous Cell Carcinoma: A Real-World Cohort Experience. *Biology* **2021**, *10*, 168. [CrossRef] [PubMed]
5. Braakhuis, B.J.; Brakenhoff, R.H.; Leemans, C.R. Treatment choice for locally advanced head and neck cancers on the basis of risk factors: Biological risk factors. *Ann. Oncol.* **2012**, *23* (Suppl. S10), x173–x177. [CrossRef]
6. Chow, L.Q.M. Head and Neck Cancer. *N. Engl. J. Med.* **2020**, *382*, 60–72. [CrossRef] [PubMed]
7. Cooper, J.S.; Pajak, T.F.; Forastiere, A.A.; Jacobs, J.; Campbell, B.H.; Saxman, S.B.; Kish, J.A.; Kim, H.E.; Cmelak, A.J.; Rotman, M.; et al. Postoperative concurrent radiotherapy and chemotherapy for high-risk squamous-cell carcinoma of the head and neck. *N. Engl. J. Med.* **2004**, *350*, 1937–1944. [CrossRef]
8. Bernier, J.; Domenge, C.; Ozsahin, M.; Matuszewska, K.; Lefèbvre, J.L.; Greiner, R.H.; Giralt, J.; Maingon, P.; Rolland, F.; Bolla, M.; et al. Postoperative irradiation with or without concomitant chemotherapy for locally advanced head and neck cancer. *N. Engl. J. Med.* **2004**, *350*, 1945–1952. [CrossRef]

9. Basheeth, N.; Patil, N. Biomarkers in Head and Neck Cancer an Update. *Indian J. Otolaryngol. Head Neck Surg.* **2019**, *71* (Suppl. S1), 1002–1011. [CrossRef]
10. Budach, V.; Tinhofer, I. Novel prognostic clinical factors and biomarkers for outcome prediction in head and neck cancer: A systematic review. *Lancet Oncol.* **2019**, *20*, e313–e326. [CrossRef]
11. Hsieh, J.C.; Wang, H.M.; Wu, M.H.; Chang, K.P.; Chang, P.H.; Liao, C.T.; Liau, C.-T. Review of emerging biomarkers in head and neck squamous cell carcinoma in the era of immunotherapy and targeted therapy. *Head Neck* **2019**, *41* (Suppl. S1), 19–45. [CrossRef] [PubMed]
12. Wang, H.C.; Yeh, T.J.; Chan, L.P.; Hsu, C.M.; Cho, S.F. Exploration of Feasible Immune Biomarkers for Immune Checkpoint Inhibitors in Head and Neck Squamous Cell Carcinoma Treatment in Real World Clinical Practice. *Int. J. Mol. Sci.* **2020**, *21*, 7621. [CrossRef] [PubMed]
13. Balkwill, F.; Mantovani, A. Inflammation and cancer: Back to Virchow? *Lancet* **2001**, *357*, 539–545. [CrossRef] [PubMed]
14. Hibino, S.; Kawazoe, T.; Kasahara, H.; Itoh, S.; Ishimoto, T.; Sakata-Yanagimoto, M.; Taniguchi, K. Inflammation-Induced Tumorigenesis and Metastasis. *Int. J. Mol. Sci.* **2021**, *22*, 5421. [CrossRef] [PubMed]
15. Hanahan, D.; Weinberg, R.A. Hallmarks of cancer: The next generation. *Cell* **2011**, *144*, 646–674. [CrossRef] [PubMed]
16. Coussens, L.M.; Werb, Z. Inflammation and cancer. *Nature* **2002**, *420*, 860–867. [CrossRef] [PubMed]
17. Greten, F.R.; Grivennikov, S.I. Inflammation and Cancer: Triggers, Mechanisms, and Consequences. *Immunity* **2019**, *51*, 27–41. [CrossRef]
18. Mantovani, A.; Allavena, P.; Sica, A.; Balkwill, F. Cancer-related inflammation. *Nature* **2008**, *454*, 436–444. [CrossRef]
19. Li, B.; Zhou, P.; Liu, Y.; Wei, H.; Yang, X.; Chen, T.; Xiao, J. Platelet-to-lymphocyte ratio in advanced Cancer: Review and meta-analysis. *Clin. Chim. Acta Int. J. Clin. Chem.* **2018**, *483*, 48–56. [CrossRef]
20. Mandaliya, H.; Jones, M.; Oldmeadow, C.; Nordman, I.I. Prognostic biomarkers in stage IV non-small cell lung cancer (NSCLC): Neutrophil to lymphocyte ratio (NLR), lymphocyte to monocyte ratio (LMR), platelet to lymphocyte ratio (PLR) and advanced lung cancer inflammation index (ALI). *Transl. Lung Cancer Res.* **2019**, *8*, 886–894. [CrossRef]
21. Hu, G.; Liu, G.; Ma, J.Y.; Hu, R.J. Lymphocyte-to-monocyte ratio in esophageal squamous cell carcinoma prognosis. *Clin. Chim. Acta Int. J. Clin. Chem.* **2018**, *486*, 44–48. [CrossRef] [PubMed]
22. Tan, D.; Fu, Y.; Tong, W.; Li, F. Prognostic significance of lymphocyte to monocyte ratio in colorectal cancer: A meta-analysis. *Int. J. Surg.* **2018**, *55*, 128–138. [CrossRef] [PubMed]
23. Li, P.; Li, H.; Ding, S.; Zhou, J. NLR, PLR, LMR and MWR as diagnostic and prognostic markers for laryngeal carcinoma. *Am. J. Transl. Res.* **2022**, *14*, 3017–3027.
24. Deans, D.A.; Tan, B.H.; Wigmore, S.J.; Ross, J.A.; de Beaux, A.C.; Paterson-Brown, S.; Fearon, K.C.H. The influence of systemic inflammation, dietary intake and stage of disease on rate of weight loss in patients with gastro-oesophageal cancer. *Br. J. Cancer* **2009**, *100*, 63–69. [CrossRef] [PubMed]
25. Templeton, A.J.; McNamara, M.G.; Šeruga, B.; Vera-Badillo, F.E.; Aneja, P.; Ocaña, A.; Leibowitz-Amit, R.; Sonpavde, G.; Knox, J.J.; Tran, B.; et al. Prognostic role of neutrophil-to-lymphocyte ratio in solid tumors: A systematic review and meta-analysis. *J. Natl. Cancer Inst.* **2014**, *106*, dju124. [CrossRef]
26. Cupp, M.A.; Cariolou, M.; Tzoulaki, I.; Aune, D.; Evangelou, E.; Berlanga-Taylor, A.J. Neutrophil to lymphocyte ratio and cancer prognosis: An umbrella review of systematic reviews and meta-analyses of observational studies. *BMC Med.* **2020**, *18*, 360. [CrossRef] [PubMed]
27. Mariani, P.; Russo, D.; Maisto, M.; Troiano, G.; Caponio, V.C.A.; Annunziata, M.; Laino, L. Pre-treatment neutrophil-to-lymphocyte ratio is an independent prognostic factor in head and neck squamous cell carcinoma: Meta-analysis and trial sequential analysis. *J. Oral Pathol. Med.* **2022**, *51*, 39–51. [CrossRef] [PubMed]
28. Mascarella, M.A.; Mannard, E.; Silva, S.D.; Zeitouni, A. Neutrophil-to-lymphocyte ratio in head and neck cancer prognosis: A systematic review and meta-analysis. *Head Neck* **2018**, *40*, 1091–1100. [CrossRef]
29. Takenaka, Y.; Oya, R.; Kitamiura, T.; Ashida, N.; Shimizu, K.; Takemura, K.; Yamamoto, Y.; Uno, A. Prognostic role of neutrophil-to-lymphocyte ratio in head and neck cancer: A meta-analysis. *Head Neck* **2018**, *40*, 647–655. [CrossRef]
30. Haddad, C.R.; Guo, L.; Clarke, S.; Guminski, A.; Back, M.; Eade, T. Neutrophil-to-lymphocyte ratio in head and neck cancer. *J. Med. Imaging Radiat. Oncol.* **2015**, *59*, 514–519. [CrossRef]
31. Yu, Y.; Wang, H.; Yan, A.; Wang, H.; Li, X.; Liu, J.; Li, W. Pretreatment neutrophil to lymphocyte ratio in determining the prognosis of head and neck cancer: A meta-analysis. *BMC Cancer* **2018**, *18*, 383. [CrossRef]
32. Rachidi, S.; Wallace, K.; Wrangle, J.M.; Day, T.A.; Alberg, A.J.; Li, Z. Neutrophil-to-lymphocyte ratio and overall survival in all sites of head and neck squamous cell carcinoma. *Head Neck* **2016**, *38* (Suppl. S1), E1068–E1074. [CrossRef] [PubMed]
33. Rosculet, N.; Zhou, X.C.; Ha, P.; Tang, M.; Levine, M.A.; Neuner, G.; Califano, J. Neutrophil-to-lymphocyte ratio: Prognostic indicator for head and neck squamous cell carcinoma. *Head Neck* **2017**, *39*, 662–667. [CrossRef] [PubMed]
34. Frampton, G.M.; Fichtenholtz, A.; Otto, G.A.; Wang, K.; Downing, S.R.; He, J.; Schnall-Levin, M.; White, J.; Sanford, E.M.; An, P.; et al. Development and validation of a clinical cancer genomic profiling test based on massively parallel DNA sequencing. *Nat. Biotechnol.* **2013**, *31*, 1023–1031. [CrossRef] [PubMed]
35. Nagai, N.; Kudo, Y.; Aki, D.; Nakagawa, H.; Taniguchi, K. Immunomodulation by Inflammation during Liver and Gastrointestinal Tumorigenesis and Aging. *Int. J. Mol. Sci.* **2021**, *22*, 2238. [CrossRef] [PubMed]
36. Wang, K.; Karin, M. Tumor-Elicited Inflammation and Colorectal Cancer. *Adv. Cancer Res.* **2015**, *128*, 173–196. [PubMed]

37. Zahorec, R. Neutrophil-to-lymphocyte ratio, past, present and future perspectives. *Bratisl. Lek. Listy* **2021**, *122*, 474–488. [CrossRef] [PubMed]
38. Zahorec, R. Ratio of neutrophil to lymphocyte counts—Rapid and simple parameter of systemic inflammation and stress in critically ill. *Bratisl. Lek. Listy* **2001**, *102*, 5–14.
39. Walsh, S.R.; Cook, E.J.; Goulder, F.; Justin, T.A.; Keeling, N.J. Neutrophil-lymphocyte ratio as a prognostic factor in colorectal cancer. *J. Surg. Oncol.* **2005**, *91*, 181–184. [CrossRef]
40. Jiang, Y.; Xu, D.; Song, H.; Qiu, B.; Tian, D.; Li, Z.; Ji, Y.; Wang, J. Inflammation and nutrition-based biomarkers in the prognosis of oesophageal cancer: A systematic review and meta-analysis. *BMJ Open* **2021**, *11*, e048324. [CrossRef]
41. Miyamoto, R.; Inagawa, S.; Sano, N.; Tadano, S.; Adachi, S.; Yamamoto, M. The neutrophil-to-lymphocyte ratio (NLR) predicts short-term and long-term outcomes in gastric cancer patients. *Eur. J. Surg. Oncol.* **2018**, *44*, 607–612. [CrossRef] [PubMed]
42. Yu, L.; Lv, C.Y.; Yuan, A.H.; Chen, W.; Wu, A.W. Significance of the preoperative neutrophil-to-lymphocyte ratio in the prognosis of patients with gastric cancer. *World J. Gastroenterol.* **2015**, *21*, 6280–6286. [CrossRef] [PubMed]
43. Cheng, H.; Luo, G.; Lu, Y.; Jin, K.; Guo, M.; Xu, J.; Long, J.; Liu, L.; Yu, X.; Liu, C. The combination of systemic inflammation-based marker NLR and circulating regulatory T cells predicts the prognosis of resectable pancreatic cancer patients. *Pancreatology* **2016**, *16*, 1080–1084. [CrossRef] [PubMed]
44. Beal, E.W.; Wei, L.; Ethun, C.G.; Black, S.M.; Dillhoff, M.; Salem, A.; Weber, S.M.; Tran, T.; Poultsides, G.; Son, A.Y.; et al. Elevated NLR in gallbladder cancer and cholangiocarcinoma—Making bad cancers even worse: Results from the US Extrahepatic Biliary Malignancy Consortium. *HPB* **2016**, *18*, 950–957. [CrossRef] [PubMed]
45. Ma, S.J.; Yu, H.; Khan, M.; Gill, J.; Santhosh, S.; Chatterjee, U.; Iovoli, A.; Farrugia, M.; Mohammadpour, H.; Wooten, K.; et al. Evaluation of Optimal Threshold of Neutrophil-Lymphocyte Ratio and Its Association With Survival Outcomes Among Patients With Head and Neck Cancer. *JAMA Netw. Open* **2022**, *5*, e227567. [CrossRef] [PubMed]
46. Chen, M.F.; Tsai, M.S.; Chen, W.C.; Chen, P.T. Predictive Value of the Pretreatment Neutrophil-to-Lymphocyte Ratio in Head and Neck Squamous Cell Carcinoma. *J. Clin. Med.* **2018**, *7*, 294. [CrossRef] [PubMed]
47. Rassouli, A.; Saliba, J.; Castano, R.; Hier, M.; Zeitouni, A.G. Systemic inflammatory markers as independent prognosticators of head and neck squamous cell carcinoma. *Head Neck* **2015**, *37*, 103–110. [CrossRef]
48. Salzano, G.; Dell'Aversana Orabona, G.; Abbate, V.; Vaira, L.A.; Committeri, U.; Bonavolontà, P.; Piombino, P.; Maglitto, F.; Russo, C.; Russo, D.; et al. The prognostic role of the pre-treatment neutrophil to lymphocyte ratio (NLR) and tumor depth of invasion (DOI) in early-stage squamous cell carcinomas of the oral tongue. *Oral Maxillofac. Surg.* **2022**, *26*, 21–32. [CrossRef]
49. Woodley, N.; Rogers, A.D.G.; Turnbull, K.; Slim, M.A.M.; Ton, T.; Montgomery, J.; Douglas, C. Prognostic scores in laryngeal cancer. *Eur. Arch. Oto-Rhino-Laryngol.* **2022**, *279*, 3705–3715. [CrossRef]
50. Kotha, N.V.; Voora, R.S.; Qian, A.S.; Kumar, A.; Qiao, E.M.; Stewart, T.F.; Rose, B.S.; Orosco, R.K. Prognostic Utility of Pretreatment Neutrophil-Lymphocyte Ratio in Advanced Larynx Cancer. *Biomark. Insights* **2021**, *16*, 11772719211049848. [CrossRef]
51. Takenaka, Y.; Oya, R.; Takemoto, N.; Inohara, H. Neutrophil-to-lymphocyte ratio as a prognostic marker for head and neck squamous cell carcinoma treated with immune checkpoint inhibitors: Meta-analysis. *Head Neck* **2022**, *44*, 1237–1245. [CrossRef]
52. Ueda, T.; Chikuie, N.; Takumida, M.; Furuie, H.; Kono, T.; Taruya, T.; Hamamoto, T.; Hattori, M.; Ishino, T.; Takeno, S. Baseline neutrophil-to-lymphocyte ratio (NLR) is associated with clinical outcome in recurrent or metastatic head and neck cancer patients treated with nivolumab. *Acta Oto-Laryngol.* **2020**, *140*, 181–187. [CrossRef]
53. Ng, S.P.; Bahig, H.; Jethanandani, A.; Sturgis, E.M.; Johnson, F.M.; Elgohari, B.; Gunn, G.B.; Ferrarotto, R.; Phan, J.; Rosenthal, D.I.; et al. Prognostic significance of pre-treatment neutrophil-to-lymphocyte ratio (NLR) in patients with oropharyngeal cancer treated with radiotherapy. *Br. J. Cancer* **2021**, *124*, 628–633. [CrossRef]
54. Yanni, A.; Buset, T.; Bouland, C.; Loeb, I.; Lechien, J.R.; Rodriguez, A.; Journe, F.; Saussez, S.; Dequanter, D. Neutrophil-to-lymphocyte ratio as a prognostic marker for head and neck cancer with lung metastasis: A retrospective study. *Eur. Arch. Oto-Rhino-Laryngol.* **2022**, *279*, 4103–4111. [CrossRef]
55. Zhang, W.; Liu, H.T. MAPK signal pathways in the regulation of cell proliferation in mammalian cells. *Cell Res.* **2002**, *12*, 9–18. [CrossRef]
56. Preston, S.P.; Doerflinger, M.; Scott, H.W.; Allison, C.C.; Horton, M.; Cooney, J.; Pellegrini, M. The role of MKK4 in T-cell development and immunity to viral infections. *Immunol. Cell Biol.* **2021**, *99*, 428–435. [CrossRef]
57. Guo, Y.J.; Pan, W.W.; Liu, S.B.; Shen, Z.F.; Xu, Y.; Hu, L.L. ERK/MAPK signalling pathway and tumorigenesis. *Exp. Ther. Med.* **2020**, *19*, 1997–2007. [CrossRef]
58. Kyriakis, J.M.; Avruch, J. Sounding the alarm: Protein kinase cascades activated by stress and inflammation. *J. Biol. Chem.* **1996**, *271*, 24313–24316. [CrossRef]
59. Lawrence, T. The nuclear factor NF-kappaB pathway in inflammation. *Cold Spring Harb. Perspect. Biol.* **2009**, *1*, a001651. [CrossRef]
60. O'Shea, J.J.; Holland, S.M.; Staudt, L.M. JAKs and STATs in immunity, immunodeficiency, and cancer. *N. Engl. J. Med.* **2013**, *368*, 161–170. [CrossRef]
61. Manzoor, Z.; Koh, Y.-S. Mitogen-activated protein kinases in inflammation. *J. Bacteriol. Virol.* **2012**, *42*, 189–195. [CrossRef]
62. Arthur, J.S.; Ley, S.C. Mitogen-activated protein kinases in innate immunity. *Nat. Rev. Immunol.* **2013**, *13*, 679–692. [CrossRef]
63. Wu, P.K.; Park, J.I. MEK1/2 Inhibitors: Molecular Activity and Resistance Mechanisms. *Semin. Oncol.* **2015**, *42*, 849–862. [CrossRef]
64. Burkhard, K.; Shapiro, P. Use of inhibitors in the study of MAP kinases. *Methods Mol. Biol.* **2010**, *661*, 107–122.

65. Wang, C.; Wang, H.; Zheng, C.; Liu, Z.; Gao, X.; Xu, F.; Niu, Y.; Zhang, L.; Xu, P. Research progress of MEK1/2 inhibitors and degraders in the treatment of cancer. *Eur. J. Med. Chem.* **2021**, *218*, 113386. [CrossRef]

66. McNamara, B.; Harold, J.; Manavella, D.; Bellone, S.; Mutlu, L.; Hartwich, T.M.P.; Zipponi, M.; Yang-Hartwich, Y.; Demirkiran, C.; Verzosa, M.S.Z.; et al. Uterine leiomyosarcomas harboring MAP2K4 gene amplification are sensitive in vivo to PLX8725, a novel MAP2K4 inhibitor. *Gynecol. Oncol.* **2023**, *172*, 65–71. [CrossRef]

67. Xue, Z.; Vis, D.J.; Bruna, A.; Sustic, T.; van Wageningen, S.; Batra, A.S.; Rueda, O.M.; Bosdriesz, E.; Caldas, C.; Wessels, L.F.A.; et al. MAP3K1 and MAP2K4 mutations are associated with sensitivity to MEK inhibitors in multiple cancer models. *Cell Res.* **2018**, *28*, 719–729. [CrossRef]

68. De Braekeleer, E.; Douet-Guilbert, N.; Rowe, D.; Bown, N.; Morel, F.; Berthou, C.; Férec, C.; De Braekeleer, M. ABL1 fusion genes in hematological malignancies: A review. *Eur. J. Haematol.* **2011**, *86*, 361–371. [CrossRef]

69. Greuber, E.K.; Smith-Pearson, P.; Wang, J.; Pendergast, A.M. Role of ABL family kinases in cancer: From leukaemia to solid tumours. *Nat. Rev. Cancer* **2013**, *13*, 559–571. [CrossRef]

70. Khatri, A.; Wang, J.; Pendergast, A.M. Multifunctional Abl kinases in health and disease. *J. Cell Sci.* **2016**, *129*, 9–16. [CrossRef]

71. Huang, T.; Zhou, F.; Yuan, X.; Yang, T.; Liang, X.; Wang, Y.; Tu, H.; Chang, J.; Nan, K.; Wei, Y. Reactive Oxygen Species Are Involved in the Development of Gastric Cancer and Gastric Cancer-Related Depression through ABL1-Mediated Inflammation Signaling Pathway. *Oxidative Med. Cell. Longev.* **2019**, *2019*, 5813985. [CrossRef]

 *biomedicines*

Review

A Comprehensive Review of Natural Products as Therapeutic or Chemopreventive Agents against Head and Neck Squamous Cell Carcinoma Cells Using Preclinical Models

Yoon Xuan Liew [1], Lee Peng Karen-Ng [1],* and Vui King Vincent-Chong [2],*

[1] Oral Cancer Research & Coordinating Centre (OCRCC), Faculty of Dentistry, University of Malaya, Kuala Lumpur 50603, Malaysia; yoonxuan1996@gmail.com

[2] Department of Oral Oncology, Roswell Park Comprehensive Cancer Center, Buffalo, NY 14263, USA

* Correspondence: karennlp@um.edu.my (L.P.K.-N.); vincentvuiking.chong@roswellpark.org or vincent_sean@hotmail.com (V.K.V.-C.)

Abstract: Head and neck squamous cell carcinoma (HNSCC) is a type of cancer that arises from the epithelium lining of the oral cavity, hypopharynx, oropharynx, and larynx. Despite the advancement of current treatments, including surgery, chemotherapy, and radiotherapy, the overall survival rate of patients afflicted with HNSCC remains poor. The reasons for these poor outcomes are due to late diagnoses and patient-acquired resistance to treatment. Natural products have been extensively explored as a safer and more acceptable alternative therapy to the current treatments, with numerous studies displaying their potential against HNSCC. This review highlights preclinical studies in the past 5 years involving natural products against HNSCC and explores the signaling pathways altered by these products. This review also addresses challenges and future directions of natural products as chemotherapeutic and chemoprevention agents against HNSCC.

Keywords: head and neck squamous cell carcinoma; natural products; phytochemicals; chemotherapeutics; chemoprevention

Citation: Liew, Y.X.; Karen-Ng, L.P.; Vincent-Chong, V.K. A Comprehensive Review of Natural Products as Therapeutic or Chemopreventive Agents against Head and Neck Squamous Cell Carcinoma Cells Using Preclinical Models. *Biomedicines* **2023**, *11*, 2359. https://doi.org/10.3390/biomedicines11092359

Academic Editor: Giovanni Pallio

Received: 16 June 2023
Revised: 9 August 2023
Accepted: 18 August 2023
Published: 23 August 2023

1. Introduction

Head and neck cancer (HNC) represents cancers occurring in the head and neck region, which includes the lip and oral cavity, nasal cavity, larynx, and pharynx [1]. More than 90% of HNCs are head and neck squamous cell carcinomas (HNSCCs), which account for approximately 300,000 deaths and 500,000 new cases worldwide annually [2]. In the United States, HNSCC has been diagnosed as one of the top 10 leading cancers in men in 2022 [2–4]. The low survival rate of HNSCC has been proposed to be associated with cancer recurrence, distant metastases, the progression of second primary cancers, and resistance to chemo/radiotherapy [5,6], making it a public health issue that compromises patients' quality of life.

Tobacco smoking, excessive alcohol drinking, betel quid chewing, and high-risk human papillomavirus (HPV) have been documented as risk factors for HNSCC [7–10]. Clinical intervention of HNSCC often takes place at advanced stages of the disease due to late diagnoses, especially among individuals with a lower socioeconomic background [11,12]. The most common clinical interventions for HNSCC are surgery, radiotherapy, chemotherapy, or combined therapy, which causes numerous side effects during the treatment of this disease [11]. However, even with a successful clinical intervention, approximately 30% of these patients treated at advanced stages of the disease develop recurrent locoregional or second primary cancers, with the onset of chemo- or radio-resistance [13–15]. Notably, these clinical interventions are only effective on a limited subgroup of HNSCC patients and often result in additional morbidities.

Cetuximab, an epidermal growth factor receptor (EGFR)-targeting monoclonal antibody, was the first molecular-targeted drug approved by the U.S. Food and Drug Administration (FDA) as a chemotherapeutic agent for HNSCC in 2006 [5,16]. A retrospective study in Japan reported a total effective rate of 57.1%, a median progression-free survival (PFS) of 5.5 months, and an overall survival (OS) of 8.0 months with cetuximab in locally advanced HNSCC, while a total effective rate of 60.0%, a PFS of 3.8 months, and an OS of 5.8 months were reported in distantly metastatic HNSCC [17]. However, the chemoresistance ability of certain mutated cancer cell types, such as EGFRvIII, has shown resistance toward cetuximab [18]. In recent years, both pembrolizumab and nivolumab, which act as anti-programmed cell death receptor 1 (PD-1) immunotherapy drugs, were approved by the FDA for recurrent or metastatic HNSCC treatment [19,20]. However, most patient previously exposed to anti-PD-1 monoclonal antibody develop acquired resistance to immunotherapeutic drugs, making it difficult to treat recurrent or metastatic cancers [21]. There is certainly an urgent need for alternative therapeutic agents to overcome the acquired resistance of HNSCC to the standard of care. Because of the challenges faced using immunotherapies, there is a great need for therapeutic agents that can work effectively as chemoprevention or to enhance the effectiveness of chemotherapeutic agents when used in combination to kill cancer cells.

Natural products are compounds naturally found in natural resources such as plants, which possess biological activities [22]. In recent years, natural products have been widely reported for their chemotherapeutic and chemoprevention properties against HNSCC due to their low cytotoxicity, efficacy against cancers, availability, and low cost [12,22]. According to U.S. National Cancer Institute, chemoprevention is defined as the use of certain drugs or other substances to help lower a person's risk of developing cancer or keep it from coming back [23]. The beneficial properties of natural products are consistent and could overcome the challenges with current treatment such as acquired chemoresistance, cytotoxicity against normal cells, and expensive therapy. Numerous studies have been carried out with various natural products on preclinical models of HNSCC, including various HNSCC cell lines and xenograft or carcinogen-induced tumor animal models (Figure 1). For example, psorachromene, a flavonoid found in Psoralea corylifolia, which has been used in traditional Chinese medicine (TCM) and Ayurveda, had shown therapeutic effects against HNSCC via regulation of the EGFR signaling pathways and other carcinogenesis-related signaling pathway, making it a strong candidate to act as a chemotherapeutic agent [24]. Other natural products, such as calcitriol, have been reported to possess chemoprevention properties against carcinogen-induced HNSCC in animal models [25]. Therefore, natural products have shown potential as candidates for further exploration as adjuvant, neoadjuvant chemotherapy, and chemoprevention agents in HNSCC. However, we acknowledge that, in many cases, these compounds are studied only at the beginning state in animal models. More studies on safety and efficacy are needed to improve the therapeutic potential in patients, which include clinical trials.

The activation of complex signaling pathways, such as phosphoinositide 3-kinase/protein kinase B/mammalian target of rapamycin (PI3K/Akt/mTOR) and mitogen-activated protein kinase/extracellular signal-regulated protein kinase (MAPK/ERK), is important for tumor development, cell survival, and angiogenesis, while epithelial to mesenchymal transition (EMT) signaling leads to tumor invasion and migration [26,27]. Thus, the inhibition of these signaling pathways could lead to cell death, inhibiting tumor development and metastasis [27]. Notwithstanding the extensive knowledge at the molecular level of tumor-associated signaling pathways, the chemotherapeutic agents targeting the oncogenic pathways are limited [28]. To date, Crooker et al. and Rahman et al. have described the use of natural products as chemoprevention agents against HNSCC [29,30]. Several natural products, such as vitamin A, green tea extract, and curcumin, which have shown promising results in preclinical studies, were reported to show toxicity with limited bioavailability in clinical studies [31–34]. Vitamin A has been shown by Shin et al. [31] and Papadimitrakopoulou et al. [33] to induce toxicity via oral mucosa and lip inflammation, conjunctivitis, skin reactions, fatigue, and joint and muscle pain. On the other hand, poor oral absorption of green tea extract and curcumin has been observed, which limits the bioavailability of both natural products [33,34]. Other emerging chemotherapeutic phytochemicals or herbal derivatives against HNC have recently been described by Aggarwal et al. [12]. However, several new phytochemicals against HNSCC, such as actein, calcitriol, and psorachromene, were not fully addressed. Therefore, in this review, the comprehensive mechanisms of various natural products showing significant preclinical results in HNSCC models for the past 5 years are discussed.

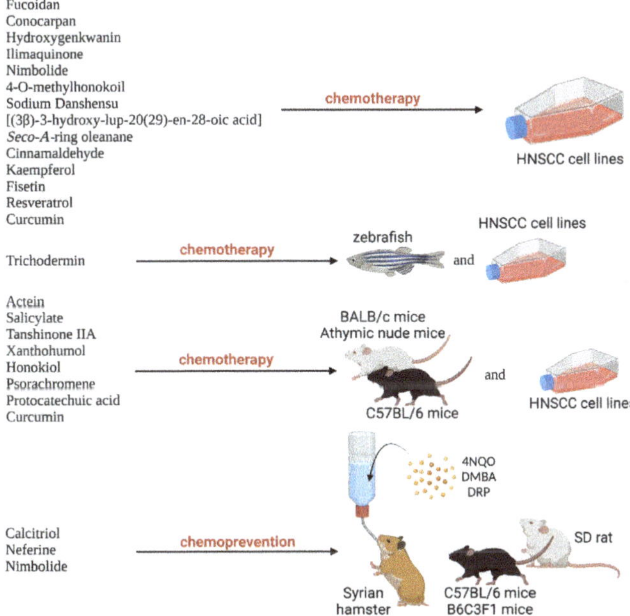

Figure 1. Natural products involved in preclinical trials involving HNSCC cell lines and animal models.

2. Chemotherapeutic Properties of Natural Products on Essential Pathways for HNSCC

2.1. PI3K/Akt/mTOR Pathway

It was previously reported that the activation of the PI3K/Akt/mTOR pathway has been observed in approximately 90% of HNSCC cases, making it a prominent target for treatments of HNSCC [35,36]. The activation of PI3K/Akt/mTOR signaling also plays an important role in HNSCC chemotherapy and radiotherapy resistances, as the inhibition of the signaling pathway has shown positive effects on tumor proliferation and radiotherapy sensitization in preclinical studies [37,38]. The PI3K/Akt/mTOR pathway is activated when a ligand-like growth factor binds with the receptor tyrosine kinase (RTK), leading to the activation of PI3K, which, in turn, partially activates Akt [39]. Then, mTORC2 is required to completely activate Akt via phosphorylation, leading to the activation of multiple proteins involved in cell proliferation and motility [40]. Inhibition of Akt formation will therefore limit the expression of oncoprotein. Various natural products, namely actein, salicylate, tanshinone IIA, xanthohumol, fucoidan, honokiol, ilimaquinone, nimbolide, and cinnamaldehyde, have led to the downregulation of Akt expression and phosphorylation of Akt (p-Akt) using HNSCC preclinical models [41–49]. The downregulation of mTOR expression in HNSCC preclinical models has been shown by salicylate, honokiol, cinnamaldehyde, and *Seco-A*-ring oleanane, with PI3K also limiting the phosphorylation of Akt [42,45,48,50]. The expression of glycogen synthase kinase-3β (GSK-3β) plays an important role in regulating growth, cell cycle progression, apoptosis, and cancer cell invasion in HNSCC [51,52]. The inactivation of GSK-3β via phosphorylation has previously shown significant inhibition in cancer cell growth and migration in HNSCC [51]. Preclinical studies with nimbolide have also shown the upregulation of phosphorylated GSK-3β, inactivating GSK-3β and thus inhibiting cell proliferation [47]. The expression of c-Myc has been associated with the upregulation of Akt-related pathways, leading to cancer cell proliferation and tumorigenesis [53]. The downregulation of c-Myc using HNSCC preclinical models has been observed with tanshinone IIA, leading to the inhibition of tumorigenesis [43].

2.2. MAPK/ERK Pathway

Activation of the MAPK/ERK signaling pathway, which includes signaling molecules such as interleukin-8 (IL-8), vascular endothelial growth factor (VEGF), mitogen-activated extracellular protein kinase (MEK), and extracellular signal-regulated protein kinase (ERK), is strongly associated with the expression of oncoproteins leading to cell proliferation and angiogenesis [54]. Honokiol, sodium Danshensu, cinnamaldehyde, and protocatechuic acid have been reported to inhibit MAPK/ERK signaling in HNSCC preclinical models, thus inhibiting cell proliferation [45,48,55,56]. Protocatechuic acid inhibits MAPK/ERK signaling via activating the c-Jun N-terminal kinase/p38 (JNK/p38) signaling pathway [56]. However, the expression of JNK plays a dual role in HNSCC, both tumor-suppressive and -progressive, due to the complex crosstalk between multiple signaling molecules and pathways [57]. Therefore, understanding of the role of JNK and p38 in HNSCC is required to ensure relevant clinical research in future. The overexpression of matrix metalloproteinases (MMPs), such as MMP-2 and MMP-9, has been associated with the progression and metastasis of HNSCC [58]. Fucoidan, sodium Danshensu, [(3β)-3-hydroxy-lup-20(29)-en-28-oic acid], and Seco-A-ring oleanane have been reported to downregulate MMP-2 [49,50,55]. Meanwhile, fucoidan and trichodermin downregulate MMP-9 in HNSCC preclinical models [49,56], inhibiting cancer cell metastasis. The expression of VEGF plays an important role in blood vessel formation, known as angiogenesis, which, in turn, promotes cancer cell growth due to the supply of nutrients via newly formed blood vessels [59,60]. It was also noticed that VEGF could be upregulated by human growth factor (HGF) through the PI3K/Akt/mTOR and MAPK signaling pathways [61]. Preclinical studies of HNSCC have indicated that cinnamaldehyde, [(3β)-3-hydroxy-lup-20(29)-en-28-oic acid], and Seco-A-ring oleanane could downregulate VEGF, thus inhibiting angiogenesis [48,50].

2.3. NF-κB and STAT3 Transcription Factors

Nuclear factor-kappa B (NF-κB), a protein complex consisting of five transcription factors, namely RelA, RelB, c-Rel, NF-κB1, and NF-κB2, has been reported to play an important role in cell proliferation and survival in HNSCC [62,63]. Preclinical studies using HNSCC models have demonstrated the inhibition of NF-κB, leading to apoptosis with neferine, trichodermin, cinnamaldehyde, and *Seco-A*-ring oleanane [48,50,64,65].

The expression of the signal transducer and activator of transcription-3 (STAT3), a transcription factor from the STAT family, plays an important role in cell proliferation, survival, and metastasis in HNSCC [66–68]. Phosphorylation of STAT3 (p-STAT3) has been proposed as a crucial reaction for the complete activation of STAT3, which is regulated by p38, EGFR, and janus kinases (JAKs) [69,70]. The activation of STAT3 would also lead to various STAT3-dependent pathways, such as the interleukin-8/STAT3 (IL-8/STAT3) and EGFR/STAT3/SRY-box transcription factor 2 (EGFR/STAT3/SOX2) pathways, which have been shown to play important roles in cancer stemness [71,72]. Preclinical studies in HNSCC models have indicated the downregulation of p-STAT3 with trichodermin and *Seco-A*-ring oleanane [50,65]. Meanwhile, psorachromene caused the downregulation of EGFR, leading to apoptosis and inhibiting metastasis [24].

Actein, a biologically active compound found in the rhizome of *Cinicifuga foetida*, has been discovered to inhibit forkhead box O1 (FoxO1) by upregulating the phosphorylation of FoxO1 (p-FoxO1), leading to the inhibition of cell proliferation via the Akt/FoxO1 signaling pathway [41]. Knockdown of FoxO1 would reverse the antiproliferation properties of actein, indicating that the Akt/FoxO1 pathway plays an important role in actein-induced effects toward HNSCC [41]. FoxO1 is one of many transcription factors among the FoxO family which regulates various biological activities in HNSCC, such as cancer cell invasion and proliferation [73].

Activated p53 via phosphorylation acts as a tumor suppressive molecule, which has been associated with induced apoptosis in HNSCC [74,75]. About half of HNSCC cases have shown a loss-of-function p53 gene mutation [76], making p53 an interesting target in improving the efficiency of HNSCC therapy through the reactivation of p53. In HPV-positive HNSCC, p53 was downregulated by the E6 oncoprotein, leading to tumorigenesis [76]. Notably, neferine and ilimaquinone have been shown to upregulate or activate p53 in preclinical HNSCC studies, leading to apoptosis [46,64].

2.4. LC3-Dependent Autophagy

Autophagy is a major process involving protein degradation followed by the turnover of cellular components, which helps maintain the intercellular homeostasis [77]. Light chain 3 (LC3) has been strongly associated with the regulation of autophagosome, a double-membrane vesicle formed during autophagy [77,78]. During the formation of an autophagosome, cytosolic LC3 (LC3-I) is converted into the activated form of intra-autophagosomal LC3 (LC3-II) via a uibiquitylation-like reaction catalyzed by Atg7 and Atg3, by conjugating with phosphatidylethanolamine [77–79]. Therefore, the increased conversion of LC3 I to LC3-II indicates higher levels of autophagosome and thus the activation of autophagy, leading to tumor suppression [80]. The accumulation of p62 was also found to be associated with the induction of autophagy in HNSCC [81]. Honokiol, ilimaquinone, and nimbolide have been reported to increase LC3-II/LC3-I ratio in HNSCC preclinical models [42,45,47].

Paradoxically, autophagy may potentially lead to tumor survival when certain conditions are met [82–85]. Hypoxia, a condition where the oxygen level is below the physiological level, is a common feature in tumor progression when the oxygen supply does not meet the demand due to the exponential growth of tumor [83,86]. Hypoxia-induced autophagy in tumors has been observed to induce tumor survival in vitro [83]. Meanwhile, the knockdown of essential autophagy proteins with a xenograft model was found to promote tumor suppression [84]. The turnover of cellular components by autophagy may essentially help with hypoxia and nutrient stress faced by exponentially growing tumors, thus enhancing tumor survival [83,84].

2.5. Bcl-2/Bax Signaling

Bcl-2 and Bax are proteins found in the mitochondrial membrane, nuclear envelop, and endoplasmic reticulum [87,88]. Bcl-2 inhibits the release of cytochrome c, which impedes tointrinsic apoptosis, while Bax reverses the reaction, leading to the induction of intrinsic apoptosis in HNSCC [89]. Cytochrome c, when released followed by a cascade of reaction involving apoptotic protease activating factor 1 (Apaf-1), Caspase-9, Caspase-7, and Caspase-3, causes apoptosis [89,90]. A preclinical study with fisetin showed that the upregulation of cytochrome c leads to apoptosis [91]. Preclinical studies in HNSCC with actein, xanthohumol, fucoidan, ilimaquinone, nimbolide, 4-O-methylhonokiol, cinnamaldehyde, kaempferol, and fisetin have indicated higher a Bax/Bcl-2 ratio (upregulating Bax and/or downregulating Bcl-2) on the mitochondrial membrane, supporting the induction of intrinsic apoptosis of HNSCC [41,44,46–49,91,92]. The activation of Caspase-8 has been associated with the activation of Caspase-7 and Caspase-3 followed by apoptosis induction, while Survivin acts as an anti-apoptotic protein [75,93,94]. In preclinical studies of HNSCC, Ilimaquinone was reported to upregulate Caspase-8, while the downregulation of Survivin has been observed with actein, xanthohumol, ilimaquinone, and seco-A-ring oleanane [41,44,46,50].

2.6. Cell Cycle Arrest by Cyclin and the CDK Signaling Pathway

The cell cycle acts as a fundamental process for cancer progression in HNSCC, including cell proliferation and differentiation [95]. It is well established that cyclins and cyclin-dependent kinases (Cdks) play an important role in the regulation and transition of cell cycles [96–98]. G0/G1 transitioning is strongly dependent on cyclin-D/Cdk4/6 complexes; S phase entry is dependent on the cyclin E/Cdk2 complex; S/G2 transitioning is dependent on the cyclin A/Cdk2 complex, followed by mitotic phase entry, which is dependent on the cyclin A/Cdk1 complex, and finally, M/G0 transitioning, which is dependent on the cyclin B/Cdk1 complex [99–103]. The inhibition of the cell cycle via the formation of cyclin-Cdk complexes would lead to cell cycle arrest and, eventually, cell death in HNSCC [104]. Anti-mitogenic signals such as p16, p21, and p53 could inhibit the cell cycle transition effectively by inhibiting the formation of cyclin-Cdk complexes, thus acting as an important target for cell cycle arrest induction [99,105,106]. Xanthohumol, neferine, fucoidan, hydroxygenkwanin, ilimaquinone, honokiol, trichodermin, cinnamaldehyde, resveratrol, and curcumin have been discovered to induce cell cycle arrest via cyclin-dependent signaling with HNSCC preclinical models [44–46,48,49,56,64,107,108].

2.7. Potential Natural Products as Therapeutic Agent for HNSCC

Actein, salicylate, honokiol, trichodermin, psorachromene, protocatechuic acid, fucoidan, hydroxygenkwanin, nimbolide, 4-O-methylhonokiol, [(3β)-3-hydroxy-lup-20(29)-en-28-oic acid], seco-A-ring oleanane, cinnamaldehyde, kaempferol, resveratrol, and curcumin induced cell cycle arrest via the regulation of cyclins and Cdks, which play an important role in coordinating the cell cycle progression [24,41,42,45,47–50,56,65,91,92,107,108]. The expressions of p21, p27, p16, and p53, acting as cyclin-Cdk complex inhibitors, have also been upregulated by actein, hydroxygenkwanin, honokiol, ilimaquinone, tanshinone IIA, resveratrol, and curcumin, which leads to induced cell cycle arrest [41,43,45,46,107,108].

Cell migration and angiogenesis have been widely studied and strongly associated with regulations of MMPs. Fucoidan, trichodermin, [(3β)-3-hydroxy-lup-20(29)-en-28-oic acid], seco-A-ring oleanane, and sodium danshensu have shown to downregulate MMPs, which inhibit cell migration and angiogenesis in HNSCC [49,50,55,65]. The VEGF and VEGF receptors play an important role in coordinating angiogenesis, while ilimaquinone, cinnamaldehyde, [(3β)-3-hydroxy-lup-20(29)-en-28-oic acid], and seco-A-ring oleanane have shown inhibition of the VEGF signaling in vitro [46,48,50].

The PI3K/Akt/mTOR and MAPK/ERK signaling pathways have been well studied and have shown a strong association with the cell proliferation and survival of HNSCC. In vitro studies with actein, salicylate, tanshinone IIA, fucoidan, honokiol, xanthohumol, ilimaquinone, cinnamaldehyde, nimbolide, and seco-A-ring oleanane have disrupted the PI3K/Akt/mTOR signaling pathway, leading to the inhibition of cell proliferation and survival [41–50]. In contrary, in vitro studies with honokiol, sodium danshensu, protocatechuic acid, and cinnamaldehyde have shown inhibition of the MAPK/ERK signaling pathway, leading to the inhibition of cell proliferation and survival [45,48,55,56].

Cell apoptosis via the mitochondrial pathway is regulated by caspases, where the proteases are activated by the Bax and Bcl-2 proteins present in the mitochondrial membrane. In vitro studies have shown that the downregulation of Bcl-2 and upregulation of Bax by fucoidan, ilimaquinone, nimbolide, 4-O-methylhonokiol, actein, and cinnamaldehyde leads to a cascade reaction of apoptosis induction [41,46–49,92]. The upregulation of caspases leading to apoptosis have also been shown by in vitro studies with hydroxygenkwanin, ilimaquinone, nimbolide, psorachromene, actein, conocarpan, trichodermin, protocatechuic acid, kaempferol, and fisetin [24,41,46,47,56,65,91,107,109]. Table 1 summarizes in vitro preclinical studies using natural products, while Table 2 summarizes in vitro and in vivo preclinical studies of natural products.

Table 1. Chemotherapeutic findings from preclinical studies with natural products involving in vitro HNSCC cell lines.

No	Author	PMID	Natural Product	Preclinical Model (Cell Lines)	Test & Dosage	Findings	Pathway Related
1	Zhang et al. [49]	31495936	Fucoidan	In vitro (SCC15, SCC25)	• Cell Proliferation; MTT assay (IC_{50} value; 2 µM, 5 µM) • Cell Cycle Arrest; flow cytometry (dosage not available) • Colony Formation; clonogenic assay (dosage not available) • Apoptosis; Annexin V assay (dosage not available) • Cell Migration and Invasion; transwell assay, wound healing assay (dosage not available)	• Reduced cell viability and inducing cell arrest (G_2 Phase) • Inducing intrinsic apoptosis via upregulation of Bcl-2 associated X protein (Bax) expression • Downregulating phosphorylation of protein kinase B (p-Akt) and cyclin-dependent kinase 1 (Cdk1) • Inhibiting cell migration and invasion via downregulation of metalloproteinases (MMP-9) and MMP-2 expression • Upregulate circRNA filament A (circFLNA) expression, mediating signaling molecules	Akt Bax/Bcl-2
2	Fonseca et al. [109]	33078660	Conocarpan	In vitro (SCC9, SCC4, SCC25)	• Cell Viability; MTT assay (1.2–300 µM) • Apoptosis; cell morphology observation	• Reduced cell viability and promote cell apoptosis • Activating Caspase-3 expression and inducing pyknotic nuclei	Caspase
3	Huang et al. [107]	31620368	Hydroxygenkwanin	In vitro (SAS & OECM1)	• Cell Proliferation; MTT assay (25, 50 & 75 µM) • Cell Cycle Arrest; flow cytometry (25, 50 & 75 µM) • Colony formation; clonogenic assay (25, 50 & 75 µM) • Cell Migration and Invasion; wound healing assay and invasion assay (25 & 50 µM)	• Reduced cell growth and colony formation • Activated p21 and inhibited Cdk2 expression, inducing cell cycle arrest • Upregulating poly (ADP-ribose) polymerase (PARP) cleavage and phosphorylated X-linked inhibitor of apoptosis protein (p-H2AX) • Induce cell apoptosis via intrinsic pathway involving Caspase-9 • Inhibition of cell invasion and migration via downregulation of Vimentin	Caspase

Table 1. *Cont.*

No	Author	PMID	Natural Product	Preclinical Model (Cell Lines)	Test & Dosage	Findings	Pathway Related
4	Lin et al. [46]	32825464	Ilimaquinone	In vitro (SCC4, SCC2095)	• Cell Viability; MTT assay (IC$_{50}$ value; 7.5 µM, 8.5 µM) • Apoptosis; Annexin V assay (5, 10, 20 & 30 µM) • Autophagy; autophagic vesicle detection (2.5, 5, 10 & 20 µM)	• Reduced cell viability • Upregulated caspase-3, caspase-8, caspase-9 and PARP cleavage leading to apoptosis • Upregulated proapoptotic protein Bax and p-p53, and downregulated anti-apoptotic protein myeloid leukemia cell differentiation protein (Mcl-1), B-cell lymphoma 2 (Bcl-2), and apoptotic inhibitor protein Survivin • Downregulated p-Akt and hypoxia-inducible factor 1-alpha (H1F-1α), mediating cell migration • Upregulated p-H2AX, as regulation due to increased reactive oxygen species (ROS) generation • Induced autophagy by upregulating light chain 3 (LC3-II) and autophagy related 5 (Atg5) expression	Akt Caspase Bax/Bcl-2
5	Sophia et al. [47]	30352996	Nimbolide	In vitro (SCC131, SCC4)	• Cell Viability; MTT assay (IC$_{50}$ value; 6 µM, 6.2 µM) • Cell Cycle Arrest; flow cytometry (6 µM, 6.2 µM) • Apoptosis; Annexin V, nuclear morphology, mitochondrial transmembrane potential (6 µM, 6.2 µM) • Autophagy; autophagic vesicle detection (6 µM, 6.2 µM)	• Reduced cell viability • Upregulated Bax and downregulated Bcl-2, leading to intrinsic apoptosis • Upregulated cleaved Caspase-9 and Caspase-3 expression • Induced conversion of LC3-I to LC3-II, inducing autophagy • Downregulating p-Akt and upregulation of phosphorylated glycogen synthase kinase 3 beta (p-GSK-3β), inhibiting phosphoinositide 3-kinase/protein kinase B/glycogen synthase kinase 3 beta (PI3K/Akt/GSK-3β)	PI3K/Akt/GSK-3β Caspase Bax/Bcl-2

Table 1. *Cont.*

No	Author	PMID	Natural Product	Preclinical Model (Cell Lines)	Test & Dosage	Findings	Pathway Related
6	Xiao et al. [92]	29332355	4-O-methylhonokoil	In vitro (PE/CA-PJ41)	• Cell Viability; MTT assay (IC$_{50}$ value; 2.5 μM) • Cell Cycle Arrest; flow cytometry (1, 2.5 & 5 μM)	• Induced G$_2$/M cell cycle arrest and apoptosis • Upregulated formation of intracellular ROS, leading to ROS-mediated reduction in mitochondrial membrane potential, inducing intrinsic apoptosis • Upregulated Bax and downregulated Bcl-2 expression, leading to apoptosis	Bax/Bcl-2
7	Kumar et al. [55]	33101201	Sodium Danshensu	In vitro (FaDu, CA9-22)	• Cell Viability; MTT assay (50 μM) • Cell Migration and Invasion; wound healing, migration, and invasion assay (25, 50 & 100 μM)	• Reduced cell motility, migration and invasion • Upregulation of E-cadherin and zonula occludens-1 (ZO-1) expression, and downregulation of MMP-2, Vimentin and N-cadherin expression, leading to anti-migratory and anti-invasive effect • Downregulate p38 phosphorylation leading to inhibition of mitogen-activated protein kinase (MAPK) signaling pathway and downregulation of extracellular signal-regulated protein kinase (ERK1/2)	MAPK/ERK
8	Aswathy et al. [50]	33860206	[(3β)-3-hydroxy-lup-20(29)-en-28-oic acid]	In vitro (SAS)	• Cell Proliferation; MTT assay (IC$_{50}$ value; 6 μM) • Colony Formation; colony-forming assay (10 & 15 μM) • Cell Cycle Arrest; flow cytometry (5, 10 & 20 μM) • Apoptosis; Annexin V (50 μM) • Cell Migration; cell migration assay (2.5 & 5 μM)	• Reduced colony formation and migration and induced apoptosis • Downregulated vascular endothelial growth factor (VEGF) and MMP-2 expression, via Akt/mTOR pathway	Akt/mTOR JAK/STAT3 VEGF NF-κB

Table 1. *Cont.*

No	Author	PMID	Natural Product	Preclinical Model (Cell Lines)	Test & Dosage	Findings	Pathway Related
8	Aswathy et al. [50]	33860206	*Seco-A*-ring oleanane	In vitro (SAS)	• Cell Proliferation; MTT assay (IC$_{50}$ value; 20 μM) • Colony Formation; colony-forming assay (10 & 15 μM) • Cell Cycle Arrest; flow cytometry (1, 3 & 5 μM) • Apoptosis; Annexin V (50 μM) • Cell Migration; cell migration assay (2.5 & 5 μM)	• Reduced colony formation and migration and induced apoptosis • Downregulated cyclooxygenase-2 (Cox-2), Survivin, MMP-2 and VEGF, via nuclear factor kappa light chain enhancer of activated B cells (NF-κB), mTOR and STAT3 pathway	
9	Aggarwal et al. [48]	35774603	Cinnamaldehyde	In vitro (SCC9, SCC25)	• Cell Viability; MTT assay (IC$_{50}$ value; 40 μM, 45 μM) • Colony formation; clonogenic assay (40 μM, 45 μM) • Cell Cycle Arrest; flow cytometry (40 μM, 45 μM) • Apoptosis; Annexin V (40 μM, 45 μM) • Cell Invasion; matrigel cell invasion assay (40 μM, 45 μM)	• Reduced cell viability and inhibited proliferation, migration and invasion • Induced cell cycle arrest at G$_2$/M and S-phase • Induces autophagy • Inhibited nuclear translocation of NF-κB from cytoplasm • Cinnamaldehyde shows binding affinity with MAPK-p38α and dihydrofolate reductase (DHFR) • Downregulating expression of NF-κB/p65, Cox-2, p110a, cyclin-D1, VEGF, Akt, mTOR, p-mTOR, and Bcl-2, and upregulating beclin-1 expression	PI3K/Akt/mTOR NF-κB MAPK

Table 1. *Cont.*

No	Author	PMID	Natural Product	Preclinical Model (Cell Lines)	Test & Dosage	Findings	Pathway Related
10	Kubina et al. [91]	37371038	Kaempferol & Fisetin	In vitro (SCC9, SCC25)	• Cell Proliferation; WST-1 assay (IC_{50} value kaempferol; 45.03 µM, 49.90 µM & fisetin; 38.85 µM, 62.34 µM) • Cell Cycle Arrest; flow cytometry (1/2 and 1/4 IC_{50} value) • Apoptosis; Annexin V Detection Kit (1/2 and 1/4 IC_{50} value) • Cell Migration; wound healing assay (1/2 and 1/4 IC_{50} value)	• Inhibited cell proliferation and migration • Induced apoptosis by activation of Caspase-3 and decreased potential of mitochondrial membrane • Downregulation of Bcl-2 • Fisetin upregulate cytochrome c • Kaempferol induced cell cycle arrest at S phase	Bax/Bcl-2
11	Bostan et al. [108]	32859062	Resveratrol & Curcumin	In vitro (PE/CA-PJ49)	• Cell Viability; MTT assay (IC_{50} value resveratrol; 46.8 µM & curcumin; 16.3 µM) • Cell Proliferation; cell proliferation assay (resveratrol; 40 µM & curcumin; 15 µM) • Cell Cycle Arrest; flow cytometry (resveratrol; 40 µM & curcumin; 15 µM) • Apoptosis; Annexin V Detection Kit (resveratrol; 40 µM & curcumin; 15 µM)	• Inhibited cell proliferation • Induced apoptosis • Amplifying effect of low concentration of cisplatin on inhibition of cell proliferation, and induction of apoptosis and cell cycle arrest • Upregulation of p21	Caspase

Table 2. Chemotherapeutic findings from preclinical studies with natural products involving in vitro HNSCC cell lines and in vivo xenograft models.

No	Author	PMID	Natural Product	Preclinical Model	Test & Dosage	Findings	Pathway Related
1	Zhao et al. [41]	34175854	Actein	In vitro (CAL27, SCC9)	• Cell Proliferation; CCK-8 Assay (IC_{50} value; 15 µM, 12 µM) • Cell Cycle Arrest; flow cytometry (7, 15 & 30 µM) • Apoptosis; Annexin V assay (7, 15 & 30 µM)	• Reduced cell viability and induced cell cycle arrest • Induced apoptosis via upregulated Bax and downregulated Bcl-2 • Downregulated Survivin and upregulated p21 and Bim • Downregulating Akt and upregulating forkhead box protein O1 (FoxO1), inhibiting Akt/FoxO1 pathway	Akt/FoxO1 Bax/Bcl-2
				In vivo (CAL27) C57BL/6 mice	• Mammary fat pad subcutaneous inoculation • Intragastrically administration of 10, 20, and 50 mg/kg dosage	• Impaired tumor growth	
2	Zhang et al. [42]	32267053	Salicylate	In vitro (SAS)	• Cell Proliferation; MTS Assay (2500, 5000 and 10,000 µM) • Cell Cycle Arrest; flow cytometry (5000 µM) • Tumorsphere Formation; tumorsphere formation assay (5000 µM)	• Combination with cisplatin enhanced the cytotoxicity and induced cell apoptosis • Downregulated p-Akt, mTOR, p-S6 and p-70S6, inhibiting Akt/mTOR signaling pathway • Downregulated Jagged-1, SRY-box transcription factor 9 (SOX9), yes-associated protein-1 (YAP-1), sonic hedgehog protein (Shh), aldehyde dehydrogenase (ALDH), oxtamer-binding transcription factor 4 (OCT4) expression	Akt/mTOR
				In vivo (SAS) Nude mice	• Subcutaneous inoculation • Oral administration of 3 mg/kg dosage	• Monotherapy impaired tumor growth • Combination with cisplatin enhanced tumor growth inhibition	

Table 2. *Cont.*

No	Author	PMID	Natural Product	Preclinical Model	Test & Dosage	Findings	Pathway Related
3	Li et al. [43]	32424132	Tanshinone IIA	In vitro (CAL27, SCC9, SCC15, SCC25)	• Cell Viability; MTS assay (2 & 5 µM)	• Reducing cell viability • Inducing intrinsic apoptosis via downregulation of p-Akt, c-Myc and hexokinase 2 (HK2) • Inhibiting glycolysis of SCC by downregulating expression of HK2 • Promoting FBW7 E3 ligase interaction with c-Myc, shortening half-life of c-Myc	Akt/c-Myc
				In vivo (CAL27, SCC15) Athymic nude mice	• Right flank subcutaneous inoculation • Intraperitoneal administration of 10 mg/kg dosage	• Reducing population of Ki-67 positive cells • Downregulating p-Akt, c-Myc and HK2 • Impaired tumor growth	
4	Li et al. [44]	32410646	Xanthohumol	In vitro (CAL27, SCC9, SCC15, SCC25)	• Cell Viability; MTS assay (1, 2 & 5 µM)	• Reducing cell viability and colony formation • Regulating Akt/Wee1/Cdk1 signaling pathway • Upregulating Bax on mitochondria, PARP cleavage and Caspase-3, leading to intrinsic apoptosis • Promote ubiquitination and degradation of Survivin by upregulating FBXL7 E3 protein	Akt/Wee1/Cdk1 Bax/Bcl-2 Caspase
				In vivo (CAL27, SCC25) Athymic nude mice	• Right flank subcutaneous inoculation • Intraperitoneal administration of 10 mg/kg dosage	• Delayed tumor development and impaired tumor growth • Reducing population of Ki-67 positive cells • Downregulating p-Akt and Survivin • Sensitizing radioresistance cells to radiotherapy	

Table 2. *Cont.*

No	Author	PMID	Natural Product	Preclinical Model	Test & Dosage	Findings	Pathway Related
5	Huang et al. [45]	29363886	Honokiol	In vitro (OC2, OCSL)	• Cell Viability; CCK-8 assay (IC$_{50}$ value; 35 μM, 22 μM) • Cell Cycle Arrest; flow cytometry (25 & 40 μM, 15 & 30 μM) • Apoptosis; Annexin V assay (25 μM, 15 μM)	• Reduced cell growth • Induced cell cycle arrest at G$_0$/G$_1$ phase via upregulating p21 and p27, accumulation of cyclin-E, and downregulating Cdk2, Cdk4 and cyclin-D1 • Inhibiting MAPK pathway by downregulating p-Akt and p-mTOR • Inducing autophagy via activation of LC3-II • Synergic therapeutic effect with Fluorouracil	Akt/mTOR MAPK
				In vivo (SAS) BALB/c nude mice, AnN.Cg-Foxn1nu/CrlNarl	• Right flank subcutaneous inoculation • Oral administration of 5 and 15 mg/kg dosage	• Inhibiting tumor growth and inducing apoptosis	
6	Chen et al. [65]	35785707	Trichodermin	In vitro (Ca922, HSC3)	• Cell Viability; MTT assay (IC$_{50}$ value; 9.65 μM, 11.49 μM) • Colony Formation; clonogenic assay (3 & 10 μM) • Cell Cycle Arrest; flow cytometry (3 & 10 μM) • Apoptosis; Annexin V assay; nuclear condensation observation (3 & 10 μM) • Cell Migration and Invasion; transwell assay (3 & 10 μM)	• Reduced cell viability, migration and invasive • Downregulation of MMP-9, inhibiting cell migration and invasion • Downregulation of cyclin A, cyclin D1, Cdk1/Cdk2 and Cdk4 expression, leading to G$_2$/M cell cycle arrest • Inducing apoptosis by upregulating Caspase-3 and cleaved PARP expression • Reduces mitochondrial membrane potential, basal respiration, ATP production, maximum respiration and proton leak, inducing intrinsic apoptosis • Downregulated expression of HDAC-2, phosphorylated STAT3 and NF-κB	HDAC-2 Caspase

Table 2. *Cont.*

No	Author	PMID	Natural Product	Preclinical Model	Test & Dosage	Findings	Pathway Related
6	Chen et al. [65]	35785707	Trichodermin	In vivo (HSC3) Zebrafish	• Embryo tumor transplantation • Embryo submerged in dosage of 3 and 10 μg/mL solution	• Inhibited tumor growth	
7	Wang et al. [24]	31750253	Psorachromene	In vitro (SAS, OECM1)	• Cell Viability; sulforhodamine B assay (25 & 50 μM) • Cell Cycle Arrest; flow cytometry (25, 50 & 75 μM) • Apoptosis; TUNEL assay (50 μM) • Cell Migration and Invasion; wound healing assay and invasion assay (25, 50 & 75 μM)	• Reduced cell growth and colony formation • Induced cell cycle arrest in G_2 phase • Induced apoptosis via activation of Caspase-9 and cleavage PARP • Inhibited cell migration and invasion, via downregulating EMT-promoting protein, Vimentin, Slug and EGFR signaling pathway • Synergic therapeutic effect with Cisplatin and Doxorubicin • Upregulating growth arrest specific 5 (GAS5), inhibiting cell growth and metastasis	EGFR EMT-related Caspase SLUG
				In vivo (SAS) BALB/c nude mice	• Right flank subcutaneous inoculation • Intraperitoneal administration of 100 μL dosage	• Downregulated EMT-promoting protein and EGFR signaling pathway • Reduced tumor growth	

Table 2. *Cont.*

No	Author	PMID	Natural Product	Preclinical Model	Test & Dosage	Findings	Pathway Related
8	Li et al. [56]	35904511	Protocatechuic acid	In vitro (HSC3, CAL27)	• Cell Proliferation; CCK-8 assay (250 & 500 μM) • Cell Cycle Arrest; flow cytometry (250 & 500 μM) • Apoptosis; TUNEL assay (250 & 500 μM) • Tumorsphere Formation; tumorsphere assay (250, 500 & 1000 μM)	• Induced cell death by interrupting Serpinb9 and granzyme B (Sb9-GrB) complex formation • Induced cell apoptosis via upregulated phosphorylation of c-Jun N-terminal kinase (JNK), p38 and cleaved Caspase-3 expression • Upregulated superoxide dismutase and nuclear factor erythroid 2-related factor 2 (Nrf2), reducing the ROS levels • Inhibiting cancer stemness	JNK/p38 Caspase
				In vivo (CAL27) Nude mice	• Right flank subcutaneous inoculation • Peritumoral administration of 100 μL dosage	• Inhibiting tumor growth • Reducing population of Ki-67 positive cell	
9	de Compos et al. [110]	28782139	Curcumin	In vitro (CAL27, SCC25, HACAT, NIH-3T3)	• Cell Proliferation; NF cell proliferation assay (2, 5, 10, 20, 30, 40 & 50 μM) • Cell Apoptosis; Annexin V assay (5 & 50 μM) • Cell Migration and Invasion; time-lapse analysis (2 & 5 μM) • Tumorsphere Formation; spheroid assay (10,20, 50 & 200 μM)	• Reducing cell proliferation • Reduced migratory rate and impairment on tumor cell directionality • Reducing cell-cell adhesion, leading to less homogenous spheroid	
				In vivo (HNSCC Biopsy) BALB/c nude mice	• Right flank subcutaneous inoculation • Treatment with 70 mg/kg dosage	• Inhibiting tumor growth • Induced a less aggressive histological phenotype	

3. Chemoprevention Properties of Natural Products against HNSCC Oral Carcinogenesis Mechanism

Oral carcinogenesis often involves the formation of abnormalities in the oral tissue, known as oral potentially malignant disorder (OPMD), before proceeding to oral squamous cell carcinoma (OSCC), a major type of HNSCC [111,112]. Previous studies have found the malignant transformation (MT) rate of OPMD to be 7.9%, while high-risk OPMD, such as erythroplakia, has shown an average MT rate of 33.1% [113]. To date, no preventive strategies including the use of drugs and/or surgical procedures has been considered the standard of care for OPMDs, thus indicating the need to investigate the chemoprevention properties of various natural products against oral carcinogenesis [114]. Several drugs, such as celecoxib, erlotinib, and metformin, have been investigated for their HNSCC prevention properties via clinical trials on oral premalignant lesions [115–117]. However, the use of erlotinib and celecoxib has shown no significant result in reducing the oral cancer-free survival rate while possessing higher toxicity [115,116]. The use of metformin has shown a low clinical response rate (17%) in terms of reduction in lesion size [117].

In vivo studies involving the use of genetically altered rodents or rodents treated with chemical carcinogens could lead to site-specific carcinogenesis, mimicking carcinogenesis in humans [118]. Moreover, 4-nitroquinoline 1-oxide (4NQO) acts as a tobacco-mimicking carcinogen, which has been widely used in carcinogen-induced HNSCC animal models, mainly due to the similarity in terms of genetic alteration and expression between 4NQO-induced mouse models and human oral carcinogenesis [119,120]. Other carcinogens, including 7,12-dimethylbenz(a)anthracene (DMBA) and dibenzo[a,l]pyrene (DBP), have also been widely used as HNSCC-inducing agents in animal models [64,121].

Several review studies have introduced various natural products as chemoprevention agents against HNSCC, where vitamin A, green tea extracts, and curcumin have shown promising results in preclinical and clinical trials [29,30]. However, vitamin A has shown toxicity [31,32], while green tea extracts and curcumin [33,34] have both shown limitations in bioavailability. In a randomized chemoprevention trial reported by Papadimitrakopoulou et al. [32], low-dose 13-cis retinoic acid (a derivative of vitamin A) could induce grade 1 (45%), 2 (37%), 3 (15%), and 4 (1%) toxicity, including cheilitis, conjunctivitis, and skin reactions. Similarly, a phase II chemoprevention trial by Shin et al. [31] with combinations of interferon-alpha, 13-cis retinoic acid, and alpha-tocopherol induced mild to moderate non-hematologic toxicity. One patient was reported with a severe throat infection due to beta-hemolytic streptococci which required an emergency tracheostomy. The patient was still able to complete the planned treatment after fully recovering from the infection [31]. Green tea extract was reported to induce adverse effects such as insomnia, nausea, nervousness, and headache, which is most likely due to the presence of caffeine in green tea extract [33]. The poor oral absorption of epigallocatechin-3-gallate, the most abundant polyphenol in green tea extract, was reported by Tsao et al. [33], leading to variability in plasma epigallocatechin-3-gallate concentrations. Similarly, a phase I chemoprevention trial by Cheng et al. [34] with curcumin also indicated the poor gastrointestinal absorption of curcumin, as the peak serum curcumin concentration was recorded at 1.77 μM with 8000 mg daily dosage. The poor bioavailability of both green tea extract and curcumin have introduced difficulties in dosage estimation as absorption varies among patients, which may lead to ineffective treatment. Therefore, in this review, we seek to provide a greater variety of promising natural products with chemoprevention properties using 4NQO, DMBA, or DBP-induced carcinogenesis animal models [25,64,121]. Table 3 summarizes preclinical studies with natural products involved in the chemoprevention of HNSCC investigated in the past 5 years.

Table 3. Chemoprevention findings from preclinical studies with natural products on induced carcinogenesis.

No	Author	PMID	Natural Product	Preclinical Model	Test & Dosage	Findings	Pathway Related
1	Vincent-Chong et al. [25]	30875566	Calcitriol	In vivo (4NQO-induced carcinogenesis) C57BL/6NCr mice	• 0.1 µg dosage, intraperitoneal administration (thrice weekly)	• Inhibition of 4-nitroquinoline-1-oxide (4NQO)-induced carcinogenesis • Reduced incidence of HNSCC induced by 4NQO • Increased Ki-67 positive dysplastic epithelium • Carcinogenesis influenced by stage of intervention and duration of exposure to calcitriol	
2	Wang et al. [64]	33156559	Neferine	In vivo (DMBA-induced carcinogenesis) Syrian hamster	• 15 mg/kg dosage, intragastric administration (thrice weekly)	• Increased body weight and suppression on formation of 7,12-Dimethylbenz[a]anthracene (DMBA) induced tumor development • Downregulated expression of NF-κB, mutant p53 and PCNA	NF-κB
3	Sophia et al. [47]	30352996	Nimbolide	In vivo (DMBA-induced carcinogenesis) Syrian hamster	• 100 µg/kg dosage, intragastric administration	• Inhibited PI3K/Akt/GSK-3β signaling pathway • Sensitized tumor to apoptosis	PI3K/Akt/GSK-3β

Preclinical in vivo studies involving neferine and nimbolide have indicated an inhibition in carcinogenesis and a reduction in the tumor growth of DMBA-induced HNSCC, making both possible chemoprevention agents [41]. Similarly, calcitriol successfully inhibited 4NQO-induced HNSCC carcinogenesis [25]. Interestingly, a study by Vincent-Chong et al. [25] with calcitriol on a 4NQO-induced animal model showed the influence of treatment stage intervention and duration of exposure to treatment in carcinogenesis. Understanding the pathways involved during carcinogenesis will allow effective treatment intervention in HNSCC patients.

Calcitriol, nimbolide, and neferine are being investigated in preclinical studies for chemoprevention, and these products have shown promising results by reducing or inhibiting carcinogenesis-related molecular mechanisms [25,47,64].

4. Limitation and Future Direction

As multiple signaling and crosstalk between pathways occur during carcinogenesis, with most recurrent or metastatic HNSCC failing the primary standard of care, an effective treatment for cancer may require a combined therapeutic approach such as the use of multiple signaling inhibitors combined with DNA-damaging drugs for the most efficient outcome [26]. Therefore, the synergic or antagonistic effects of natural products with standards of care such as chemotherapy (cisplatin, cetuximab, pembrolizumab and nivolumab) and radiotherapy should be analyzed using preclinical models. However, fewer than 10% of the reviewed studies reported the combination effects of natural products with the standard of care. For instance, xanthohumol, psorachromene, honokiol, calcitriol, and salicylate were shown to provide synergistic effects with standard-of-care treatment such as chemotherapy and radiotherapy using HNSCC preclinical models [24,25,42,44,45].

The major reason for the low survival rate of HNSCC is due to late diagnoses and risk factors associated with HNSCC progression, which lead to the risk of recurrent or metastatic SCC [5,12]. Chemoprevention therapy could potentially act as an important barrier to lower the risk of recurrent or metastatic SCC and the malignant transformation of OPMD; therefore, chemoprevention should be widely studied in the future. The prevention of HNSCC involving single-agent chemotherapy, such as retinoids and isotretinoin, possesses high toxicity and low efficacy, indicating the need for the development of new chemoprevention agents either as alternative or adjunctive agents for HNSCC prevention. In the current review, only 5 out of 37 studies explored the potential usage of natural products as chemoprevention therapy using 4NQO/DMBA/DBP-induced oral carcinogenesis [25,47,64,121]. Furthermore, the stage and duration of natural products' intervention on oral carcinogenesis should also be extensively explored given the [25] strong association with the progression of carcinogenesis. All six natural products (Table 3) are strongly encouraged to proceed with clinical trials for high-risk OPMD patients, as previous trials with celecoxib, erlotinib, and metformin on mild to advanced OPMD showed no significant clinical improvements [115–117].

A preclinical study reported by Dai et al. [122] using various cancer lines (melanoma, breast, colon, and liver cancer cell lines (H1299, BT549, MDA-MB-231, MDA-MB-468, SW620, MHCC97H, and B16F10)) indicated the crucial role of immunomodulatory roles of natural products in anti-cancer treatments, including the involvement of CD3+ CD8+ T lymphocytes in a co-culture system together with cancer cell lines. Similarly, Cattanaeo et al. [123] and Neal et al. [124] have proposed the use of co-culture organoid-tumor reactive T lymphocyte system to investigate the role of anti-PD-1 drugs on T lymphocyte activity. Other studies have also used similar systems to screen natural product-derived drugs and epigenetic inhibitors for the non-cytotoxic T lymphocyte immunomodulating effects [122,125]. Apart from the in vitro co-culture system proposed, Verma et al. [126] implemented an in vivo RPMOC1 synergic HNSCC animal model to investigate the effects of a non-oncological drug, calcitriol, on the immunomodulation of T lymphocytes, which was the only preclinical study involving synergic HNSCC in an in vivo model present at the time of preparing this review. The initiation of this review was prompted by a recent publication by Crooker et al. [29] which provided informaion on the use of natural products in HNSCC preclinical models in 2018. Over the past five years, there have been limited efforts to publish reviews related to natural products, with one notable publication by Aggarwal et al. [12] focusing solely on HPV-related HNSCC. To address the paucity of information, we explored the possibilities of using the latest research of natural products with in vitro and in vivo studies, listed in Tables 1–3, that has not been discussed in previous review papers, particularly targeting HNSCC, as some signaling pathways are common among different subtypes. The review also highlights various challenges in the field that hinder the clinical translation of natural products.

Among the reviewed natural products, actein, salicylate, tanshinone IIA, xanthohumol, honokiol, trichodermin, psorachromene, and protocatechuic acid have been investigated as chemotherapeutic agents with HNSCC cell lines and xenografted animal models, and these products have shown promising targeted molecular mechanisms against HNSCC, making them ideal candidates for further clinical trials for safety and efficacy analyses [24,41–45,56,65]. Furthermore, salicylate and xanthohumol have been investigated to provide synergistic effect with cisplatin and radiotherapy, respectively, making them the strongest candidates for future clinical trials for HNSCC or OPMD patients [42,44]. Finally, calcitriol, nimbolide, and neferine were the major natural products investigated in this review; these products showed promising chemoprevention properties against induced-carcinogenesis animal models. We encourage further investigation into the safety and efficacy of these products with human trials, as well as in high-risk OPMD such as erythroplakia and leukoplakia patients [25,47,64]. Nevertheless, it is crucial that the results of preclinical studies presented in this manuscript serve only as a principle for further investigations and should not be directly extrapolated into clinical practices without additional detailed evaluation. Preclinical studies, either in vitro or in vivo, provide valuable insights on the potential of natural products as therapeutic agents; however, their efficacy, toxicity profile, and applicability in human patients should be widely explored. Finally, caution must be exercised when interpreting the findings of preclinical studies and their potential implication for clinical trials.

5. Conclusions

Both chemotherapeutic and chemoprevention approaches showed promising anticancer effects through preclinical studies involving various HNSCC cell lines and animal models via various pathways (Figure 2). However, the lack of extensive molecular mechanisms of natural products and lack of combination of natural products with the current standard of care limits the use of natural products as new chemotherapeutic drugs. The only study that used calcitriol to determine the immunomodulatory effect of natural product in the context of HNSCC highlighted the effort to investigate the role of these natural products since the immune checkpoint inhibitor has been recognized as one of the standards of care for HNSCC patients. The low intrinsic toxicity of natural products in normal cells and significant therapeutic effects toward cancer cells have sparked an interest in oncology studies in recent years. The synergistic effect of natural products with standard of cares, including chemotherapy, radiotherapy, and immunotherapy, has showed promising properties as both alternative and adjunctive chemotherapeutic or chemoprevention agents in cancer treatment or prevention, respectively.

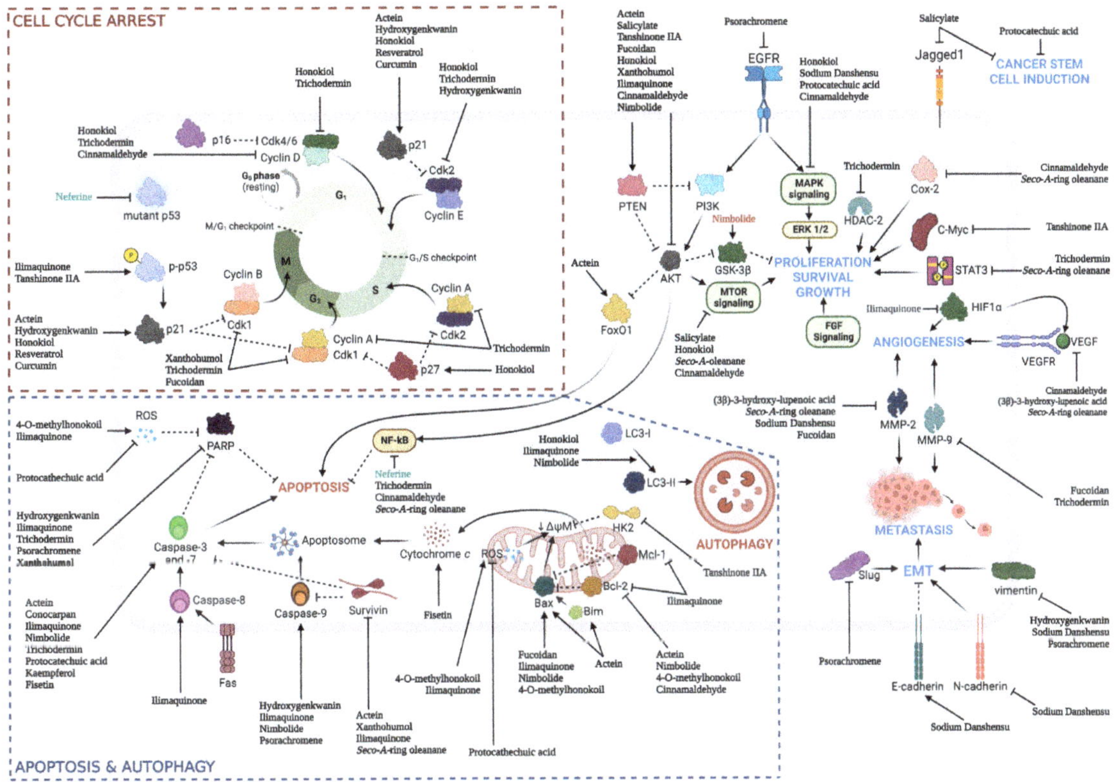

Figure 2. Various natural products and their respective molecular mechanisms against HNSCC: chemoprevention (natural product name in green), chemotherapeutic (natural product name in black).

As a conclusion, we attempted to provide a comprehensive database of natural products used for in vivo and in vitro preclinical studies involving HNSCC in the recent years, which would facilitate the identification of effective natural products that show promising chemotherapeutic and chemoprevention properties against HNSCC while possessing low toxicity. Natural products have shown to be promising molecular targets for therapy and prevention against HNSCC, making them great alternatives and adjunctive agents in cancer treatment. However, further toxicity profiles should be analyzed for natural products in future clinical trials to ensure the safety and efficacy of natural products against HNSCC.

Author Contributions: Conceptualization, V.K.V.-C.; Investigation, Y.X.L., L.P.K.-N. and V.K.V.-C.; Visualization, Y.X.L.; Writing—Original Draft Preparation, Y.X.L.; Writing—Review and Editing, L.P.K.-N. and V.K.V.-C. All authors have read and agreed to the published version of the manuscript.

Funding: This research received no external funding.

Data Availability Statement: The data that support the findings of this study are available from the corresponding author upon reasonable request.

Conflicts of Interest: The authors declare no conflict of interest.

References

1. Johnson, D.E.; Burtness, B.; Leemans, C.R.; Lui, V.W.Y.; Bauman, J.E.; Grandis, J.R. Head and Neck Squamous Cell Carcinoma. *Nat. Rev. Dis. Prim.* **2020**, *6*, 92. [CrossRef] [PubMed]
2. Sung, H.; Ferlay, J.; Siegel, R.L.; Laversanne, M.; Soerjomataram, I.; Jemal, A.; Bray, F. Global Cancer Statistics 2020: GLOBOCAN Estimates of Incidence and Mortality Worldwide for 36 Cancers in 185 Countries. *CA Cancer J. Clin.* **2021**, *71*, 209–249. [CrossRef] [PubMed]
3. Magnes, T.; Wagner, S.M.; Melchardt, T.; Weiss, L.; Rinnerthaler, G.; Huemer, F.; Kopp, M.; Gampenrieder, S.P.; Mayrbäurl, B.; Füreder, T.; et al. Postoperative Chemoradiotherapy with Cisplatin Is Superior to Radioimmunotherapy with Cetuximab and Radiotherapy Alone: Analysis of the Austrian Head and Neck Cancer Registry of the AGMT. *Wien. Klin. Wochenschr.* **2021**, *133*, 1131–1136. [CrossRef]
4. Siegel, R.L.; Miller, K.D.; Fuchs, H.E.; Jemal, A. Cancer Statistics, 2022. *CA Cancer J. Clin.* **2022**, *72*, 7–33. [CrossRef] [PubMed]
5. Bonner, J.A.; Harari, P.M.; Giralt, J.; Cohen, R.B.; Jones, C.U.; Sur, R.K.; Raben, D.; Baselga, J.; Spencer, S.A.; Zhu, J.; et al. Radiotherapy plus Cetuximab for Locoregionally Advanced Head and Neck Cancer: 5-Year Survival Data from a Phase 3 Randomised Trial, and Relation between Cetuximab-Induced Rash and Survival. *Lancet Oncol.* **2010**, *11*, 21–28. [CrossRef]
6. Ferris, R.L.; Blumenschein, G.; Fayette, J.; Guigay, J.; Colevas, A.D.; Licitra, L.; Harrington, K.; Kasper, S.; Vokes, E.E.; Even, C.; et al. Nivolumab for Recurrent Squamous-Cell Carcinoma of the Head and Neck. *N. Engl. J. Med.* **2016**, *375*, 1856–1867. [CrossRef]
7. Talamini, R.; Bosetti, C.; La Vecchia, C.; Dal Maso, L.; Levi, F.; Bidoli, E.; Negri, E.; Pasche, C.; Vaccarella, S.; Barzan, L.; et al. Combined Effect of Tobacco and Alcohol on Laryngeal Cancer Risk: A Case-Control Study. *Cancer Causes Control* **2002**, *13*, 957–964. [CrossRef]
8. Lewin, F.; Norell, S.E.; Johansson, H.; Gustavsson, P.; Wennerberg, J.; Biörklund, A.; Rutqvist, L.E. Smoking Tobacco, Oral Snuff, and Alcohol in the Etiology of Squamous Cell Carcinoma of the Head and Neck A Population-Based Case-Referent Study in Sweden. *Cancer* **1998**, *82*, 1367–1375. [CrossRef]
9. Hansson, B.G.; Rosenquist, K.; Antonsson, A.; Wennerberg, J.; Schildt, E.B.; Bladström, A.; Andersson, G. Strong Association between Infection with Human Papillomavirus and Oral and Oropharyngeal Squamous Cell Carcinoma: A Population-Based Case-Control Study in Southern Sweden. *Acta Otolaryngol.* **2005**, *125*, 1337–1344. [CrossRef]
10. Van Wyk, C.W.; Stander, I.; Padayachee, A.; Grobler-Rabie, A.F. The Areca Nut Chewing Habit and Oral Squamous Cell Carcinoma in South African Indians. A Retrospective Study. *S. Afr. Med. J.* **1993**, *83*, 425–429.
11. Haddad, R.I.; Shin, D.M. Recent Advances in Head and Neck Cancer Reconstruction. *N. Engl. J. Med.* **2008**, *359*, 1143–1154. [CrossRef] [PubMed]
12. Aggarwal, N.; Yadav, J.; Chhakara, S.; Janjua, D.; Tripathi, T.; Chaudhary, A.; Chhokar, A.; Thakur, K.; Singh, T.; Bharti, A.C. Phytochemicals as Potential Chemopreventive and Chemotherapeutic Agents for Emerging Human Papillomavirus–Driven Head and Neck Cancer: Current Evidence and Future Prospects. *Front. Pharmacol.* **2021**, *12*, 699044. [CrossRef] [PubMed]
13. Posner, M.R.; Hershock, D.M.; Blajman, C.R.; Mickiewicz, E.; Winquist, E.; Gorbounova, V.; Tjulandin, S.; Shin, D.M.; Cullen, K.; Ervin, T.J.; et al. Cisplatin and Fluorouracil Alone or with Docetaxel in Head and Neck Cancer Marshall. *N. Engl. J. Med.* **2007**, *357*, 1705–1715. [CrossRef] [PubMed]
14. Hashim, D.; Genden, E.; Posner, M.; Hashibe, M.; Boffetta, P. Head and Neck Cancer Prevention: From Primary Prevention to Impact of Clinicians on Reducing Burden. *Ann. Oncol.* **2019**, *30*, 744–756. [CrossRef]
15. Brockstein, B.; Haraf, D.J.; Rademaker, A.W.; Kies, M.S.; Stenson, K.M.; Rosen, F.; Mittal, B.B.; Pelzer, H.; Fung, B.B.; Witt, M.E.; et al. Patterns of Failure, Prognostic Factors and Survival in Locoregionally Advanced Head and Neck Cancer Treated with Concomitant Chemoradiotherapy: A 9-Year, 337-Patient, Multi-Institutional Experience. *Ann. Oncol.* **2004**, *15*, 1179–1186. [CrossRef]
16. Bonner, J.A.; Harari, P.M.; Giralt, J.; Azarnia, N.; Shin, D.M.; Cohen, R.B.; Jones, C.U.; Sur, R.; Raben, D.; Jassem, J.; et al. Results of Radiotherapy plus Cetuximab for Squamous Cell Carcinoma of the Head and Neck. *N. Engl. J. Med.* **2006**, *354*, 567–578. [CrossRef]
17. Naruse, T.; Yanamoto, S.; Matsushita, Y.; Sakamoto, Y.; Morishita, K.; Ohba, S.; Shiraishi, T.; Yamada, S.-I.; Asahina, I.; Umeda, M. Cetuximab for the Treatment of Locally Advanced and Recurrent/Metastatic Oral Cancer: An Investigation of Distant Metastasis. *Mol. Clin. Oncol.* **2016**, *5*, 246–252. [CrossRef]
18. Sok, J.C.; Coppelli, F.M.; Thomas, S.M.; Lango, M.N.; Xi, S.; Hunt, J.L.; Freilino, M.L.; Graner, M.W.; Wikstrand, C.J.; Bigner, D.D.; et al. Mutant Epidermal Growth Factor Receptor (EGFRvIII) Contributes to Head and Neck Cancer Growth and Resistance to EGFR Targeting. *Clin. Cancer Res.* **2006**, *12*, 5064–5073. [CrossRef]
19. US Food and Drug Administration. FDA Approves Pembrolizumab for First-Line Treatment of Head and Neck Squamous Cell Carcinoma. Available online: https://www.fda.gov/drugs/resources-information-approved-drugs/fda-approves-pembrolizumab-first-line-treatment-head-and-neck-squamous-cell-carcinoma (accessed on 1 February 2023).
20. US Food and Drug Administration. Nivolumab for SCCHN. Available online: https://www.fda.gov/drugs/resources-information-approved-drugs/nivolumab-scchn (accessed on 1 February 2023).
21. dos Santos, L.V.; Abrahão, C.M.; William, W.N. Overcoming Resistance to Immune Checkpoint Inhibitors in Head and Neck Squamous Cell Carcinomas. *Front. Oncol.* **2021**, *11*, 596290. [CrossRef]
22. Newman, D.J.; Cragg, G.M. Natural Products as Sources of New Drugs over the Nearly Four Decades from 01/1981 to 09/2019. *J. Nat. Prod.* **2020**, *83*, 770–803. [CrossRef]

23. National Cancer Institute. Dictionary of Cancer Terms. Available online: https://www.cancer.gov/publications/dictionaries/cancer-terms (accessed on 26 September 2022).
24. Wang, T.H.; Leu, Y.L.; Chen, C.C.; Shieh, T.M.; Lian, J.H.; Chen, C.Y. Psorachromene Suppresses Oral Squamous Cell Carcinoma Progression by Inhibiting Long Non-Coding RNA GAS5 Mediated Epithelial-Mesenchymal Transition. *Front. Oncol.* **2019**, *9*, 1168. [CrossRef]
25. Vincent-Chong, V.K.; DeJong, H.; Attwood, K.; Hershberger, P.A.; Seshadri, M. Preclinical Prevention Trial of Calcitriol: Impact of Stage of Intervention and Duration of Treatment on Oral Carcinogenesis. *Neoplasia* **2019**, *21*, 376–388. [CrossRef] [PubMed]
26. Hashem, S.; Ali, T.A.; Akhtar, S.; Nisar, S.; Sageena, G.; Ali, S.; Al-Mannai, S.; Therachiyil, L.; Mir, R.; Elfaki, I.; et al. Targeting Cancer Signaling Pathways by Natural Products: Exploring Promising Anti-Cancer Agents. *Biomed. Pharmacother.* **2022**, *150*, 113054. [CrossRef]
27. Sever, R.; Brugge, J.S. Signal Transduction in Cancer. *Cold Spring Harb. Perspect. Med.* **2015**, *5*, a006098. [CrossRef]
28. Sanchez-Vega, F.; Mina, M.; Armenia, J.; Chatila, W.K.; Luna, A.; La, K.C.; Dimitriadoy, S.; Liu, D.L.; Kantheti, H.S.; Saghafinia, S.; et al. Oncogenic Signaling Pathways in The Cancer Genome Atlas. *Cell* **2019**, *173*, 321–337. [CrossRef] [PubMed]
29. Crooker, K.; Aliani, R.; Ananth, M.; Arnold, L.; Anant, S.; Thomas, S.M. A Review of Promising Natural Chemopreventive Agents for Head and Neck Cancer. *Cancer Prev. Res.* **2018**, *11*, 441–450, ISBN 1913588467. [CrossRef]
30. Rahman, M.A.; Amin, A.R.M.R.; Shin, D.M. Chemopreventive Potential of Natural Compounds in Head and Neck Cancer. *Nutr. Cancer* **2010**, *62*, 973–987. [CrossRef] [PubMed]
31. Shin, D.M.; Khuri, F.R.; Murphy, B.; Garden, A.S.; Clayman, G.; Francisco, M.; Liu, D.; Glisson, B.S.; Ginsberg, L.; Papadimitrakopoulou, V.; et al. Combined Interferon-Alfa, 13-Cis-Retinoic Acid, and Alpha-Tocopherol in Locally Advanced Head and Neck Squamous Cell Carcinoma: Novel Bioadjuvant Phase II Trial. *J. Clin. Oncol.* **2001**, *19*, 3010–3017. [CrossRef]
32. Papadimitrakopoulou, V.A.; Lee, J.J.; William, W.N.; Martin, J.W.; Thomas, M.; Kim, E.S.; Khuri, F.R.; Shin, D.M.; Feng, L.; Waun, K.H.; et al. Randomized Trial of 13-Cis Retinoic Acid Compared with Retinyl Palmitate with or without Beta-Carotene in Oral Premalignancy. *J. Clin. Oncol.* **2009**, *27*, 599–604. [CrossRef]
33. Tsao, A.S.; Liu, D.; Martin, J.; Tang, X.M.; Lee, J.J.; El-Naggar, A.K.; Wistuba, I.; Culotta, K.S.; Mao, L.; Gillenwater, A.; et al. Phase II Randomized, Placebo-Controlled Trial of Green Tea Extract in Patients with High-Risk Oral Premalignant Lesions. *Cancer Prev. Res.* **2009**, *2*, 931–941. [CrossRef]
34. Cheng, A.L.; Hsu, C.H.; Lin, J.K.; Hse, M.W.; Ho, Y.F.; Shen, T.S.; Ko, J.Y.; Lin, J.T.; Lin, B.R.; Wu, M.S.; et al. Phase I Clinical Trial of Curcumin, a Chemopreventive Agent, in Patients with High-Risk or Pre-Malignant Lesions. *Anticancer Res.* **2001**, *21*, 2895–2900. [PubMed]
35. Molinolo, A.A.; Hewitt, S.M.; Amornphimoltham, P.; Keelawat, S.; Rangdaeng, S.; García, A.M.; Raimondi, A.R.; Jufe, R.; Itoiz, M.; Gao, Y.; et al. Dissecting the Akt/Mammalian Target of Rapamycin Signaling Network: Emerging Results from the Head and Neck Cancer Tissue Array Initiative. *Clin. Cancer Res.* **2007**, *13*, 4964–4973. [CrossRef] [PubMed]
36. Freudlsperger, C.; Horn, D.; Weißfuß, S.; Weichert, W.; Weber, K.J.; Saure, D.; Sharma, S.; Dyckhoff, G.; Grabe, N.; Plinkert, P.; et al. Phosphorylation of AKT(Ser473) Serves as an Independent Prognostic Marker for Radiosensitivity in Advanced Head and Neck Squamous Cell Carcinoma. *Int. J. Cancer* **2015**, *136*, 2775–2785. [CrossRef] [PubMed]
37. Zumsteg, Z.S.; Morse, N.; Krigsfeld, G.; Gupta, G.; Higginson, D.S.; Lee, N.Y.; Morris, L.; Ganly, I.; Shiao, S.L.; Powell, S.N.; et al. Taselisib (GDC-0032), a Potent β-Sparing Small Molecule Inhibitor of PI3K, Radiosensitizes Head and Neck Squamous Carcinomas Containing Activating PIK3CA Alterations. *Clin. Cancer Res.* **2016**, *22*, 2009–2019. [CrossRef] [PubMed]
38. Iwase, M.; Yoshiba, S.; Uchid, M.; Takaoka, S.; Kurihara, Y.; Ito, D.; Hatori, M.; Shintani, S. Enhanced Susceptibility to Apoptosis of Oral Squamous Cell Carcinoma Cells Subjected to Combined Treatment with Anticancer Drugs and Phosphatidylinositol 3-Kinase Inhibitors. *Int. J. Oncol.* **2007**, *31*, 1141–1147. [PubMed]
39. Cantley, L.C. The Phosphoinositide 3-Kinase Pathway. *Science* **2002**, *296*, 1655–1657. [CrossRef]
40. Sarbassov, D.D.; Guertin, D.A.; Ali, S.M.; Sabatini, D.M. Phosphorylation and Regulation of Akt/PKB by the Rictor-MTOR Complex. *Science* **2005**, *307*, 1098–1101. [CrossRef]
41. Zhao, C.; Zhang, Z.; Dai, X.; Wang, J.; Liu, H.; Ma, H. Actein Antagonizes Oral Squamous Cell Carcinoma Proliferation through Activating FoxO1. *Pharmacology* **2021**, *106*, 551–563. [CrossRef]
42. Zhang, X.; Wang, F.; Zeng, Y.; Zhu, X.; Peng, L.; Zhang, L.; Gu, J.; Han, H.; Yi, X.; Shi, J. Salicylate Sensitizes Oral Squamous Cell Carcinoma to Chemotherapy through Targeting MTOR Pathway. *Oral Dis.* **2020**, *26*, 1131–1140. [CrossRef]
43. Li, M.; Gao, F.; Zhao, Q.; Zuo, H.; Liu, W.; Li, W. Tanshinone IIA Inhibits Oral Squamous Cell Carcinoma via Reducing Akt-c-Myc Signaling-Mediated Aerobic Glycolysis. *Cell Death Dis.* **2020**, *11*, 381. [CrossRef]
44. Li, M.; Gao, F.; Yu, X.; Zhao, Q.; Zhou, L.; Liu, W.; Li, W. Promotion of Ubiquitination-Dependent Survivin Destruction Contributes to Xanthohumol-Mediated Tumor Suppression and Overcomes Radioresistance in Human Oral Squamous Cell Carcinoma. *J. Exp. Clin. Cancer Res.* **2020**, *39*, 88. [CrossRef]
45. Huang, K.J.; Kuo, C.H.; Chen, S.H.; Lin, C.Y.; Lee, Y.R. Honokiol Inhibits in Vitro and in Vivo Growth of Oral Squamous Cell Carcinoma through Induction of Apoptosis, Cell Cycle Arrest and Autophagy. *J. Cell. Mol. Med.* **2018**, *22*, 1894–1908. [CrossRef]
46. Lin, C.W.; Bai, L.Y.; Su, J.H.; Chiu, C.F.; Lin, W.Y.; Huang, W.T.; Shih, M.C.; Huang, Y.T.; Hu, J.L.; Weng, J.R. Ilimaquinone Induces Apoptosis and Autophagy in Human Oral Squamous Cell Carcinoma Cells. *Biomedicines* **2020**, *8*, 296. [CrossRef]

47. Sophia, J.; Kowshik, J.; Dwivedi, A.; Bhutia, S.K.; Manavathi, B.; Mishra, R.; Nagini, S. Nimbolide, a Neem Limonoid Inhibits Cytoprotective Autophagy to Activate Apoptosis via Modulation of the PI3K/Akt/GSK-3β Signalling Pathway in Oral Cancer. *Cell Death Dis.* **2018**, *9*, 1087. [CrossRef] [PubMed]
48. Aggarwal, S.; Bhadana, K.; Singh, B.; Rawat, M.; Mohammad, T.; Al-Keridis, L.A.; Alshammari, N.; Hassan, M.I.; Das, S.N. Cinnamomum Zeylanicum Extract and Its Bioactive Component Cinnamaldehyde Show Anti-Tumor Effects via Inhibition of Multiple Cellular Pathways. *Front. Pharmacol.* **2022**, *13*, 918479. [CrossRef] [PubMed]
49. Zhang, N.; Gao, L.; Ren, W.; Li, S.; Zhang, D.; Song, X.; Zhao, C.; Zhi, K. Fucoidan Affects Oral Squamous Cell Carcinoma Cell Functions in Vitro by Regulating FLNA-Derived Circular RNA. *Ann. N. Y. Acad. Sci.* **2020**, *1462*, 65–78. [CrossRef] [PubMed]
50. Aswathy, M.; Banik, K.; Parama, D.; Sasikumar, P.; Harsha, C.; Joseph, A.G.; Sherin, D.R.; Thanathu, M.K.; Kunnumakkara, A.B.; Vasu, R.K. Exploring the Cytotoxic Effects of the Extracts and Bioactive Triterpenoids from Dillenia Indica against Oral Squamous Cell Carcinoma: A Scientific Interpretation and Validation of Indigenous Knowledge. *ACS Pharmacol. Transl. Sci.* **2021**, *4*, 834–847. [CrossRef]
51. Schulz, L.; Pries, R.; Lanka, A.S.; Drenckhan, M.; Rades, D.; Wollenberg, B. Inhibition of GSK3α/β Impairs the Progression of HNSCC. *Oncotarget* **2018**, *9*, 27630–27644. [CrossRef]
52. Ugolkov, A.V.; Matsangou, M.; Taxter, T.J.; O'halloran, T.V.; Cryns, V.L.; Giles, F.J.; Mazar, A.P. Aberrant Expression of Glycogen Synthase Kinase-3β in Human Breast and Head and Neck Cancer. *Oncol. Lett.* **2018**, *16*, 6437–6444. [CrossRef]
53. Marconi, G.D.; Della Rocca, Y.; Fonticoli, L.; Melfi, F.; Rajan, T.S.; Carradori, S.; Pizzicannella, J.; Trubiani, O.; Diomede, F. C-Myc Expression in Oral Squamous Cell Carcinoma: Molecular Mechanisms in Cell Survival and Cancer Progression. *Pharmaceuticals* **2022**, *15*, 890. [CrossRef]
54. Bancroft, C.C.; Chen, Z.; Dong, G.; Sunwoo, J.B.; Yeh, N.; Park, C.; Van Waes, C. Coexpression of Proangiogenic Factors IL-8 and VEGF by Human Head and Neck Squamous Cell Carcinoma Involves Coactivation by MEK-MAPK and IKK-NF-KB Signal Pathways. *Clin. Cancer Res.* **2001**, *7*, 435–442. [PubMed]
55. Kumar, V.B.; Lin, S.H.; Mahalakshmi, B.; Lo, Y.S.; Lin, C.C.; Chuang, Y.C.; Hsieh, M.J.; Chen, M.K. Sodium Danshensu Inhibits Oral Cancer Cell Migration and Invasion by Modulating P38 Signaling Pathway. *Front. Endocrinol.* **2020**, *11*, 568436. [CrossRef] [PubMed]
56. Li, Z.; Cao, L.; Yang, C.; Liu, T.; Zhao, H.; Luo, X.; Chen, Q. Protocatechuic Acid-Based Supramolecular Hydrogel Targets SerpinB9 to Achieve Local Chemotherapy for OSCC. *ACS Appl. Mater. Interfaces* **2022**, *14*, 32. [CrossRef] [PubMed]
57. Gkouveris, I.; Nikitakis, N.; Karanikou, M.; Rassidakis, G.; Sklavounou, A. JNK1/2 Expression and Modulation of STAT3 Signaling in Oral Cancer. *Oncol. Lett.* **2016**, *12*, 699–706. [CrossRef]
58. charoenrat, P.; Rhys-Evans, P.H.; Eccles, S.A. Expression of Matrix Metalloproteinases and Their Inhibitors Correlates with Invasion and Metastasis in Squamous Cell Carcinoma of the Head and Neck. *Arch. Otolaryngol. Head Neck Surg.* **2001**, *127*, 813–820.
59. Carmeliet, P. VEGF as a Key Mediator of Angiogenesis in Cancer. *Oncology* **2005**, *69*, 4–10. [CrossRef] [PubMed]
60. Pepper, M.S.; Ferrara, N.; Orci, L.; Montesano, R. Potent Synergism between Vascular Endothelial Growth Factor and Basic Fibroblast Growth Factor in the Induction of Angiogenesis in Vitro. *Biochem. Biophys. Res. Commun.* **1992**, *189*, 824–831. [CrossRef]
61. Zhang, Y.H.; Wei, W.; Xu, H.; Wang, Y.Y.; Wu, W.X. Inducing Effects of Hepatocyte Growth Factor on the Expression of Vascular Endothelial Growth Factor in Human Colorectal Carcinoma Cells through MEK and PI3K Signaling Pathways. *Chin. Med. J.* **2007**, *120*, 743–748. [CrossRef]
62. Lun, M.; Zhang, P.L.; Pellitteri, P.K.; Law, A.; Kennedy, T.L.; Brown, R.E. Nuclear Factor-KappaB Pathway as a Therapeutic Target in Head and Neck Squamous Cell Carcinoma: Pharmaceutical and Molecular Validation in Human Cell Lines Using Velcade and SiRNA/NF-KB. *Ann. Clin. Lab. Sci.* **2005**, *35*, 251–258.
63. Oeckinghaus, A.; Ghosh, S. The NF-KB Family of Transcription Factors and Its Regulation. *Cold Spring Harb. Perspect. Biol.* **2009**, *1*, a000034. [CrossRef]
64. Wang, J.; Hu, Y.; Yuan, J.; Zhang, Y.; Wang, Y.; Yang, Y.; Alahmadi, T.A.; Ali Alharbi, S.; Zhuang, Z.; Wu, F. Chemomodulatory Effect of Neferine on DMBA-Induced Squamous Cell Carcinogenesis: Biochemical and Molecular Approach. *Environ. Toxicol.* **2021**, *36*, 460–471. [CrossRef] [PubMed]
65. Chen, H.; Lo, Y.; Lin, C.; Lee, T.; Leung, W.; Wang, S.; Lin, I.; Lin, M.; Lee, C. Biomedicine & Pharmacotherapy Trichodermin Inhibits the Growth of Oral Cancer through Apoptosis-Induced Mitochondrial Dysfunction and HDAC-2-Mediated Signaling. *Biomed. Pharmacother.* **2022**, *153*, 113351. [CrossRef] [PubMed]
66. Ihle, J.N. The Stat Family in Cytokine Signaling. *Curr. Opin. Cell Biol.* **2001**, *13*, 211–217. [CrossRef] [PubMed]
67. Sun, S.S.; Zhou, X.; Huang, Y.Y.; Kong, L.P.; Mei, M.; Guo, W.Y.; Zhao, M.H.; Ren, Y.; Shen, Q.; Zhang, L. Targeting STAT3/MiR-21 Axis Inhibits Epithelial-Mesenchymal Transition via Regulating CDK5 in Head and Neck Squamous Cell Carcinoma. *Mol. Cancer* **2015**, *14*, 213. [CrossRef]
68. Zhou, X.; Ren, Y.; Liu, A.; Han, L.; Zhang, K.; Li, S.; Li, P.; Li, P.; Kang, C.; Wang, X.; et al. STAT3 Inhibitor WP1066 Attenuates MiRNA-21 to Suppress Human Oral Squamous Cell Carcinoma Growth in Vitro and in Vivo. *Oncol. Rep.* **2014**, *31*, 2173–2180. [CrossRef]
69. Riebe, C.; Pries, R.; Schroeder, K.N.; Wollenberg, B. Phosphorylation of STAT3 in Head and Neck Cancer Requires P38 MAPKinase, Whereas Phosphorylation of STAT1 Occurs via a Different Signaling Pathway. *Anticancer Res.* **2011**, *31*, 3819–3825.

70. Zhao, C.; Yang, L.; Zhou, F.; Yu, Y.; Du, X.; Xiang, Y.; Li, C.; Huang, X.; Xie, C.; Liu, Z.; et al. Feedback Activation of EGFR Is the Main Cause for STAT3 Inhibition-Irresponsiveness in Pancreatic Cancer Cells. *Oncogene* **2020**, *39*, 3997–4013. [CrossRef] [PubMed]
71. Ning, Y.; Cui, Y.; Li, X.; Cao, X.; Chen, A.; Xu, C.; Cao, J.; Luo, X. Co-Culture of Ovarian Cancer Stem-like Cells with Macrophages Induced SKOV3 Cells Stemness via IL-8/STAT3 Signaling. *Biomed. Pharmacother.* **2018**, *103*, 262–271. [CrossRef]
72. Yang, J.; Liao, D.; Chen, C.; Liu, Y.; Chuang, T.H.; Xiang, R.; Markowitz, D.; Reisfeld, R.A.; Luo, Y. Tumor-Associated Macrophages Regulate Murine Breast Cancer Stem Cells through a Novel Paracrine Egfr/Stat3/Sox-2 Signaling Pathway. *Stem Cells* **2013**, *31*, 248–258. [CrossRef]
73. Chan, C.Y.; Huang, S.Y.; Sheu, J.J.C.; Roth, M.M.; Chou, I.T.; Lien, C.H.; Lee, M.F.; Huang, C.Y. Transcription Factor HBP1 Is a Direct Anti-Cancer Target of Transcription Factor FOXO1 in Invasive Oral Cancer. *Oncotarget* **2017**, *8*, 14537–14548. [CrossRef]
74. Roh, J.L.; Kang, S.K.; Minn, I.; Califano, J.A.; Sidransky, D.; Koch, W.M. P53-Reactivating Small Molecules Induce Apoptosis and Enhance Chemotherapeutic Cytotoxicity in Head and Neck Squamous Cell Carcinoma. *Oral Oncol.* **2011**, *47*, 8–15. [CrossRef] [PubMed]
75. Khan, Z.; Tiwari, R.P.; Mulherker, R.; Sah, N.K.; Prasad, G.B.; Shrivastava, B.R.; Bisen, P.S. Detection of Survivin and P53 in Human Oral Cancer: Correlation with Clinicopathological Findings. *Head Neck* **2009**, *31*, 1039–1048. [CrossRef] [PubMed]
76. The Cancer Genome Atlas Network. Comprehensive Genomic Characterization of Head and Neck Squamous Cell Carcinomas. *Nature* **2015**, *517*, 576–582. [CrossRef]
77. Tanida, I.; Ueno, T.; Kominami, E. LC3 and Autophagy. *Methods Mol. Biol.* **2008**, *445*, 77–88. [CrossRef] [PubMed]
78. Kabeya, Y.; Mizushima, N.; Ueno, T.; Yamamoto, A.; Kirisako, T.; Noda, T.; Kominami, E.; Ohsumi, Y.; Yoshimori, T. LC3, a Mammalian Homologue of Yeast Apg8p, Is Localized in Autophagosome Membranes after Processing. *EMBO J.* **2000**, *19*, 5720–5728. [CrossRef] [PubMed]
79. Yun, C.W.; Lee, S.H. The Roles of Autophagy in Cancer. *Int. J. Mol. Sci.* **2018**, *19*, 3466. [CrossRef]
80. Khan, T.; Relitti, N.; Brindisi, M.; Magnano, S.; Zisterer, D.; Gemma, S.; Butini, S.; Campiani, G. Autophagy Modulators for the Treatment of Oral and Esophageal Squamous Cell Carcinomas. *Med. Res. Rev.* **2020**, *40*, 1002–1060. [CrossRef]
81. Lee, M.; Nam, H.Y.; Kang, H.B.; Lee, W.H.; Lee, G.H.; Sung, G.J.; Han, M.W.; Cho, K.J.; Chang, E.J.; Choi, K.C.; et al. Epigenetic Regulation of P62/SQSTM1 Overcomes the Radioresistance of Head and Neck Cancer Cells via Autophagy-Dependent Senescence Induction. *Cell Death Dis.* **2021**, *12*, 250. [CrossRef]
82. Eisenberg-Lerner, A.; Kimchi, A. The Paradox of Autophagy and Its Implication in Cancer Etiology and Therapy. *Apoptosis* **2009**, *14*, 376–391. [CrossRef]
83. Garcia-Medina, R.; Gounon, P.; Chiche, J.; Pouysse, J.; Mazure, N.M. Hypoxia-Induced Autophagy Is Mediated through Hypoxia-Inducible Factor Induction of BNIP3 and BNIP3L via Their BH3 Domains. *Mol. Cell. Biol.* **2009**, *29*, 2570–2581. [CrossRef]
84. Wei, H.; Wei, S.; Gan, B.; Peng, X.; Zou, W.; Guan, J. Suppression of Autophagy by FIP200 Deletion Inhibits Mammary Tumorigenesis. *Genes Dev.* **2011**, *37*, 1510–1527. [CrossRef] [PubMed]
85. Singh, S.S.; Vats, S.; Chia, A.Y.; Zea, T.; Shuo, T.; Mei, D.; Ong, S.; Arfuso, F.; Yap, C.T.; Cher, B.; et al. Dual Role of Autophagy in Hallmarks of Cancer. *Oncogene* **2018**, *37*, 1142–1158. [CrossRef] [PubMed]
86. Muz, B.; Azab, A.K. The Role of Hypoxia in Cancer Progression, Angiogenesis, Metastasis, and Resistance to Therapy. *Hypoxia* **2015**, *3*, 83–92. [CrossRef] [PubMed]
87. Oltval, Z.N.; Milliman, C.L.; Korsmeyer, S.J. Bcl-2 Heterodimerizes in Vivo with a Conserved Homolog, Bax, That Accelerates Programed Cell Death. *Cell* **1993**, *74*, 609–619. [CrossRef]
88. Hockenbery, D.; Nuñez, G.; Milliman, C.; Schreiber, R.D.; Korsmeyer, S.J. Bcl-2 Is an Inner Mitochondrial Membrane Protein That Blocks Programmed Cell Death. *Nature* **1990**, *348*, 334–336. [CrossRef]
89. Loro, L.L.; Vintermyr, O.K.; Liavaag, P.G.; Jonsson, R.; Johannessen, A.C. Oral Squamous Cell Carcinoma Is Associated with Decreased Bcl-2/Bax Expression Ratio and Increased Apoptosis. *Hum. Pathol.* **1999**, *30*, 1097–1105. [CrossRef]
90. Zou, H.; Henzel, W.J.; Liu, X. Apaf-1, a Human Protein Homologous to C. Elegans CED-4, Participates in cytochrome c-dependent activation of caspase-3. *Cell* **1997**, *90*, 405–413. [CrossRef]
91. Kubina, R.; Krzykawski, K.; Dziedzic, A. Kaempferol and Fisetin-Related Signaling Pathways Induce Apoptosis in Head and Neck Cancer Cells. *Cells* **2023**, *12*, 1568. [CrossRef]
92. Xiao, S.; Chen, F.; Gao, C. Antitumor Activity of 4-O-Methylhonokiol in Human Oral Cancer Cells Is Mediated via ROS Generation, Disruption of Mitochondrial Potential, Cell Cycle Arrest and Modulation of Bcl-2/Bax Proteins. *J. BUON* **2017**, *22*, 1577–1581.
93. Liu, J.; Uematsu, H.; Tsuchida, N.; Ikeda, M.A. Essential Role of Caspase-8 in P53/P73-Dependent Apoptosis Induced by Etoposide in Head and Neck Carcinoma Cells. *Mol. Cancer* **2011**, *10*, 95. [CrossRef]
94. Liu, J.; Uematsu, H.; Tsuchida, N.; Ikeda, M.A. Association of Caspase-8 Mutation with Chemoresistance to Cisplatin in HOC313 Head and Neck Squamous Cell Carcinoma Cells. *Biochem. Biophys. Res. Commun.* **2009**, *390*, 989–994. [CrossRef] [PubMed]
95. Coleman, S.C.; Stewart, Z.A.; Day, T.A.; Netterville, J.L.; Burkey, B.B.; Pietenpol, J.A. Analysis of Cell-Cycle Checkpoint Pathways in Head and Neck Cancer Cell Lines: Implications for Therapeutic Strategies. *Arch. Otolaryngol.-Head Neck Surg.* **2002**, *128*, 167–176. [CrossRef] [PubMed]
96. Norbury, C.; Nurse, P. Animal Cell Cycles and Their Control. *Annu. Rev. Biochem.* **1992**, *61*, 441–470. [CrossRef] [PubMed]
97. Xiong, Y.; Connolly, T.; Futcher, B.; Beach, D. Human D-Type Cyclin. *Cell* **1991**, *65*, 691–699. [CrossRef]
98. Matsushime, H.; Ewen, M.E.; Strom, D.K.; Kato, J.Y.; Hanks, S.K.; Roussel, M.F.; Sherr, C.J. Identification and Properties of an Atypical Catalytic Subunit (P34PSK-J3/Cdk4) for Mammalian D Type G1 Cyclins. *Cell* **1992**, *71*, 323–334. [CrossRef]

99. Lea, N.C.; Orr, S.J.; Stoeber, K.; Williams, G.H.; Lam, E.W.-F.; Ibrahim, M.A.A.; Mufti, G.J.; Thomas, N.S.B. Commitment Point during $G_0 \rightarrow G_1$ That Controls Entry into the Cell Cycle. *Mol. Cell. Biol.* **2003**, *23*, 2351–2361. [CrossRef]
100. Duronio, R.J.; Brook, A.; Dyson, N.; O'Farrell, P.H. E2F-Induced S Phase Requires Cyclin E. *Genes Dev.* **1996**, *10*, 2505–2513. [CrossRef]
101. Oakes, V.; Wang, W.; Harrington, B.; Lee, W.J.; Beamish, H.; Chia, K.M.; Pinder, A.; Goto, H.; Inagaki, M.; Pavey, S.; et al. Cyclin A/Cdk2 Regulates Cdh1 and Claspin during Late S/G2 Phase of the Cell Cycle. *Cell Cycle* **2014**, *13*, 3302–3311. [CrossRef]
102. Vigneron, S.; Sundermann, L.; Labbé, J.C.; Pintard, L.; Radulescu, O.; Castro, A.; Lorca, T. Cyclin A-Cdk1-Dependent Phosphorylation of Bora Is the Triggering Factor Promoting Mitotic Entry. *Dev. Cell* **2018**, *45*, 637–650.e7. [CrossRef]
103. Gavet, O.; Pines, J. Activation of Cyclin B1-Cdk1 Synchronizes Events in the Nucleus and the Cytoplasm at Mitosis. *J. Cell Biol.* **2010**, *189*, 247–259. [CrossRef]
104. Li, B.; Zhou, P.; Xu, K.; Chen, T.; Jiao, J.; Wei, H.; Yang, X.; Xu, W.; Wan, W.; Xiao, J. Metformin Induces Cell Cycle Arrest, Apoptosis and Autophagy through ROS/JNK Signaling Pathway in Human Osteosarcoma. *Int. J. Biol. Sci.* **2020**, *16*, 74–84. [CrossRef]
105. McConnell, B.B.; Gregory, F.J.; Stott, F.J.; Hara, E.; Peters, G. Induced Expression of P16 INK4a Inhibits Both CDK4- and CDK2-Associated Kinase Activity by Reassortment of Cyclin-CDK-Inhibitor Complexes. *Mol. Cell. Biol.* **1999**, *19*, 1981–1989. [CrossRef] [PubMed]
106. Qin, X.Q.; Livingston, D.M.; Kaelin, W.G.; Adams, P.D. Deregulated Transcription Factor E2F-1 Expression Leads to S-Phase Entry and P53-Mediated Apoptosis. *Proc. Natl. Acad. Sci. USA* **1994**, *91*, 10918–10922. [CrossRef] [PubMed]
107. Huang, Y.C.; Lee, P.C.; Wang, J.J.; Hsu, Y.C. Anticancer Effect and Mechanism of Hydroxygenkwanin in Oral Squamous Cell Carcinoma. *Front. Oncol.* **2019**, *9*, 911. [CrossRef] [PubMed]
108. Bostan, M.; Petrică-Matei, G.G.; Radu, N.; Hainarosie, R.; Stefanescu, C.D.; Diaconu, C.C.; Roman, V. The Effect of Resveratrol or Curcumin on Head and Neck Cancer Cells Sensitivity to the Cytotoxic Effects of Cisplatin. *Nutrients* **2020**, *12*, 2596. [CrossRef]
109. Da Fonseca, A.C.C.; de Queiroz, L.N.; Sales Felisberto, J.; Jessé Ramos, Y.; Mesquita Marques, A.; Wermelinger, G.F.; Pontes, B.; de Lima Moreira, D.; Robbs, B.K. Cytotoxic Effect of Pure Compounds from Piper Rivinoides Kunth against Oral Squamous Cell Carcinoma. *Nat. Prod. Res.* **2021**, *35*, 6163–6167. [CrossRef]
110. De Campos, P.S.; Matte, B.F.; Diel, L.F.; Jesus, L.H.; Bernardi, L.; Alves, A.M.; Rados, P.V.; Lamers, M.L. Low Doses of Curcuma Longa Modulates Cell Migration and Cell–Cell Adhesion. *Phyther. Res.* **2017**, *31*, 1433–1440. [CrossRef]
111. Chourasia, N.R.; Borle, R.M.; Vastani, A. Concomitant Association of Oral Submucous Fibrosis and Oral Squamous Cell Carcinoma and Incidence of Malignant Transformation of Oral Submucous Fibrosis in a Population of Central India: A Retrospective Study. *J. Maxillofac. Oral Surg.* **2015**, *14*, 902–906. [CrossRef]
112. Evren, I.; Brouns, E.R.; Wils, L.J.; Poell, J.B.; Peeters, C.F.W.; Brakenhoff, R.H.; Bloemena, E.; de Visscher, J.G.A.M. Annual Malignant Transformation Rate of Oral Leukoplakia Remains Consistent: A Long-Term Follow-up Study. *Oral Oncol.* **2020**, *110*, 105014. [CrossRef]
113. Iocca, O.; Sollecito, T.P.; Alawi, F.; Weinstein, G.S.; Newman, J.G.; De Virgilio, A.; Di Maio, P.; Spriano, G.; Pardiñas López, S.; Shanti, R.M. Potentially Malignant Disorders of the Oral Cavity and Oral Dysplasia: A Systematic Review and Meta-Analysis of Malignant Transformation Rate by Subtype. *Head Neck* **2020**, *42*, 539–555. [CrossRef]
114. Lodi, G.; Franchini, R.; Warnakulasuriya, S.; Varoni, E.M.; Sardella, A.; Kerr, A.R.; Carrassi, A.; MacDonald, L.C.; Worthington, H.V.; Mauleffinch, L.F. Interventions for Treating Oral Leukoplakia to Prevent Oral Cancer (Review). *Cochrane Database Syst. Rev.* **2016**, *7*, CD001829. [CrossRef] [PubMed]
115. Papadimitrakopoulou, V.A.; William, W.N.; Dannenberg, A.J.; Lippman, S.M.; Lee, J.J.; Ondrey, F.G.; Peterson, D.E.; Feng, L.; Atwell, A.; El-Naggar, A.K.; et al. Pilot Randomized Phase II Study of Celecoxib in Oral Premalignant Lesions. *Clin. Cancer Res.* **2008**, *14*, 2095–2101. [CrossRef] [PubMed]
116. William, W.N.; Papadimitrakopoulou, V.; Lee, J.J.; Mao, L.; Cohen, E.E.W.; Lin, H.Y.; Gillenwater, A.M.; Martin, J.W.; Lingen, M.W.; Boyle, J.O.; et al. Erlotinib and the Risk of Oral Cancer the Erlotinib Prevention of Oral Cancer (EPOC) Randomized Clinical Trial. *JAMA Oncol.* **2016**, *2*, 209–216. [CrossRef]
117. Gutkind, J.S.; Molinolo, A.A.; Wu, X.; Wang, Z.; Nachmanson, D.; Harismendy, O.; Alexandrov, L.B.; Wuertz, B.R.; Ondrey, F.G.; Laronde, D.; et al. Inhibition of MTOR Signaling and Clinical Activity of Metformin in Oral Premalignant Lesions. *JCI Insight* **2021**, *6*, 147096. [CrossRef] [PubMed]
118. Tian, H.; Lyu, Y.; Yang, Y.G.; Hu, Z. Humanized Rodent Models for Cancer Research. *Front. Oncol.* **2020**, *10*, 1696. [CrossRef]
119. Bouaoud, J.; De Souza, G.; Darido, C.; Tortereau, A.; Elkabets, M.; Bertolus, C.; Saintigny, P. The 4-NQO Mouse Model: An Update on a Well-Established in Vivo Model of Oral Carcinogenesis. *Methods Cell Biol.* **2021**, *163*, 197–229. [CrossRef]
120. Schoop, R.A.L.; Noteborn, M.H.M.; Baatenburg De Jong, R.J. A Mouse Model for Oral Squamous Cell Carcinoma. *J. Mol. Histol.* **2009**, *40*, 177–181. [CrossRef]
121. Wang, J.; Wang, S.; Wang, Y.; Wang, L.; Xia, Q.; Tian, Z.; Guan, X. Chemopreventive Effect of Modified Zengshengping on Oral Cancer in a Hamster Model and Assessment of Its Effect on Liver. *J. Ethnopharmacol.* **2020**, *255*, 112774. [CrossRef]
122. Dai, Z.; Zhu, P.F.; Liu, H.; Li, X.C.; Zhu, Y.Y.; Liu, Y.Y.; Shi, X.L.; Chen, W.D.; Liu, Y.P.; Zhao, Y.L.; et al. Discovery of Potent Immune-Modulating Molecule Taccaoside A against Cancers from Structures-Active Relationships of Natural Steroidal Saponins. *Phytomedicine* **2022**, *104*, 154335. [CrossRef]
123. Cattaneo, C.M.; Dijkstra, K.K.; Fanchi, L.F.; Kelderman, S.; Kaing, S.; van Rooij, N.; van den Brink, S.; Schumacher, T.N.; Voest, E.E. Tumor Organoid–T-Cell Coculture Systems. *Nat. Protoc.* **2020**, *15*, 15–39. [CrossRef]

124. Neal, J.T.; Li, X.; Zhu, J.; Giangarra, V.; Grzeskowiak, C.L.; Ju, J.; Liu, I.H.; Chiou, S.H.; Salahudeen, A.A.; Smith, A.R.; et al. Organoid Modeling of the Tumor Immune Microenvironment. *Cell* **2018**, *175*, 1972–1988.e16. [CrossRef] [PubMed]
125. Zhou, Z.; Van der Jeught, K.; Fang, Y.; Yu, T.; Li, Y.; Ao, Z.; Liu, S.; Zhang, L.; Yang, Y.; Eyvani, H.; et al. An Organoid-Based Screen for Epigenetic Inhibitors That Stimulate Antigen Presentation and Potentiate T-Cell-Mediated Cytotoxicity. *Nat. Biomed. Eng.* **2021**, *5*, 1320–1335. [CrossRef] [PubMed]
126. Verma, A.; Vincent-chong, V.K.; Dejong, H.; Hershberger, P.A.; Seshadri, M. Impact of Dietary Vitamin D on Initiation and Progression of Oral Cancer. *J. Steroid Biochem. Mol. Biol.* **2020**, *199*, 105603. [CrossRef] [PubMed]

Table 2. Summary of studies evaluating the association between head and neck cancers with TERT promoter mutations.

Author, Country (Year)	Case Numbers	Cancer Sites	Prevalence of TERT Promoter Mutations	Special Findings	The Association with Survival
Killela, USA (2013) [25]	70	31 Oral cavity 23 Oropharynx 4 Supraglottic 12 Others	Total: 17.1% (12/70) C228T: 14.8% C250T: 2.8%	Highest frequency in tongues (47.8%, 11/23)	N/A
Schwaederle, USA (2018) [32]	28	28 HNC	Total: 28.6% (8/28)	N/A	Trend toward shorter survival
Cheng, USA (2015) [93]	12	12 HNSCC	Total: 16.67% (2/12) C228T: 16.67% C250T: 0%	No significant correlation was observed.	N/A
Barczak, USA (2017) [15]	61	25 Mouth 25 Voice box 5 Nose/sinuses 6 Throat	C250T homozygous T/T allele: 36% heterozygous C/T allele: 26%	Homozygous T/T mutation is associated with the grade of the tumor.	N/A
Yu, USA (2021) [29]	117	74 Oral cavity 24 Larynx 5 Hypopharynx 14 HPV (-) oropharynx	Total: 53.8% (63/117) C228T: 33.3% C250T: 9.4% C250T, C254T: 6% C228A: 4.3% CC434TT: 0.9%	Highest frequency in the oral cavity (81.1%, 60/74)	Increased risk of locoregional failure, but not distant failure or OS.
Morris, USA (2017) [97]	53	20 Oral cavity 18 Oropharynx 7 Larynx 2 Hypopharynx 6 Others (4 sinonasal cavity)	Total: 32.1% (17/53) C228T: 20.8% C250T: 5.7% C228A: 1.9%	TERT mutation and HPV infection may represent parallel mechanisms.	N/A
Boscolo-Rizzo, Italy (2020) [3]	101	27 Oral cavity 23 Oropharynx 15 Hypopharynx 36 Larynx	Total: 11.9% (12/101) C228T: 9.9% C250T: 2%	Highest frequency in the oral cavity (37%) TERT levels did not significantly differ according to the mutational status of TERT promoter.	No significant association between TERT promoter status and OS. Higher TERT levels, worse OS (43.6% vs. 60.1%)
Annunziata, Italy (2018) [96]	24	15 Oral cavity 9 Oropharynx	Total: 37.5% (9/24) C228T: 8.3% C250T: 12.5% Other: 16.7%	No mutation in oropharynx cancer. Mutations were independent of HPV status.	N/A
Yilmaz, Turkey (2020) [4]	189	102 Oral cavity 22 Oropharynx 6 Hypopharynx 59 Larynx	Total: 43.9% (83/189) C228T: 29.6% C250T: 11.6% C228A: 2.6%	Highest frequency in the oral cavity (75.5%, 77/102). TERT mutations are associated with younger age, female gender, and an inverse relationship to smoking and alcohol consumption.	No difference
Arantes, Brazil (2020) [13]	88	69 Oral cavity 11 Larynx 8 Pharynx	Total: 27.3% (24/88) C228T: 6.8% C250T: 20.5%	94.4% C250T were alcohol consumers. 66.7% C228T were not alcohol consumers	Decreased 5-year DFS and OS in C228T
Vinothkumar, India (2016) [26]	41	41 Oral cavity	Total: 31.7% (13/41) C228T: 21.9% C250T: 9.7%	No significant correlation was observed.	N/A
Chang, Taiwan (2017) [28]	201	201 Oral cavity	Total: 64.7% (130/201) C228T: 51.7% C250T: 12.9%	C228T mutation was associated with betel nut chewing.	No difference
Qu, China (2014) [27]	235	235 Laryngeal	Total: 27% (64/235) C250T: 23.8% C228T: 3.4%	Not significantly correlate with any clinicopathological variables	Poor survival, especially C250T mutation

Killela et al. surveyed 70 oral cavity cancers and identified TERT promoter mutations in 12 of the tumors (17.1%) [25]. Schwaederle et al. analyzed 423 cases of TERT promoter alterations using next-generation sequencing (NGS). Only 28 patients (6.6%) had HNCs. The incidence of TERT promoter alternations was 14.4% (61 of 423) in the overall population and 28.6% (8 of 28) in HNCs [32]. Cheng et al. collected 84 cases of SCC from different sites, including 12 HNC and C228T mutations, which were detected in 16.67% (2 of 12) [93].

Barczak et al. analyzed 61 HNC patients to determine the prevalence of the hTERT promoter C250T mutation. High-resolution melting mutation analysis was used to identify the C250T hTERT promoter mutation, followed by sequencing verification in 10% of the samples. The prevalence of the hTERT promoter C250T mutation was 36% [15]. Yu et al. identified TERT promoter mutations in 117 patients with SCC of the oral cavity (N = 74), larynx (N = 24), hypopharynx (N = 5), and HPV-negative SCC of the oropharynx (N = 14) using NGS. Overall, 63 patients (53.8%) had TERT promoter alterations, and the most common mutations were C228T and C250T [29]. Morris et al. collected 53 patients and 20 oral cavities, 18 oropharynges, 7 larynges, and 2 hypopharynges. Overall, the frequency of the TERT alteration was 32.1% (17 of 53), yet it was much higher in HPV-negative tumors (53% vs. 4.3%). Remarkably, 91% (10 of 11) of the HPV-negative tongue SCCs possessed TERT mutations [97].

In Italy, Boscolo-Rizzo et al. analyzed cancer tissue and adjacent mucosa specimens from 101 patients with HNCs and evaluated the prevalence of the TERT promoter mutations by Sanger sequencing. The tumor subsites in the HNCs included the oral cavity (N = 27), oropharynx (N = 23), hypopharynx (N = 15), and larynx (N = 36). The TERT promoter harbored mutations in 12 tumors (11.9%), with C228T and C250T, which accounted for 83.3% and 16.7%, respectively. They also evaluated the TERT mRNA level and found no significant difference between the TERT mRNA level and the mutational status of the TERT promoter [3]. Annunziata et al. analyzed tumor biopsies from 15 oral SCCs and nine oropharyngeal SCCs. The frequency of TERT promoter mutations was 60% (9 of 15) in oral SCCs and was absent in oropharyngeal SCCs. There were five hotspot mutations (three C228T and two C250T) and four other mutations. They also investigated the TERT mRNA levels and identified that the TERT mRNA levels were comparable to those detected in peri-tumor tissues. However, these data were from six oropharyngeal SCCs and illustrated that they all lacked mutations in the TERT promoter [96].

In Turkey, Yilmaz et al. collected a total of 189 patients with HNCs, including 102 oral cavities, 22 oropharynges, 6 hypopharynges, and 59 larynges. The TERT gene expression was examined by polymerase chain reaction (PCR)-based direct sequencing. TERT promoter mutations were detected in 43.9% (83 of 189) of the cases. Three TERT promoter region mutations were detected: C228T (56 of 83; 67.5%), C250T (22 of 83; 26.5%), and C228A (5 of 83; 6%). The frequency of the C228T mutation was almost twice that of the C250T and C228A mutations [4]. In Brazil, Arantes et al. collected 88 HNC patients and analyzed the TERT promoter mutations C228T and C250T using pyrosequencing. The overall prevalence of the TERT hotspot mutations is 27.3% (6.8% at locus C228T and 20.5% at C250T) [13]. In India, Vinothkumar et al. analyzed 181 primary tumors of the uterine cervix and oral cavity using PCR amplification and sequencing. A high frequency of TERT hotspot mutations was observed in both cervical (30 of 140, 21.4%) and oral (13 of 41, 31.7%) SCCs. Among the oral cancer samples, the TERT promoter hotspot mutations were frequent, while the C228T mutation (69.2%) was twice as frequent as the C250T (30.8%) [26].

In Taiwan, Chang et al. included 201 oral cavity SCC tumors and adjacent normal tissues to detect two TERT promoter mutations (C228T and C250T) using Sanger sequencing. Overall, the TERT hotspot promoter mutations occurred at a high frequency (64.7%) in patients with oral cavity SCCs. There were 52.5% (104 of 201) and 12.9% (26 of 201) oral cavity SCC tumor tissues containing that contained the C228T and C250T mutations, respectively [28].

In China, Qu et al. obtained 235 laryngeal cancer tissues using a pyrosequencing assay to detect the TERT promoter mutations C228T and C250T. The TERT promoter hotspot mutations were present in 27% (64 of 235) of the samples. The TERT C250T mutations were more common (56 of 235) than the C228T mutations (8 of 235) [27]. In Figure 1, we summarized the reported frequencies of the TERT promoter mutations in HNCs from the various studies mentioned above.

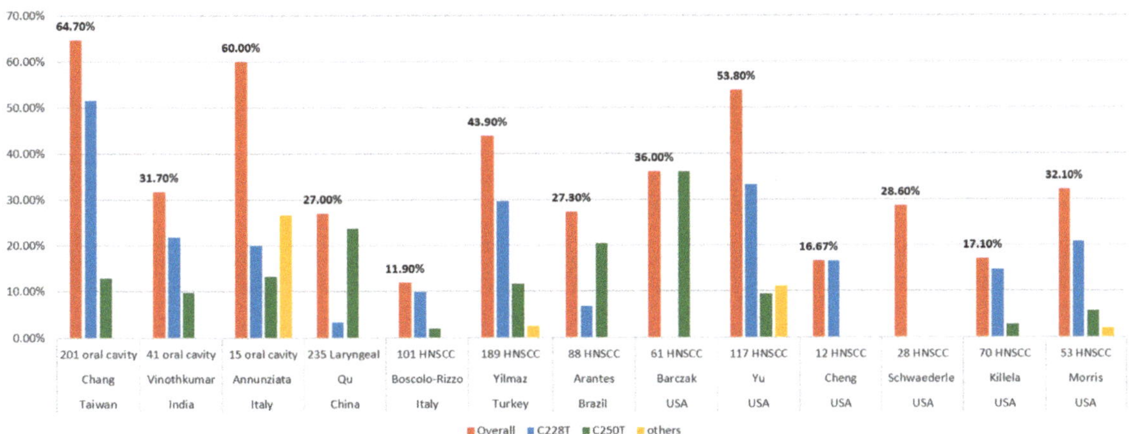

Figure 1. Frequencies of TERT promoter mutations in head and neck cancers from different studies.

5.2. TERT Promoter Mutations in Different Anatomic Distribution

HNC is a heterogeneous group of tumors involving distinct anatomical sites and sub-sites with varying etiological factors. Yu et al. showed that TERT promoter mutations were more abundant in oral cavity SCCs than in laryngopharyngeal cancers (81.1% vs. 7.0%) [29]. Boscolo-Rizzo et al. demonstrated that the prevalence of TERT hotspot promoter mutations is significantly higher in oral cavity SCCs (37%) [3]. Annunziata et al. also showed that TERT promoter mutations were predominant in oral SCCs (60%), yet absent in oropharyngeal SCCs [96]. Arantes et al. reported that 92% of the mutation cases were located in the oral cavity [13]. Finally, Yilmaz et al. showed that the frequency of the TERT promoter mutations in oral SCCs (75.5%) was significantly higher than in the other locations [4]. The anatomic distribution of cases is strongly associated with TERT promoter mutations, and the highest frequency is in oral cavity cancers.

As for the subsites in oral cavity SCCs, Arantes et al. noticed that 92% of the mutated cases were mainly in the tongue [13]. Killela et al. also revealed that 11 out of the 12 cancers with TERT promoter mutations were in the oral tongue, although only 23 of the 70 oral cavity cancers originated in the oral tongue [25]. However, Yilmaz et al. demonstrated that the highest rate was related to the buccal location and the lowest to the floor of the mouth (82.35% and 61.53%, respectively), although the difference was not statistically significant [4].

5.3. TERT Promoter Mutation and Human Papillomavirus Status

An association between HPV infection and oropharyngeal SCC has been proven. It was also clear that the molecular landscape and clinical pattern were different between HPV-positive and HPV-negative oropharyngeal cancers [10]. Only two studies have investigated the association between HPV status and TERT promoter mutations.

In a cohort of 53 patients with advanced HNCs, performed by Morris et al., a very high TERT alternation rate (53%, 16 of 30) was present in 30 HPV-negative tumors, however, there was only one TERT alternation (4.3%), which was a TERT amplification rather than a hotspot mutation, in 23 HPV-positive tumors. HPV-negative tongue SCCs showed the highest TERT mutation rate (91%). This demonstrated that TERT mutations and HPV infection may represent parallel mechanisms of telomerase activation in HNCs [97]. In another cohort study conducted by Annunziata et al., among the 9 patients with TERT promoter mutations in 15 oral SCC patients, 7 were HPV-negative and 2 were HPV-positive ($p = 0.486$). The frequency of TERT mutations was also independent of HPV tumor status in oral cancer [96].

5.4. TERT Promoter Mutation and Tobacco, Alcohol, and Betel Quid

Aside from HPV infection, tobacco smoking, alcohol consumption, and betel quid chewing are the other three main etiological factors of HNC [4,5]. Until now, the relationship between the TERT promoter mutations and these three factors remains inconclusive.

In a Brazilian cohort of 88 patients with HNC conducted by Arantes et al., the frequency of the C250T mutation appeared to be higher in alcohol consumers. Of the patients harboring the TERT promoter mutation C250T, 94.4% were alcohol consumers, and 66.7% of the patients harboring the TERT promoter mutation C228T did not consume alcohol [13]. In a Chinese cohort of 235 laryngeal cancer cases reported by Qu et al., hotspot mutations were not significantly correlated with any clinicopathological variables. However, TERT promoter mutations, particularly the C250T mutation, were more frequent in smoking patients (47 of 130) than in non-smoking patients (9 of 49), although no statistical significance was noted [27]. In a cohort of 201 patients with oral cavity SCC performed by Chang et al. in Taiwan, the C228T mutation was significantly associated with betel nut chewing [28]. In contrast, in a Turkish cohort of 189 HNC patients performed by Yilmaz et al., TERT promoter region mutations in HNC were inversely related to smoking and alcohol consumption [4].

5.5. TERT Promoter Mutation and Other Factors

Schwaederle et al. demonstrated that TERT promoter alterations are more frequent in men. They were also associated with brain cancers, skin/melanoma, head, and neck tumors, and increased median numbers of alterations in the univariate analysis. However, this association in head and neck tumors was not found in further multivariate analyses [32]. Yilmaz et al. reported that TERT promoter region mutations in HNCs are associated with younger age and female genders in a cohort from Turkey [4]. Barczak et al. demonstrated a significant association between the frequency of the homozygous C250T mutation and tumor grade (T1 = 27%, T2 = 36%, T3 = 35%, T4 = 46%, $p \leq 0.0001$) [15]. However, in a cohort of 41 patients with oral SCCs, performed by Vinothkumar et al. in India, no significant correlation was observed between any of the genotypes and the clinicopathological characteristics [26].

5.6. TERT Promoter Mutation and Survival

TERT promoter mutations in various reports of different cancers have been associated with aggressive characteristics, poor outcomes, and shorter survival [98–101]. In HNC, Qu et al. showed that TERT promoter mutations significantly affected the overall survival of laryngeal cancer patients, particularly those with the C250T mutation. TERT promoter mutations were significant predictors of poor prognosis in patients with laryngeal cancer, as an independent variable, with respect to age, tumor localization, TNM stage, tumor invasion, lymph node metastasis, and smoking history [27]. Schwaederle et al. also demonstrated a significantly shorter overall survival in patients harboring the TERT promoter alterations in the overall population in a univariate analysis. Subanalyses of the three tumor types with the highest prevalence of TERT alterations consistently showed a trend toward shorter survival for patients with altered TERT promoters in brain tumors, head, and neck cancers, and melanoma/skin tumors [32]. Arantes et al. demonstrated no statistically significant association between the presence of hotspot mutations (C228T and C250T) and survival. However, the presence of the C228T mutation impacted patient outcomes, with a significant decrease in 5-year disease-free survival (20.0 vs. 63.0%) and 5-year overall survival (16.7 vs. 45.1%) [13].

Similar results were reported by Yu et al. [29]. They reported that the TERT promoter mutations were associated with locoregional failure (LRF) in the overall cohort and in oral cavity SCCs. This increased risk for LRF is independent of the oral cavity primary site, TP53 mutation status, extracapsular extension, and positive surgical margins suggesting that the TERT promoter mutations are an independent biomarker of LRF rather than a surrogate for OSCCs, or other known prognostic markers. The cumulative incidence of LRF was similar between the two types of TERT promoter mutations (C250T and C228A/T groups), and both were associated with a higher cumulative incidence of LRF compared to wildtype tumors. Overall, they demonstrated that TERT promoter mutations were associated with an increased risk of LRF, although not with distant failure or overall survival [29].

In contrast, Yilmaz et al. did not find a significant association between the presence of TERT mutations and OS, despite patients with HNCs harboring TERT mutations exhibiting a slightly shorter median OS [4]. Boscolo-Rizzo et al. showed no significant association between the TERT promoter status and overall survival, although the TERT mRNA level had an impact on clinical outcomes [3]. Chang et al. also reported that there was no significant difference in overall survival, disease-specific survival, and disease-free survival between TERT promoter mutations and the wildtype [28].

6. Anti-Telomerase Therapeutics

The unique feature of overexpression in most cancer cells, although absent or with low expression in somatic cells, makes telomerase and other telomere components a target for the development of therapeutics [30]. Several therapeutic strategies have been proposed to target telomerase, and some have already been evaluated in clinical trials against various cancer types [30,35,44,102]. However, the development of successful clinical therapies is hampered by significant challenges [35].

6.1. Direct Telomerase Inhibition

Direct telomerase inhibition by small molecules or oligonucleotides that directly bind to the TERT or TERC template region suppresses telomere extension.

The first-in-class modified oligonucleotide, GRN163L (Imetelstat), was developed in 2003 [103]. Imetelstat is a lipidated 13-mer thiophosphoramidate oligonucleotide complementary to the TERC template region, which competitively inhibits telomerase activity and suppresses cancer cell viability [103]. After showing activity and efficacy against multiple cancer cell lines and in mouse xenograft models, Imetelstat has moved to early clinical trials against solid tumor malignancies (such as breast cancer, non-small-cell lung cancer, brain tumor, and melanoma) and hematologic diseases (such as multiple myeloma, myelodysplastic syndrome, and myeloproliferative neoplasms) [35,44,102]. Although Imetelstat did not meet its efficacy endpoints in trials on non-small cell lung cancer and breast cancers [104,105], it showed robust response rates in patients with lower-risk myelodysplastic syndromes, myelofibrosis, and essential thrombocythemia [106–109], and further late-stage clinical trials are underway [110].

BIBR1532, 2-[[(E)-3-naphthalen-2-ylbut-2-enoyl]amino]benzoic acid, inhibits telomerase by non-competitively binding to the TERT active site [111]. BIBR1532 has generated promising preclinical results [112–118]. For example, it enhances the radiosensitivity of non-small cell lung cancer by increasing telomere dysfunction and ATM/CHK1 inhibition [117]. However, this has not yet progressed to clinical testing. Some natural compounds have also been reported to act as telomerase inhibitors, such as allicin (from garlic), curcumin (from turmeric), silibinin (from thistle), and epigallocatechin gallate (EGCG, from tea), and the EGCG's derivative, MST-312 [30,35,119].

6.2. G-Quadruplex Stabilizers

G-quadruplexes are tetrad planar structures formed in guanine-rich DNA or RNA sequences, including telomeres [120]. Compounds that would stabilize telomeric G-quadruplex secondary structures can disrupt telomere extension via telomerase, triggering a DNA damage response and cell death. Several G-quadruplex stabilizers, including telomestatin, BRACO-19, RHPS4, TMPyP4, CX-3543 (Quarfloxin), CX-5461 (Pidnarulex) and AS1411, have been tested in preclinical studies and some already progressed to clinical trials [30,102,121–130].

6.3. Nucleoside Analogues

Nucleoside analogs mimic the presence of uncapped telomeres and induce DNA damage response, apoptosis, and autophagy [102]. To date, several nucleotide analogs, including T-oligo, 6-thio-2′-deoxyguanosine (6-thio-dG), and 5-fluoro-2′-deoxyuridine (5-FdU) triphosphate, are under investigation, although have not yet advanced to clinical trials [131–134].

6.4. Telomerase-Based Cancer Vaccines

Telomerase-based therapeutic cancer vaccines aim to induce T cells that target a tumor antigen, leading to improved antitumor immune responses and cancer cell death [135]. TERT is an appropriate tumor-associated antigen. To date, telomerase vaccinations, including peptide vaccines (such as GV1001, GX301, UV1, and Vx-001), dendritic cell-based vaccines (such as GRNVAC1), and DNA vaccines (such as INVAC-1) have been evaluated in many clinical trials spanning almost two decades [35,44,102,135,136]. Clinical studies on hTERT have been applied to both solid tumors and hematologic malignancies, and some of them have already moved to the later stages of trials [136]. However, the efficacy of the TERT vaccines is insufficient [137,138]. Furthermore, therapeutic TERT-based vaccines can mediate specific T cell responses in a high proportion of cancer patients [35]. A more robust antitumor activity was observed when combining immune checkpoint blockade with TERT-based vaccines in preclinical research, proving the synergistic effect between these two drugs [139].

6.5. TERT or TERC Promoter-Driven Therapy

Owing to the hallmark role of TERT promoter mutation-induced TERT expression in tumorigenesis, correction of this mutation and reduction of TERT expression has become a therapeutic method, by using recently developed gene editing techniques, including oncolytic virus and suicide gene therapy [102]. Telomelysin (OBP-301), a telomerase-specific replication-component adenovirus with an hTERT promoter element, has shown strong antitumor effects in various human cancer cells, including HNCs [140,141]. Phase I trials for solid tumors [142] and advanced hepatocellular carcinoma [143] have already been completed. Further phase 2 trials on gastric/gastroesophageal junction cancers (NCT03921021), head and neck cancers (NCT04685499), and esophageal cancers (NCT03213054) are currently ongoing.

6.6. Other Therapeutics Strategies

In addition to the strategies mentioned above, there are other anti-telomerase therapies. For example, telomerase interference by altered TERC templates is introduced by lentiviral infection [144], CRISPR genome editing targeting TERT gene expression [145], inhibition of oncogenic signaling MAPK pathways that impinge on TERT transcription [35], epigenetic mechanisms using histone deacetylase [30,146], and Tankyrase inhibitors for telomere length regulation [102,147]. Furthermore, strategies relying on telomere attrition in the setting of adjuvant or maintenance therapies rather than frontline therapy are another consideration [35].

7. Conclusions

Telomeres shorten with each cell division and result in cellular senescence. Telomere maintenance via telomerase reactivation plays a critical role in tumorigenesis (Figure 2). Several mechanisms could increase telomerase activity and TERT promoter, C228T and C250T, mutations are the most well-known alternations, which have already been reported in several malignancies, including HNCs. The frequency of the TERT promoter mutations in HNCs is quite high, ranging from 11.9% to 64.7%, and is more enriched in oral cavity SCCs than in other subsites. In addition, several reports have demonstrated an association between TERT promoter mutations and poor survival. Over the past 10 to 20 years, several anti-telomerase therapeutic strategies have been developed. However, only a few drugs have been used in clinical trials, and the results are only passable. Studies on immunotherapies targeting telomerase, such as cancer vaccines and oncolytic viruses, are very promising, however, trials are still ongoing. In addition, the potential synergy between TERT vaccines and immune checkpoint blockade may be another way to maximize anti-telomerase therapy.

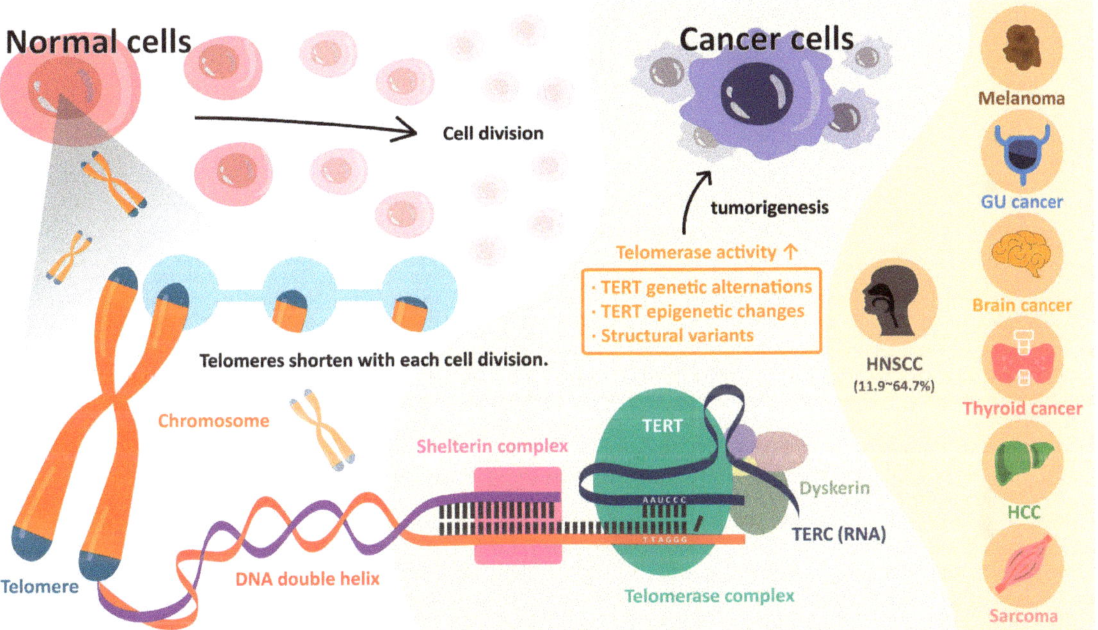

Figure 2. Telomere and telomerase play important roles in cellular biology and tumorigenesis. Telomeres are specialized structures that are located at the ends of chromosomes. They are composed of DNA repeating sequences (TTAGGG). Shelterin complexes are specific proteins known to protect chromosomes and regulate telomere length. Telomerase is a reverse transcriptase that synthesizes telomeric DNA sequences to maintain telomere length. Telomerase comprises two major components: the telomeric RNA component (TERC) and the telomerase reverse transcriptase (TERT). Other proteins, such as dyskerin, are also found in a complex with TERC. Telomeres shorten with each round of cell division and this mechanism limits the proliferation of cells to a finite number of cell divisions. Unlike normal cells, cancer cells are characterized by high telomerase activity, which could be achieved via mechanisms including TERT genetic alterations, TERT epigenetic change, or structural variants. TERT promoter mutations are the most common alternation and have been reported in several malignancies such as melanoma, genitourinary cancers, CNS tumors, hepatocellular carcinoma, thyroid cancers, sarcomas, and HNCs. In HNCs, the frequency of TERT promoter mutations is high (11.9–64.7% based on different studies).

Author Contributions: Conceptualization, T.-J.Y., S.-F.C. and H.-C.W.; writing—original draft preparation, T.-J.Y., J.-S.D. and Y.-C.G.; writing—review and editing, C.-W.L., T.-M.C., S.-F.C., Y.-C.L., H.-H.H., L.-T.C., H.-C.W. and S.-H.M.; visualization, T.-J.Y., C.-T.H. and M.-H.W.; supervision, M.-R.P., H.-C.W. and S.-H.M. All authors have read and agreed to the published version of the manuscript.

Funding: We acknowledge the support from the following grants: (1) KMUH110-0T04 and KMUH111-1M14 from the Kaohsiung Medical University Hospital; (2) 111-2321-B-037 -002 -, 110-2321-B-037-002, 109-2314-B-037-019, 109-2314-B-037-132- and 109-2314-B-037-036-MY3 from the Ministry of Science and Technology, Taiwan; (3) KMU-KI110001, KMU-TC109A03, KMU-TC111A04, KMU-TC108A03-6, and KMU-TC109B05 from Kaohsiung Medical University Research Center Grant.

Institutional Review Board Statement: Not applicable.

Informed Consent Statement: Not applicable.

Data Availability Statement: The data that support the findings of this study are available upon reasonable request (e.g., for research purposes) from the authors.

Acknowledgments: The authors wish to thank Center for Cancer Research at Kaohsiung Medical University and Cancer Center at Kaohsiung Medical University Hospital for their support.

Conflicts of Interest: The authors declare no conflict of interest.

References

1. Siegel, R.L.; Miller, K.D.; Jemal, A. Cancer statistics, 2019. *CA Cancer J. Clin.* **2019**, *69*, 7–34. [CrossRef] [PubMed]
2. Bray, F.; Ferlay, J.; Soerjomataram, I.; Siegel, R.L.; Torre, L.A.; Jemal, A. Global cancer statistics 2018: GLOBOCAN estimates of incidence and mortality worldwide for 36 cancers in 185 countries. *CA Cancer J. Clin.* **2018**, *68*, 394–424. [CrossRef] [PubMed]
3. Boscolo-Rizzo, P.; Giunco, S.; Rampazzo, E.; Brutti, M.; Spinato, G.; Menegaldo, A.; Stellin, M.; Mantovani, M.; Bandolin, L.; Rossi, M.; et al. TERT promoter hotspot mutations and their relationship with TERT levels and telomere erosion in patients with head and neck squamous cell carcinoma. *J. Cancer Res. Clin. Oncol.* **2020**, *146*, 381–389. [CrossRef] [PubMed]
4. Yilmaz, I.; Erkul, B.E.; Ozturk Sari, S.; Issin, G.; Tural, E.; Terzi Kaya Terzi, N.; Karatay, H.; Celik, M.; Ulusan, M.; Bilgic, B. Promoter region mutations of the telomerase reverse transcriptase (TERT) gene in head and neck squamous cell carcinoma. *Oral Surg. Oral Med. Oral Pathol. Oral Radiol.* **2020**, *130*, 63–70. [CrossRef] [PubMed]
5. Chen, Y.J.; Chang, J.T.; Liao, C.T.; Wang, H.M.; Yen, T.C.; Chiu, C.C.; Lu, Y.C.; Li, H.F.; Cheng, A.J. Head and neck cancer in the betel quid chewing area: Recent advances in molecular carcinogenesis. *Cancer Sci.* **2008**, *99*, 1507–1514. [CrossRef] [PubMed]
6. Gooi, Z.; Chan, J.Y.; Fakhry, C. The epidemiology of the human papillomavirus related to oropharyngeal head and neck cancer. *Laryngoscope* **2016**, *126*, 894–900. [CrossRef]
7. Mallen-St Clair, J.; Alani, M.; Wang, M.B.; Srivatsan, E.S. Human papillomavirus in oropharyngeal cancer: The changing face of a disease. *Biochim. Biophys. Acta* **2016**, *1866*, 141–150. [CrossRef]
8. Lee, Y.A.; Li, S.; Chen, Y.; Li, Q.; Chen, C.J.; Hsu, W.L.; Lou, P.J.; Zhu, C.; Pan, J.; Shen, H.; et al. Tobacco smoking, alcohol drinking, betel quid chewing, and the risk of head and neck cancer in an East Asian population. *Head Neck* **2019**, *41*, 92–102. [CrossRef]
9. Hashibe, M.; Brennan, P.; Benhamou, S.; Castellsague, X.; Chen, C.; Curado, M.P.; Dal Maso, L.; Daudt, A.W.; Fabianova, E.; Fernandez, L.; et al. Alcohol drinking in never users of tobacco, cigarette smoking in never drinkers, and the risk of head and neck cancer: Pooled analysis in the International Head and Neck Cancer Epidemiology Consortium. *J. Natl. Cancer Inst.* **2007**, *99*, 777–789. [CrossRef]
10. Leemans, C.R.; Snijders, P.J.F.; Brakenhoff, R.H. The molecular landscape of head and neck cancer. *Nat. Rev. Cancer* **2018**, *18*, 269–282. [CrossRef]
11. Boscolo-Rizzo, P.; Del Mistro, A.; Bussu, F.; Lupato, V.; Baboci, L.; Almadori, G.; MC, D.A.M.; Paludetti, G. New insights into human papillomavirus-associated head and neck squamous cell carcinoma. *Acta Otorhinolaryngol. Ital. Organo Uff. Della Soc. Ital. Di Otorinolaringol. E Chir. Cervico-Facciale* **2013**, *33*, 77–87.
12. Boscolo-Rizzo, P.; Da Mosto, M.C.; Rampazzo, E.; Giunco, S.; Del Mistro, A.; Menegaldo, A.; Baboci, L.; Mantovani, M.; Tirelli, G.; De Rossi, A. Telomeres and telomerase in head and neck squamous cell carcinoma: From pathogenesis to clinical implications. *Cancer Metastasis Rev.* **2016**, *35*, 457–474. [CrossRef] [PubMed]
13. Arantes, L.; Cruvinel-Carloni, A.; de Carvalho, A.C.; Sorroche, B.P.; Carvalho, A.L.; Scapulatempo-Neto, C.; Reis, R.M. TERT Promoter Mutation C228T Increases Risk for Tumor Recurrence and Death in Head and Neck Cancer Patients. *Front. Oncol.* **2020**, *10*, 1275. [CrossRef]
14. Carvalho, A.L.; Nishimoto, I.N.; Califano, J.A.; Kowalski, L.P. Trends in incidence and prognosis for head and neck cancer in the United States: A site-specific analysis of the SEER database. *Int. J. Cancer* **2005**, *114*, 806–816. [CrossRef] [PubMed]
15. Barczak, W.; Suchorska, W.M.; Sobecka, A.; Bednarowicz, K.; Machczynski, P.; Golusinski, P.; Rubis, B.; Masternak, M.M.; Golusinski, W. hTERT C250T promoter mutation and telomere length as a molecular markers of cancer progression in patients with head and neck cancer. *Mol. Med. Rep.* **2017**, *16*, 441–446. [CrossRef] [PubMed]
16. Kang, H.; Kiess, A.; Chung, C.H. Emerging biomarkers in head and neck cancer in the era of genomics. *Nat. Rev. Clin. Oncol.* **2015**, *12*, 11–26. [CrossRef]
17. The Cancer Genome Atlas Network. Comprehensive genomic characterization of head and neck squamous cell carcinomas. *Nature* **2015**, *517*, 576–582. [CrossRef]
18. Juodzbalys, G.; Kasradze, D.; Cicciù, M.; Sudeikis, A.; Banys, L.; Galindo-Moreno, P.; Guobis, Z. Modern molecular biomarkers of head and neck cancer. Part I. Epigenetic diagnostics and prognostics: Systematic review. *Cancer Biomark. Sect. A Dis. Markers* **2016**, *17*, 487–502. [CrossRef]

19. Ausoni, S.; Boscolo-Rizzo, P.; Singh, B.; Da Mosto, M.C.; Spinato, G.; Tirelli, G.; Spinato, R.; Azzarello, G. Targeting cellular and molecular drivers of head and neck squamous cell carcinoma: Current options and emerging perspectives. *Cancer Metastasis Rev.* **2016**, *35*, 413–426. [CrossRef]

20. Agrawal, N.; Frederick, M.J.; Pickering, C.R.; Bettegowda, C.; Chang, K.; Li, R.J.; Fakhry, C.; Xie, T.X.; Zhang, J.; Wang, J.; et al. Exome sequencing of head and neck squamous cell carcinoma reveals inactivating mutations in NOTCH1. *Science* **2011**, *333*, 1154–1157. [CrossRef]

21. Stransky, N.; Egloff, A.M.; Tward, A.D.; Kostic, A.D.; Cibulskis, K.; Sivachenko, A.; Kryukov, G.V.; Lawrence, M.S.; Sougnez, C.; McKenna, A.; et al. The mutational landscape of head and neck squamous cell carcinoma. *Science* **2011**, *333*, 1157–1160. [CrossRef] [PubMed]

22. Parfenov, M.; Pedamallu, C.S.; Gehlenborg, N.; Freeman, S.S.; Danilova, L.; Bristow, C.A.; Lee, S.; Hadjipanayis, A.G.; Ivanova, E.V.; Wilkerson, M.D.; et al. Characterization of HPV and host genome interactions in primary head and neck cancers. *Proc. Natl. Acad. Sci. USA* **2014**, *111*, 15544–15549. [CrossRef]

23. Seiwert, T.Y.; Zuo, Z.; Keck, M.K.; Khattri, A.; Pedamallu, C.S.; Stricker, T.; Brown, C.; Pugh, T.J.; Stojanov, P.; Cho, J.; et al. Integrative and comparative genomic analysis of HPV-positive and HPV-negative head and neck squamous cell carcinomas. *Clin. Cancer Res.* **2015**, *21*, 632–641. [CrossRef] [PubMed]

24. Lechner, M.; Frampton, G.M.; Fenton, T.; Feber, A.; Palmer, G.; Jay, A.; Pillay, N.; Forster, M.; Cronin, M.T.; Lipson, D.; et al. Targeted next-generation sequencing of head and neck squamous cell carcinoma identifies novel genetic alterations in HPV+ and HPV-tumors. *Genome Med.* **2013**, *5*, 49. [CrossRef] [PubMed]

25. Killela, P.J.; Reitman, Z.J.; Jiao, Y.; Bettegowda, C.; Agrawal, N.; Diaz, L.A., Jr.; Friedman, A.H.; Friedman, H.; Gallia, G.L.; Giovanella, B.C.; et al. TERT promoter mutations occur frequently in gliomas and a subset of tumors derived from cells with low rates of self-renewal. *Proc. Natl. Acad. Sci. USA* **2013**, *110*, 6021–6026. [CrossRef] [PubMed]

26. Vinothkumar, V.; Arunkumar, G.; Revathidevi, S.; Arun, K.; Manikandan, M.; Rao, A.K.; Rajkumar, K.S.; Ajay, C.; Rajaraman, R.; Ramani, R.; et al. TERT promoter hot spot mutations are frequent in Indian cervical and oral squamous cell carcinomas. *Tumour Biol. J. Int. Soc. Oncodevelopmental Biol. Med.* **2016**, *37*, 7907–7913. [CrossRef] [PubMed]

27. Qu, Y.; Dang, S.; Wu, K.; Shao, Y.; Yang, Q.; Ji, M.; Shi, B.; Hou, P. TERT promoter mutations predict worse survival in laryngeal cancer patients. *Int. J. Cancer* **2014**, *135*, 1008–1010. [CrossRef]

28. Chang, K.P.; Wang, C.I.; Pickering, C.R.; Huang, Y.; Tsai, C.N.; Tsang, N.M.; Kao, H.K.; Cheng, M.H.; Myers, J.N. Prevalence of promoter mutations in the TERT gene in oral cavity squamous cell carcinoma. *Head Neck* **2017**, *39*, 1131–1137. [CrossRef]

29. Yu, Y.; Fan, D.; Song, X.; Zakeri, K.; Chen, L.; Kang, J.; McBride, S.; Tsai, C.J.; Dunn, L.; Sherman, E.; et al. *TERT* Promoter Mutations Are Enriched in Oral Cavity Cancers and Associated With Locoregional Recurrence. *JCO Precis. Oncol.* **2021**, *5*, 1259–1269. [CrossRef]

30. Dratwa, M.; Wysoczańska, B.; Łacina, P.; Kubik, T.; Bogunia-Kubik, K. TERT-Regulation and Roles in Cancer Formation. *Front. Immunol.* **2020**, *11*, 589929. [CrossRef]

31. Giardini, M.A.; Segatto, M.; da Silva, M.S.; Nunes, V.S.; Cano, M.I. Telomere and telomerase biology. *Prog. Mol. Biol. Transl. Sci.* **2014**, *125*, 1–40. [CrossRef] [PubMed]

32. Schwaederle, M.; Krishnamurthy, N.; Daniels, G.A.; Piccioni, D.E.; Kesari, S.; Fanta, P.T.; Schwab, R.B.; Patel, S.P.; Parker, B.A.; Kurzrock, R. Telomerase reverse transcriptase promoter alterations across cancer types as detected by next-generation sequencing: A clinical and molecular analysis of 423 patients. *Cancer* **2018**, *124*, 1288–1296. [CrossRef] [PubMed]

33. Bhari, V.K.; Kumar, D.; Kumar, S.; Mishra, R. Shelterin complex gene: Prognosis and therapeutic vulnerability in cancer. *Biochem. Biophys. Rep.* **2021**, *26*, 100937. [CrossRef] [PubMed]

34. Rhodes, D.; Fairall, L.; Simonsson, T.; Court, R.; Chapman, L. Telomere architecture. *EMBO Rep.* **2002**, *3*, 1139–1145. [CrossRef]

35. Guterres, A.N.; Villanueva, J. Targeting telomerase for cancer therapy. *Oncogene* **2020**, *39*, 5811–5824. [CrossRef]

36. Cong, Y.S.; Wright, W.E.; Shay, J.W. Human telomerase and its regulation. *Microbiol. Mol. Biol. Rev. MMBR* **2002**, *66*, 407–425. [CrossRef]

37. Cao, Y.; Bryan, T.M.; Reddel, R.R. Increased copy number of the TERT and TERC telomerase subunit genes in cancer cells. *Cancer Sci.* **2008**, *99*, 1092–1099. [CrossRef]

38. Lu, W.; Zhang, Y.; Liu, D.; Songyang, Z.; Wan, M. Telomeres-structure, function, and regulation. *Exp. Cell Res.* **2013**, *319*, 133–141. [CrossRef]

39. McKelvey, B.A.; Umbricht, C.B.; Zeiger, M.A. Telomerase Reverse Transcriptase (TERT) Regulation in Thyroid Cancer: A Review. *Front. Endocrinol.* **2020**, *11*, 485. [CrossRef]

40. Greider, C.W.; Blackburn, E.H. The telomere terminal transferase of Tetrahymena is a ribonucleoprotein enzyme with two kinds of primer specificity. *Cell* **1987**, *51*, 887–898. [CrossRef]

41. Shammas, M.A. Telomeres, lifestyle, cancer, and aging. *Curr. Opin. Clin. Nutr. Metab. Care* **2011**, *14*, 28–34. [CrossRef] [PubMed]

42. Yuan, X.; Larsson, C.; Xu, D. Mechanisms underlying the activation of TERT transcription and telomerase activity in human cancer: Old actors and new players. *Oncogene* **2019**, *38*, 6172–6183. [CrossRef] [PubMed]

43. Jacobs, J.J.; de Lange, T. Significant role for p16INK4a in p53-independent telomere-directed senescence. *Curr. Biol. CB* **2004**, *14*, 2302–2308. [CrossRef] [PubMed]

44. Jafri, M.A.; Ansari, S.A.; Alqahtani, M.H.; Shay, J.W. Roles of telomeres and telomerase in cancer, and advances in telomerase-targeted therapies. *Genome Med.* **2016**, *8*, 69. [CrossRef]

45. Hanahan, D.; Weinberg, R.A. Hallmarks of cancer: The next generation. *Cell* **2011**, *144*, 646–674. [CrossRef]
46. Hanahan, D. Hallmarks of Cancer: New Dimensions. *Cancer Discov.* **2022**, *12*, 31–46. [CrossRef]
47. Chen, C.H.; Chen, R.J. Prevalence of telomerase activity in human cancer. *J. Formos. Med. Assoc.* **2011**, *110*, 275–289. [CrossRef]
48. Cesare, A.J.; Reddel, R.R. Alternative lengthening of telomeres: Models, mechanisms and implications. *Nat. Rev. Genet.* **2010**, *11*, 319–330. [CrossRef]
49. Shay, J.W. Are short telomeres predictive of advanced cancer? *Cancer Discov.* **2013**, *3*, 1096–1098. [CrossRef]
50. Leão, R.; Apolónio, J.D.; Lee, D.; Figueiredo, A.; Tabori, U.; Castelo-Branco, P. Mechanisms of human telomerase reverse transcriptase (hTERT) regulation: Clinical impacts in cancer. *J. Biomed. Sci.* **2018**, *25*, 22. [CrossRef]
51. Barthel, F.P.; Wei, W.; Tang, M.; Martinez-Ledesma, E.; Hu, X.; Amin, S.B.; Akdemir, K.C.; Seth, S.; Song, X.; Wang, Q.; et al. Systematic analysis of telomere length and somatic alterations in 31 cancer types. *Nat. Genet.* **2017**, *49*, 349–357. [CrossRef] [PubMed]
52. Masutomi, K.; Possemato, R.; Wong, J.M.; Currier, J.L.; Tothova, Z.; Manola, J.B.; Ganesan, S.; Lansdorp, P.M.; Collins, K.; Hahn, W.C. The telomerase reverse transcriptase regulates chromatin state and DNA damage responses. *Proc. Natl. Acad. Sci. USA* **2005**, *102*, 8222–8227. [CrossRef] [PubMed]
53. Nitta, E.; Yamashita, M.; Hosokawa, K.; Xian, M.; Takubo, K.; Arai, F.; Nakada, S.; Suda, T. Telomerase reverse transcriptase protects ATM-deficient hematopoietic stem cells from ROS-induced apoptosis through a telomere-independent mechanism. *Blood* **2011**, *117*, 4169–4180. [CrossRef] [PubMed]
54. Koh, C.M.; Khattar, E.; Leow, S.C.; Liu, C.Y.; Muller, J.; Ang, W.X.; Li, Y.; Franzoso, G.; Li, S.; Guccione, E.; et al. Telomerase regulates MYC-driven oncogenesis independent of its reverse transcriptase activity. *J. Clin. Investig.* **2015**, *125*, 2109–2122. [CrossRef]
55. Stewart, S.A.; Hahn, W.C.; O'Connor, B.F.; Banner, E.N.; Lundberg, A.S.; Modha, P.; Mizuno, H.; Brooks, M.W.; Fleming, M.; Zimonjic, D.B.; et al. Telomerase contributes to tumorigenesis by a telomere length-independent mechanism. *Proc. Natl. Acad. Sci. USA* **2002**, *99*, 12606–12611. [CrossRef]
56. Romaniuk, A.; Paszel-Jaworska, A.; Totoń, E.; Lisiak, N.; Hołysz, H.; Królak, A.; Grodecka-Gazdecka, S.; Rubiś, B. The non-canonical functions of telomerase: To turn off or not to turn off. *Mol. Biol. Rep.* **2019**, *46*, 1401–1411. [CrossRef]
57. Nassir, N.; Hyde, G.J.; Baskar, R. A telomerase with novel non-canonical roles: TERT controls cellular aggregation and tissue size in Dictyostelium. *PLoS Genet.* **2019**, *15*, e1008188. [CrossRef]
58. Saretzki, G. Extra-telomeric functions of human telomerase: Cancer, mitochondria and oxidative stress. *Curr. Pharm. Des.* **2014**, *20*, 6386–6403. [CrossRef]
59. Weinhold, N.; Jacobsen, A.; Schultz, N.; Sander, C.; Lee, W. Genome-wide analysis of noncoding regulatory mutations in cancer. *Nat. Genet.* **2014**, *46*, 1160–1165. [CrossRef]
60. Bell, R.J.; Rube, H.T.; Xavier-Magalhães, A.; Costa, B.M.; Mancini, A.; Song, J.S.; Costello, J.F. Understanding TERT Promoter Mutations: A Common Path to Immortality. *Mol. Cancer Res. MCR* **2016**, *14*, 315–323. [CrossRef]
61. Vinagre, J.; Almeida, A.; Pópulo, H.; Batista, R.; Lyra, J.; Pinto, V.; Coelho, R.; Celestino, R.; Prazeres, H.; Lima, L.; et al. Frequency of TERT promoter mutations in human cancers. *Nat. Commun.* **2013**, *4*, 2185. [CrossRef] [PubMed]
62. Huang, F.W.; Hodis, E.; Xu, M.J.; Kryukov, G.V.; Chin, L.; Garraway, L.A. Highly recurrent TERT promoter mutations in human melanoma. *Science* **2013**, *339*, 957–959. [CrossRef] [PubMed]
63. Horn, S.; Figl, A.; Rachakonda, P.S.; Fischer, C.; Sucker, A.; Gast, A.; Kadel, S.; Moll, I.; Nagore, E.; Hemminki, K. TERT promoter mutations in familial and sporadic melanoma. *Sciense* **2013**, *339*, 959–961. [CrossRef] [PubMed]
64. Heidenreich, B.; Rachakonda, P.S.; Hemminki, K.; Kumar, R. TERT promoter mutations in cancer development. *Curr. Opin. Genet. Dev.* **2014**, *24*, 30–37. [CrossRef]
65. Sharma, S.; Chowdhury, S. Emerging mechanisms of telomerase reactivation in cancer. *Trends Cancer* **2022**, *8*, 632–641. [CrossRef]
66. Min, J.; Shay, J.W. TERT Promoter Mutations Enhance Telomerase Activation by Long-Range Chromatin Interactions. *Cancer Discov.* **2016**, *6*, 1212–1214. [CrossRef]
67. Bell, R.J.; Rube, H.T.; Kreig, A.; Mancini, A.; Fouse, S.D.; Nagarajan, R.P.; Choi, S.; Hong, C.; He, D.; Pekmezci, M.; et al. Cancer. The transcription factor GABP selectively binds and activates the mutant TERT promoter in cancer. *Science* **2015**, *348*, 1036–1039. [CrossRef]
68. Stern, J.L.; Theodorescu, D.; Vogelstein, B.; Papadopoulos, N.; Cech, T.R. Mutation of the TERT promoter, switch to active chromatin, and monoallelic TERT expression in multiple cancers. *Genes Dev.* **2015**, *29*, 2219–2224. [CrossRef]
69. Borah, S.; Xi, L.; Zaug, A.J.; Powell, N.M.; Dancik, G.M.; Cohen, S.B.; Costello, J.C.; Theodorescu, D.; Cech, T.R. Cancer. TERT promoter mutations and telomerase reactivation in urothelial cancer. *Science* **2015**, *347*, 1006–1010. [CrossRef]
70. Koelsche, C.; Sahm, F.; Capper, D.; Reuss, D.; Sturm, D.; Jones, D.T.; Kool, M.; Northcott, P.A.; Wiestler, B.; Böhmer, K.; et al. Distribution of TERT promoter mutations in pediatric and adult tumors of the nervous system. *Acta Neuropathol.* **2013**, *126*, 907–915. [CrossRef]
71. Roh, M.R.; Park, K.H.; Chung, K.Y.; Shin, S.J.; Rha, S.Y.; Tsao, H. Telomerase reverse transcriptase (TERT) promoter mutations in Korean melanoma patients. *Am. J. Cancer Res.* **2017**, *7*, 134–138.
72. Gandini, S.; Zanna, I.; De Angelis, S.; Palli, D.; Raimondi, S.; Ribero, S.; Masala, G.; Suppa, M.; Bellerba, F.; Corso, F.; et al. TERT promoter mutations and melanoma survival: A comprehensive literature review and meta-analysis. *Crit. Rev. Oncol./Hematol.* **2021**, *160*, 103288. [CrossRef]

73. Liu, X.; Wu, G.; Shan, Y.; Hartmann, C.; von Deimling, A.; Xing, M. Highly prevalent *TERT* promoter mutations in bladder cancer and glioblastoma. *Cell Cycle* **2013**, *12*, 1637–1638. [CrossRef] [PubMed]
74. Hurst, C.D.; Platt, F.M.; Knowles, M.A. Comprehensive mutation analysis of the TERT promoter in bladder cancer and detection of mutations in voided urine. *Eur. Urol.* **2014**, *65*, 367–369. [CrossRef] [PubMed]
75. Allory, Y.; Beukers, W.; Sagrera, A.; Flández, M.; Marqués, M.; Márquez, M.; van der Keur, K.A.; Dyrskjot, L.; Lurkin, I.; Vermeij, M.; et al. Telomerase reverse transcriptase promoter mutations in bladder cancer: High frequency across stages, detection in urine, and lack of association with outcome. *Eur. Urol.* **2014**, *65*, 360–366. [CrossRef] [PubMed]
76. Rachakonda, P.S.; Hosen, I.; de Verdier, P.J.; Fallah, M.; Heidenreich, B.; Ryk, C.; Wiklund, N.P.; Steineck, G.; Schadendorf, D.; Hemminki, K.; et al. TERT promoter mutations in bladder cancer affect patient survival and disease recurrence through modification by a common polymorphism. *Proc. Natl. Acad. Sci. USA* **2013**, *110*, 17426–17431. [CrossRef] [PubMed]
77. Wang, K.; Liu, T.; Ge, N.; Liu, L.; Yuan, X.; Liu, J.; Kong, F.; Wang, C.; Ren, H.; Yan, K.; et al. TERT promoter mutations are associated with distant metastases in upper tract urothelial carcinomas and serve as urinary biomarkers detected by a sensitive castPCR. *Oncotarget* **2014**, *5*, 12428–12439. [CrossRef]
78. Huang, D.S.; Wang, Z.; He, X.J.; Diplas, B.H.; Yang, R.; Killela, P.J.; Meng, Q.; Ye, Z.Y.; Wang, W.; Jiang, X.T.; et al. Recurrent TERT promoter mutations identified in a large-scale study of multiple tumour types are associated with increased TERT expression and telomerase activation. *Eur. J. Cancer* **2015**, *51*, 969–976. [CrossRef]
79. Stoehr, R.; Taubert, H.; Zinnall, U.; Giedl, J.; Gaisa, N.T.; Burger, M.; Ruemmele, P.; Hurst, C.D.; Knowles, M.A.; Wullich, B.; et al. Frequency of TERT Promoter Mutations in Prostate Cancer. *Pathobiol. J. Immunopathol. Mol. Cell. Biol.* **2015**, *82*, 53–57. [CrossRef]
80. Nonoguchi, N.; Ohta, T.; Oh, J.E.; Kim, Y.H.; Kleihues, P.; Ohgaki, H. TERT promoter mutations in primary and secondary glioblastomas. *Acta Neuropathol.* **2013**, *126*, 931–937. [CrossRef]
81. Pezzuto, F.; Buonaguro, L.; Buonaguro, F.M.; Tornesello, M.L. Frequency and geographic distribution of TERT promoter mutations in primary hepatocellular carcinoma. *Infect. Agents Cancer* **2017**, *12*, 27. [CrossRef] [PubMed]
82. Lombardo, D.; Saitta, C.; Giosa, D.; Di Tocco, F.C.; Musolino, C.; Caminiti, G.; Chines, V.; Franzè, M.S.; Alibrandi, A.; Navarra, G.; et al. Frequency of somatic mutations in TERT promoter, TP53 and CTNNB1 genes in patients with hepatocellular carcinoma from Southern Italy. *Oncol. Lett.* **2020**, *19*, 2368–2374. [CrossRef] [PubMed]
83. Nault, J.C.; Mallet, M.; Pilati, C.; Calderaro, J.; Bioulac-Sage, P.; Laurent, C.; Laurent, A.; Cherqui, D.; Balabaud, C.; Zucman-Rossi, J. High frequency of telomerase reverse-transcriptase promoter somatic mutations in hepatocellular carcinoma and preneoplastic lesions. *Nat. Commun.* **2013**, *4*, 2218. [CrossRef] [PubMed]
84. Cevik, D.; Yildiz, G.; Ozturk, M. Common telomerase reverse transcriptase promoter mutations in hepatocellular carcinomas from different geographical locations. *World J. Gastroenterol.* **2015**, *21*, 311–317. [CrossRef] [PubMed]
85. Liu, X.; Bishop, J.; Shan, Y.; Pai, S.; Liu, D.; Murugan, A.K.; Sun, H.; El-Naggar, A.K.; Xing, M. Highly prevalent TERT promoter mutations in aggressive thyroid cancers. *Endocr.-Relat. Cancer* **2013**, *20*, 603–610. [CrossRef]
86. Yang, H.; Park, H.; Ryu, H.J.; Heo, J.; Kim, J.S.; Oh, Y.L.; Choe, J.H.; Kim, J.H.; Kim, J.S.; Jang, H.W.; et al. Frequency of TERT Promoter Mutations in Real-World Analysis of 2,092 Thyroid Carcinoma Patients. *Endocrinol. Metab.* **2022**, *37*, 652–663. [CrossRef]
87. Alzahrani, A.S.; Alsaadi, R.; Murugan, A.K.; Sadiq, B.B. TERT Promoter Mutations in Thyroid Cancer. *Horm. Cancer* **2016**, *7*, 165–177. [CrossRef]
88. Campanella, N.C.; Celestino, R.; Pestana, A.; Scapulatempo-Neto, C.; de Oliveira, A.T.; Brito, M.J.; Gouveia, A.; Lopes, J.M.; Guimarães, D.P.; Soares, P.; et al. Low frequency of TERT promoter mutations in gastrointestinal stromal tumors (GISTs). *Eur. J. Hum. Genet. EJHG* **2015**, *23*, 877–879. [CrossRef]
89. Tallet, A.; Nault, J.C.; Renier, A.; Hysi, I.; Galateau-Sallé, F.; Cazes, A.; Copin, M.C.; Hofman, P.; Andujar, P.; Le Pimpec-Barthes, F.; et al. Overexpression and promoter mutation of the TERT gene in malignant pleural mesothelioma. *Oncogene* **2014**, *33*, 3748–3752. [CrossRef]
90. Griewank, K.G.; Schilling, B.; Murali, R.; Bielefeld, N.; Schwamborn, M.; Sucker, A.; Zimmer, L.; Hillen, U.; Schaller, J.; Brenn, T.; et al. TERT promoter mutations are frequent in atypical fibroxanthomas and pleomorphic dermal sarcomas. *Mod. Pathol.* **2014**, *27*, 502–508. [CrossRef]
91. Koelsche, C.; Renner, M.; Hartmann, W.; Brandt, R.; Lehner, B.; Waldburger, N.; Alldinger, I.; Schmitt, T.; Egerer, G.; Penzel, R.; et al. TERT promoter hotspot mutations are recurrent in myxoid liposarcomas but rare in other soft tissue sarcoma entities. *J. Exp. Clin. Cancer Res. CR* **2014**, *33*, 33. [CrossRef]
92. Scott, G.A.; Laughlin, T.S.; Rothberg, P.G. Mutations of the TERT promoter are common in basal cell carcinoma and squamous cell carcinoma. *Mod. Pathol.* **2014**, *27*, 516–523. [CrossRef]
93. Cheng, K.A.; Kurtis, B.; Babayeva, S.; Zhuge, J.; Tantchou, I.; Cai, D.; Lafaro, R.J.; Fallon, J.T.; Zhong, M. Heterogeneity of TERT promoter mutations status in squamous cell carcinomas of different anatomical sites. *Ann. Diagn. Pathol.* **2015**, *19*, 146–148. [CrossRef] [PubMed]
94. Zhao, Y.; Gao, Y.; Chen, Z.; Hu, X.; Zhou, F.; He, J. Low frequency of TERT promoter somatic mutation in 313 sporadic esophageal squamous cell carcinomas. *Int. J. Cancer* **2014**, *134*, 493–494. [CrossRef] [PubMed]
95. Kim, S.K.; Kim, J.H.; Han, J.H.; Cho, N.H.; Kim, S.J.; Kim, S.I.; Choo, S.H.; Kim, J.S.; Park, B.; Kwon, J.E. TERT promoter mutations in penile squamous cell carcinoma: High frequency in non-HPV-related type and association with favorable clinicopathologic features. *J. Cancer Res. Clin. Oncol.* **2021**, *147*, 1125–1135. [CrossRef] [PubMed]

96. Annunziata, C.; Pezzuto, F.; Greggi, S.; Ionna, F.; Losito, S.; Botti, G.; Buonaguro, L.; Buonaguro, F.M.; Tornesello, M.L. Distinct profiles of TERT promoter mutations and telomerase expression in head and neck cancer and cervical carcinoma. *Int. J. Cancer* **2018**, *143*, 1153–1161. [CrossRef]

97. Morris, L.G.T.; Chandramohan, R.; West, L.; Zehir, A.; Chakravarty, D.; Pfister, D.G.; Wong, R.J.; Lee, N.Y.; Sherman, E.J.; Baxi, S.S.; et al. The Molecular Landscape of Recurrent and Metastatic Head and Neck Cancers: Insights From a Precision Oncology Sequencing Platform. *JAMA Oncol.* **2017**, *3*, 244–255. [CrossRef]

98. Liu, R.; Xing, M. TERT promoter mutations in thyroid cancer. *Endocr.-Relat. Cancer* **2016**, *23*, R143–R155. [CrossRef]

99. Andrés-Lencina, J.J.; Rachakonda, S.; García-Casado, Z.; Srinivas, N.; Skorokhod, A.; Requena, C.; Soriano, V.; Kumar, R.; Nagore, E. TERT promoter mutation subtypes and survival in stage I and II melanoma patients. *Int. J. Cancer* **2019**, *144*, 1027–1036. [CrossRef]

100. Heidenreich, B.; Kumar, R. Altered TERT promoter and other genomic regulatory elements: Occurrence and impact. *Int. J. Cancer* **2017**, *141*, 867–876. [CrossRef]

101. Heidenreich, B.; Kumar, R. TERT promoter mutations in telomere biology. *Mutat. Res. Rev. Mutat. Res.* **2017**, *771*, 15–31. [CrossRef] [PubMed]

102. Xu, Y.; Goldkorn, A. Telomere and Telomerase Therapeutics in Cancer. *Genes* **2016**, *7*, 22. [CrossRef] [PubMed]

103. Asai, A.; Oshima, Y.; Yamamoto, Y.; Uochi, T.A.; Kusaka, H.; Akinaga, S.; Yamashita, Y.; Pongracz, K.; Pruzan, R.; Wunder, E.; et al. A novel telomerase template antagonist (GRN163) as a potential anticancer agent. *Cancer Res.* **2003**, *63*, 3931–3939. [PubMed]

104. Chiappori, A.A.; Kolevska, T.; Spigel, D.R.; Hager, S.; Rarick, M.; Gadgeel, S.; Blais, N.; Von Pawel, J.; Hart, L.; Reck, M.; et al. A randomized phase II study of the telomerase inhibitor imetelstat as maintenance therapy for advanced non-small-cell lung cancer. *Ann. Oncol.* **2015**, *26*, 354–362. [CrossRef]

105. Kozloff, M.; Sledge, G.; Benedetti, F.; Starr, A.; Wallace, J.; Stuart, M.; Gruver, D.; Miller, K. Phase I study of imetelstat (GRN163L) in combination with paclitaxel (P) and bevacizumab (B) in patients (pts) with locally recurrent or metastatic breast cancer (MBC). *J. Clin. Oncol.* **2010**, *28*, 2598. [CrossRef]

106. Tefferi, A.; Lasho, T.L.; Begna, K.H.; Patnaik, M.M.; Zblewski, D.L.; Finke, C.M.; Laborde, R.R.; Wassie, E.; Schimek, L.; Hanson, C.A.; et al. A Pilot Study of the Telomerase Inhibitor Imetelstat for Myelofibrosis. *N. Engl. J. Med.* **2015**, *373*, 908–919. [CrossRef]

107. Baerlocher, G.M.; Oppliger Leibundgut, E.; Ottmann, O.G.; Spitzer, G.; Odenike, O.; McDevitt, M.A.; Röth, A.; Daskalakis, M.; Burington, B.; Stuart, M.; et al. Telomerase Inhibitor Imetelstat in Patients with Essential Thrombocythemia. *N. Engl. J. Med.* **2015**, *373*, 920–928. [CrossRef]

108. Tefferi, A.; Begna, K.; Laborde, R.R.; Patnaik, M.M.; Lasho, T.L.; Zblewski, D.; Finke, C.; Schimek, L.; LaPlant, B.R.; Hanson, C.A. *Imetelstat, a Telomerase Inhibitor, Induces Morphologic and Molecular Remissions in Myelofibrosis and Reversal of Bone Marrow Fibrosis*; American Society of Hematology: Washington, DC, USA, 2013.

109. Steensma, D.P.; Fenaux, P.; Van Eygen, K.; Raza, A.; Santini, V.; Germing, U.; Font, P.; Diez-Campelo, M.; Thepot, S.; Vellenga, E.; et al. Imetelstat Achieves Meaningful and Durable Transfusion Independence in High Transfusion-Burden Patients With Lower-Risk Myelodysplastic Syndromes in a Phase II Study. *J. Clin. Oncol.* **2021**, *39*, 48–56. [CrossRef]

110. Mascarenhas, J.; Harrison, C.N.; Kiladjian, J.J.; Komrokji, R.S.; Koschmieder, S.; Vannucchi, A.M.; Berry, T.; Redding, D.; Sherman, L.; Dougherty, S.; et al. Imetelstat in intermediate-2 or high-risk myelofibrosis refractory to JAK inhibitor: IMpactMF phase III study design. *Future Oncol.* **2022**, *18*, 2393–2402. [CrossRef]

111. Damm, K.; Hemmann, U.; Garin-Chesa, P.; Hauel, N.; Kauffmann, I.; Priepke, H.; Niestroj, C.; Daiber, C.; Enenkel, B.; Guilliard, B.; et al. A highly selective telomerase inhibitor limiting human cancer cell proliferation. *EMBO J.* **2001**, *20*, 6958–6968. [CrossRef]

112. Nasrollahzadeh, A.; Bashash, D.; Kabuli, M.; Zandi, Z.; Kashani, B.; Zaghal, A.; Mousavi, S.A.; Ghaffari, S.H. Arsenic trioxide and BIBR1532 synergistically inhibit breast cancer cell proliferation through attenuation of NF-κB signaling pathway. *Life Sci.* **2020**, *257*, 118060. [CrossRef] [PubMed]

113. Tawfik, H.O.; El-Hamaky, A.A.; El-Bastawissy, E.A.; Shcherbakov, K.A.; Veselovsky, A.V.; Gladilina, Y.A.; Zhdanov, D.D.; El-Hamamsy, M.H. New Genetic Bomb Trigger: Design, Synthesis, Molecular Dynamics Simulation, and Biological Evaluation of Novel BIBR1532-Related Analogs Targeting Telomerase against Non-Small Cell Lung Cancer. *Pharmaceuticals* **2022**, *15*, 481. [CrossRef] [PubMed]

114. Doğan, F.; Özateş, N.P.; Bağca, B.G.; Abbaszadeh, Z.; Söğütlü, F.; Gasımlı, R.; Gündüz, C.; Biray Avcı, Ç. Investigation of the effect of telomerase inhibitor BIBR1532 on breast cancer and breast cancer stem cells. *J. Cell. Biochem.* **2019**, *120*, 1282–1293. [CrossRef] [PubMed]

115. Bashash, D.; Zareii, M.; Safaroghli-Azar, A.; Omrani, M.D.; Ghaffari, S.H. Inhibition of telomerase using BIBR1532 enhances doxorubicin-induced apoptosis in pre-B acute lymphoblastic leukemia cells. *Hematology* **2017**, *22*, 330–340. [CrossRef]

116. Shi, Y.; Sun, L.; Chen, G.; Zheng, D.; Li, L.; Wei, W. A combination of the telomerase inhibitor, BIBR1532, and paclitaxel synergistically inhibit cell proliferation in breast cancer cell lines. *Target. Oncol.* **2015**, *10*, 565–573. [CrossRef]

117. Ding, X.; Cheng, J.; Pang, Q.; Wei, X.; Zhang, X.; Wang, P.; Yuan, Z.; Qian, D. BIBR1532, a Selective Telomerase Inhibitor, Enhances Radiosensitivity of Non-Small Cell Lung Cancer Through Increasing Telomere Dysfunction and ATM/CHK1 Inhibition. *Int. J. Radiat. Oncol. Biol. Phys.* **2019**, *105*, 861–874. [CrossRef]

118. Altamura, G.; Degli Uberti, B.; Galiero, G.; De Luca, G.; Power, K.; Licenziato, L.; Maiolino, P.; Borzacchiello, G. The Small Molecule BIBR1532 Exerts Potential Anti-cancer Activities in Preclinical Models of Feline Oral Squamous Cell Carcinoma Through Inhibition of Telomerase Activity and Down-Regulation of TERT. *Front. Vet. Sci.* **2020**, *7*, 620776. [CrossRef]

119. Ameri, Z.; Ghiasi, S.; Farsinejad, A.; Hassanshahi, G.; Ehsan, M.; Fatemi, A. Telomerase inhibitor MST-312 induces apoptosis of multiple myeloma cells and down-regulation of anti-apoptotic, proliferative and inflammatory genes. *Life Sci.* **2019**, *228*, 66–71. [CrossRef]

120. Biffi, G.; Tannahill, D.; McCafferty, J.; Balasubramanian, S. Quantitative visualization of DNA G-quadruplex structures in human cells. *Nat. Chem.* **2013**, *5*, 182–186. [CrossRef]

121. Konieczna, N.; Romaniuk-Drapała, A.; Lisiak, N.; Totoń, E.; Paszel-Jaworska, A.; Kaczmarek, M.; Rubiś, B. Telomerase Inhibitor TMPyP4 Alters Adhesion and Migration of Breast-Cancer Cells MCF7 and MDA-MB-231. *Int. J. Mol. Sci.* **2019**, *20*, 2670. [CrossRef]

122. Leonetti, C.; Scarsella, M.; Riggio, G.; Rizzo, A.; Salvati, E.; D'Incalci, M.; Staszewsky, L.; Frapolli, R.; Stevens, M.F.; Stoppacciaro, A.; et al. G-quadruplex ligand RHPS4 potentiates the antitumor activity of camptothecins in preclinical models of solid tumors. *Clin. Cancer Res.* **2008**, *14*, 7284–7291. [CrossRef] [PubMed]

123. Fujimori, J.; Matsuo, T.; Shimose, S.; Kubo, T.; Ishikawa, M.; Yasunaga, Y.; Ochi, M. Antitumor effects of telomerase inhibitor TMPyP4 in osteosarcoma cell lines. *J. Orthop. Res.* **2011**, *29*, 1707–1711. [CrossRef]

124. Mikami-Terao, Y.; Akiyama, M.; Yuza, Y.; Yanagisawa, T.; Yamada, O.; Yamada, H. Antitumor activity of G-quadruplex-interactive agent TMPyP4 in K562 leukemic cells. *Cancer Lett.* **2008**, *261*, 226–234. [CrossRef] [PubMed]

125. Zhou, G.; Liu, X.; Li, Y.; Xu, S.; Ma, C.; Wu, X.; Cheng, Y.; Yu, Z.; Zhao, G.; Chen, Y. Telomere targeting with a novel G-quadruplex-interactive ligand BRACO-19 induces T-loop disassembly and telomerase displacement in human glioblastoma cells. *Oncotarget* **2016**, *7*, 14925–14939. [CrossRef] [PubMed]

126. Kim, M.Y.; Vankayalapati, H.; Shin-Ya, K.; Wierzba, K.; Hurley, L.H. Telomestatin, a potent telomerase inhibitor that interacts quite specifically with the human telomeric intramolecular g-quadruplex. *J. Am. Chem. Soc.* **2002**, *124*, 2098–2099. [CrossRef] [PubMed]

127. Hasegawa, D.; Okabe, S.; Okamoto, K.; Nakano, I.; Shin-ya, K.; Seimiya, H. G quadruplex ligand-induced DNA damage response coupled with telomere dysfunction and replication stress in glioma stem cells. *Biochem. Biophys. Res. Commun.* **2016**, *471*, 75–81. [CrossRef] [PubMed]

128. Yao, Y.X.; Xu, B.H.; Zhang, Y. CX-3543 Promotes Cell Apoptosis through Downregulation of CCAT1 in Colon Cancer Cells. *BioMed Res. Int.* **2018**, *2018*, 9701957. [CrossRef] [PubMed]

129. Xu, H.; Di Antonio, M.; McKinney, S.; Mathew, V.; Ho, B.; O'Neil, N.J.; Santos, N.D.; Silvester, J.; Wei, V.; Garcia, J.; et al. CX-5461 is a DNA G-quadruplex stabilizer with selective lethality in BRCA1/2 deficient tumours. *Nat. Commun.* **2017**, *8*, 14432. [CrossRef]

130. Carvalho, J.; Mergny, J.L.; Salgado, G.F.; Queiroz, J.A.; Cruz, C. G-quadruplex, Friend or Foe: The Role of the G-quartet in Anticancer Strategies. *Trends Mol. Med.* **2020**, *26*, 848–861. [CrossRef]

131. Mender, I.; Gryaznov, S.; Dikmen, Z.G.; Wright, W.E.; Shay, J.W. Induction of telomere dysfunction mediated by the telomerase substrate precursor 6-thio-2′-deoxyguanosine. *Cancer Discov.* **2015**, *5*, 82–95. [CrossRef]

132. Zeng, X.; Hernandez-Sanchez, W.; Xu, M.; Whited, T.L.; Baus, D.; Zhang, J.; Berdis, A.J.; Taylor, D.J. Administration of a Nucleoside Analog Promotes Cancer Cell Death in a Telomerase-Dependent Manner. *Cell Rep.* **2018**, *23*, 3031–3041. [CrossRef] [PubMed]

133. Sarkar, S.; Faller, D.V. T-oligos inhibit growth and induce apoptosis in human ovarian cancer cells. *Oligonucleotides* **2011**, *21*, 47–53. [CrossRef] [PubMed]

134. Pitman, R.T.; Wojdyla, L.; Puri, N. Mechanism of DNA damage responses induced by exposure to an oligonucleotide homologous to the telomere overhang in melanoma. *Oncotarget* **2013**, *4*, 761–771. [CrossRef] [PubMed]

135. Ellingsen, E.B.; Mangsbo, S.M.; Hovig, E.; Gaudernack, G. Telomerase as a Target for Therapeutic Cancer Vaccines and Considerations for Optimizing Their Clinical Potential. *Front. Immunol.* **2021**, *12*, 682492. [CrossRef]

136. Negrini, S.; De Palma, R.; Filaci, G. Anti-cancer Immunotherapies Targeting Telomerase. *Cancers* **2020**, *12*, 2260. [CrossRef]

137. Middleton, G.; Silcocks, P.; Cox, T.; Valle, J.; Wadsley, J.; Propper, D.; Coxon, F.; Ross, P.; Madhusudan, S.; Roques, T.; et al. Gemcitabine and capecitabine with or without telomerase peptide vaccine GV1001 in patients with locally advanced or metastatic pancreatic cancer (TeloVac): An open-label, randomised, phase 3 trial. *Lancet. Oncol.* **2014**, *15*, 829–840. [CrossRef]

138. Zanetti, M. A second chance for telomerase reverse transcriptase in anticancer immunotherapy. *Nat. Rev. Clin. Oncol.* **2017**, *14*, 115–128. [CrossRef]

139. Duperret, E.K.; Wise, M.C.; Trautz, A.; Villarreal, D.O.; Ferraro, B.; Walters, J.; Yan, J.; Khan, A.; Masteller, E.; Humeau, L.; et al. Synergy of Immune Checkpoint Blockade with a Novel Synthetic Consensus DNA Vaccine Targeting TERT. *Mol. Ther.* **2018**, *26*, 435–445. [CrossRef]

140. Fujita, K.; Kimura, M.; Kondo, N.; Sakakibara, A.; Sano, D.; Ishiguro, Y.; Tsukuda, M. Anti-tumor effects of telomelysin for head and neck squamous cell carcinoma. *Oncol. Rep.* **2008**, *20*, 1363–1368. [CrossRef]

141. Kondo, N.; Tsukuda, M.; Kimura, M.; Fujita, K.; Sakakibara, A.; Takahashi, H.; Ishiguro, Y.; Toth, G.; Matsuda, H. Antitumor effects of telomelysin in combination with paclitaxel or cisplatin on head and neck squamous cell carcinoma. *Oncol. Rep.* **2010**, *23*, 355–363. [CrossRef]

142. Nemunaitis, J.; Tong, A.W.; Nemunaitis, M.; Senzer, N.; Phadke, A.P.; Bedell, C.; Adams, N.; Zhang, Y.A.; Maples, P.B.; Chen, S.; et al. A phase I study of telomerase-specific replication competent oncolytic adenovirus (telomelysin) for various solid tumors. *Mol. Ther.* **2010**, *18*, 429–434. [CrossRef] [PubMed]

143. Heo, J.; Liang, J.-D.; Kim, C.W.; Woo, H.Y.; Shih, I.-L.; Su, T.-H.; Lin, Z.-Z.; Chang, S.; Urata, Y.; Chen, P.-J. *Safety and Dose-Escalation Study of a Targeted Oncolytic Adenovirus, Suratadenoturev (OBP-301), in Patients with Refractory Advanced Liver Cancer: Phase I Clinical Trial*; American Society of Clinical Oncology: Alexandria, Virginia, 2022.
144. Li, S.; Rosenberg, J.E.; Donjacour, A.A.; Botchkina, I.L.; Hom, Y.K.; Cunha, G.R.; Blackburn, E.H. Rapid inhibition of cancer cell growth induced by lentiviral delivery and expression of mutant-template telomerase RNA and anti-telomerase short-interfering RNA. *Cancer Res.* **2004**, *64*, 4833–4840. [CrossRef] [PubMed]
145. Akıncılar, S.C.; Khattar, E.; Boon, P.L.; Unal, B.; Fullwood, M.J.; Tergaonkar, V. Long-Range Chromatin Interactions Drive Mutant TERT Promoter Activation. *Cancer Discov.* **2016**, *6*, 1276–1291. [CrossRef] [PubMed]
146. Wu, Y.F.; Ou, C.C.; Chien, P.J.; Chang, H.Y.; Ko, J.L.; Wang, B.Y. Chidamide-induced ROS accumulation and miR-129-3p-dependent cell cycle arrest in non-small lung cancer cells. *Phytomedicine* **2019**, *56*, 94–102. [CrossRef]
147. Kim, M.K. Novel insight into the function of tankyrase. *Oncol. Lett.* **2018**, *16*, 6895–6902. [CrossRef] [PubMed]

Article

Prognostic Significance of β-Catenin in Relation to the Tumor Immune Microenvironment in Oral Cancer

Paloma Lequerica-Fernández [1,2], Tania Rodríguez-Santamarta [2,3], Eduardo García-García [3], Verónica Blanco-Lorenzo [4], Héctor E. Torres-Rivas [4], Juan P. Rodrigo [2,5,6,7], Faustino J. Suárez-Sánchez [8], Juana M. García-Pedrero [2,5,7,*] and Juan Carlos De Vicente [2,3,6,*]

[1] Department of Biochemistry, Hospital Universitario Central de Asturias (HUCA), Carretera de Rubín, 33011 Oviedo, Spain; palomalequerica@gmail.com

[2] Instituto de Investigación Sanitaria del Principado de Asturias (ISPA), Instituto Universitario de Oncología del Principado de Asturias (IUOPA), Universidad de Oviedo, Carretera de Rubín, 33011 Oviedo, Spain; taniasantamarta@gmail.com (T.R.-S.); jprodrigo@uniovi.es (J.P.R.)

[3] Department of Oral and Maxillofacial Surgery, Hospital Universitario Central de Asturias (HUCA), Carretera de Rubín, 33011 Oviedo, Spain; edu.gargar.95@gmail.com

[4] Department of Pathology, Hospital Universitario Central de Asturias (HUCA), Carretera de Rubín, 33011 Oviedo, Spain; veronica.blanco@sespa.es (V.B.-L.); ress_444@yahoo.com (H.E.T.-R.)

[5] Department of Otolaryngology, Hospital Universitario Central de Asturias (HUCA), Carretera de Rubín, 33011 Oviedo, Spain

[6] Department of Surgery, University of Oviedo, 33011 Oviedo, Spain

[7] Centro de Investigación Biomédica en Red de Cáncer (CIBERONC), Instituto de Salud Carlos III, Av. Monforte de Lemos, 28029 Madrid, Spain

[8] Department of Pathology, Hospital Universitario de Cabueñes, Prados, 33394 Gijon, Spain; faustinosuarezsanchez@gmail.com

[*] Correspondence: juanagp.finba@gmail.com (J.M.G.-P.); jvicente@uniovi.es (J.C.D.V.)

Citation: Lequerica-Fernández, P.; Rodríguez-Santamarta, T.; García-García, E.; Blanco-Lorenzo, V.; Torres-Rivas, H.E.; Rodrigo, J.P.; Suárez-Sánchez, F.J.; García-Pedrero, J.M.; De Vicente, J.C. Prognostic Significance of β-Catenin in Relation to the Tumor Immune Microenvironment in Oral Cancer. *Biomedicines* **2023**, *11*, 2675. https://doi.org/10.3390/biomedicines11102675

Academic Editor: Vui King Vincent-Chong

Received: 6 September 2023
Revised: 26 September 2023
Accepted: 28 September 2023
Published: 29 September 2023

Abstract: The aim of this study was to investigate the prognostic relevance of β-catenin expression in oral squamous cell carcinoma (OSCC) and to explore relationships with the tumor immune microenvironment. Expression of β-catenin and PD-L1, as well as lymphocyte and macrophage densities, were evaluated by immunohistochemistry in 125 OSCC patient specimens. Membranous β-catenin expression was detected in 102 (81.6%) and nuclear β-catenin in 2 (1.6%) tumors. There was an association between β-catenin expression, tumoral, and stromal CD8$^+$ T-cell infiltration (TIL) and also the type of tumor immune microenvironment (TIME). Tumors harboring nuclear β-catenin were associated with a type II TIME (i.e., immune ignorance defined by a negative PD-L1 expression and low CD8$^+$ TIL density), whereas tumors with membranous β-catenin expression were predominantly type IV (i.e., immune tolerance defined by negative PD-L1 and high CD8$^+$ TIL density). Combined, but not individual, high stromal CD8$^+$ TILs and membranous β-catenin expression was independently associated with better disease-specific survival (HR = 0.48, *p* = 0.019). Taken together, a combination of high stromal CD8$^+$ T-cell infiltration and membranous β-catenin in the tumor emerges as an independent predictor of better survival in OSCC patients.

Keywords: oral squamous cell carcinoma; β-catenin; PD-L1; CD8$^+$ lymphocytes; tumor microenvironment; prognosis

1. Introduction

Oral squamous cell carcinoma (OSCC) represents more than 90% of all oral malignancies, with over 377,713 new cases reported worldwide in the year 2020. It is characterized by poor prognosis, with a five-year mortality rate close to 50% [1], despite recent advances in therapy. It is therefore necessary to possess a deeper knowledge of tumor biology in order to improve OSCC treatment and patient survival. Immunotherapy is commonly used nowadays for the treatment of many cancers, and the PD1/PD-L1 axis has emerged in recent years as a key complex to maintain a balance between immune tolerance and immunopathology. Programmed cell death protein 1 (PD-1) is an inhibitory immune checkpoint expressed on the surface of T cells [2]. PD-L1 acts as main ligand of the PD-1 receptor, and it is expressed in activated T cells, B cells, dendritic cells, macrophages, and certain tumor types, including OSCC [2]. Tumor PD-L1 binding to PD-1 on T cells inhibits the CD8[+] T-cell activation/functions and promotes the induction of regulatory or suppressor T cells (Tregs), contributing to cancer immune evasion [3,4]. In turn, anti-PD-1/PD-L1 treatments block the interaction between PD-1 and its ligand, thereby restoring T-cell activity and the anti-tumor immune response. However, around 66–85% of patients do not respond to immunotherapy or show any significant clinical benefit [5].

Tumors interact continually with the surrounding microenvironment (TME), composed of diverse immune cells, fibroblasts, blood vessels, signaling molecules, and the extracellular matrix [6]. A lymphocytic reaction characterized by high T-cell infiltration (defined as a T cell-inflamed TME) is commonly associated with a favorable clinical outcome in OSCC patients, thus supporting the importance of T cell-mediated immunity in tumor clearance [7]. Nevertheless, it still remains unclear whether the TME plays a direct role in PD-L1 transcription regulation and tumor immune evasion [8]. Recently, a TCGA database study demonstrated that most cancers are inversely associated with a T cell-inflamed gene expression signature, hence emerging Wnt/β-catenin signaling activation as a potential causal pathway [9].

The Wnt/β-catenin signaling pathway encompasses two major categories: the canonical and the non-canonical Wnt pathway. The key elements involved in the canonical Wnt/β-catenin signaling pathway include Wnt, Frizzled (Fz) receptors, low-density lipoprotein-related protein 5/6 (LRP5/6) co-receptors, destruction complex components [adenomatous polyposis coli protein (APC), Axin, glycogen synthetase 3 (GSK3β), and casein protein kinase 1α (CK1 α)], Dishevelled (DVL), β-catenin, and T-cell factor (TCF)/lymphoid enhancer factor (LEF) transcription factors [10]. When Fz receptors are unoccupied, cytoplasmic β-catenin is degraded by the destruction complex after its sequential phosphorylation and subsequent ubiquitylation [11]. The first phosphorylation is at Ser45 by CK1α, and then at Thr41, Ser37, and Ser 33 by GSK3β. However, when the Wnt ligand binds to the Fz receptor and its LRP5/LRP6 co-receptor, Wnt signaling is activated [10], which results in Axin translocation and DVL phosphorylation, and ultimately destruction complex dissociation [12]. Subsequently, Axin, GSK3β, and CK1 migrate from the cytoplasm to the cell membrane, thus resulting in β-catenin stabilization through dephosphorylation. Stable β-catenin translocates into the cell nucleus where it interacts with TCF/LEF transcription factors to induce the expression of Wnt target genes, such as Axin-2, c-Myc, cyclin D1, ITF-2, Lgr5, MMP-1, MMP-7, and PPAR-δ [11,13], thereby altering cellular processes such as proliferation, differentiation, and stemness, and promoting cancer cell proliferation and survival [4,10]. In addition, there are two non-canonical Wnt pathway categories: the Wnt/PCP (planar cell polarity) that participates in cell polarity and migration and the Wnt/Ca^{2+} pathway that is crucial for cell adhesion and motility during gastrulation [4].

The Wnt/β-catenin pathway is aberrantly activated in numerous tumor types [4], including OSCC [14]. The altered expression of β-catenin correlates with oral tumorigenesis, and it has also been linked to a poor prognosis in OSCC [15]. Wnt/β-catenin has been considered the most important pathway in OSCC [16]. In a recent meta-analysis including 41 studies and 2746 OSCC patients, the loss of membrane expression, cytoplasmic expression, and/or nuclear expression of β-catenin was found a poor prognostic factor [17]. Furthermore, a novel role of β-catenin has been described, regulating PD-L1 transcription [8]. β-catenin activation increased PD-L1 transcription, limited CD8[+] T-cell activation, and promoted tumor growth. Conversely, β-catenin depletion reduced PD-L1 expression levels in tumor cells, enhanced CD8[+] T-cell infiltration, and inhibited tumor growth [8].

In this study, we investigated the significance of the β-catenin expression pattern in a large cohort of OSCC patients using tissue microarray immunohistochemistry, its association with the type of tumor immune microenvironment (TIME), and the potential prognostic relevance in OSCC.

2. Materials and Methods

2.1. Patients and Tissue Specimens

A cohort of 125 OSCC patients who received surgical treatment at the Hospital Universitario Central de Asturias (HUCA) between 1996 and 2007 was retrospectively selected. This study was conducted following the ethical criteria of the Declaration of Helsinki and approved by the HUCA Ethics Committee and also by the Regional CEIC from Principado de Asturias (date of approval 14 May 2019; approval number 136/19, for the project PI19/01255). Written informed consent was obtained from all patients. We retrieved clinical and histopathologic information from the patients' files, and pathology reports, which are summarized in Table 1. The clinical staging was determined according to the 8th edition of the AJCC TNM classification [18] and the histological grading according to the WHO classification [19].

Table 1. Clinical and pathological characteristics of 125 patients with oral squamous cell carcinoma.

Variable	Number (%)
Age (year) (mean ± SD; median; range)	58.69 ± 14.34; 57; 28–91
Gender	
Men	82 (66)
Women	43 (34)
Tobacco use	
Smoker	84 (67)
Non-smoker	41 (33)
Alcohol use	
Drinker	69 (55)
Non-drinker	56 (45)
Location of oral squamous oral cell carcinoma	
Tongue	51 (41)
Floor of the mouth	37 (30)
Gum	22 (18)
Buccal	7 (5)
Retromolar	6 (5)
Palate	2 (1)
Tumor status	
pT1	27 (22)
pT2	54 (43)
pT3	16 (13)
pT4	28 (22)
Nodal status	
pN0	76 (61)
pN1	25 (20)
pN2	24 (19)
Clinical stage	
Stage I	20 (16)
Stage II	32 (26)
Stage III	26 (21)
Stage IV	47 (37)
G status	
G1	80 (64)
G2	41 (33)
G3	4 (3)

Formalin-fixed paraffin-embedded (FFPE) tissue samples were sourced from the Principado de Asturias BioBank (PT20/0161) and processed following standard operating procedures. Clinicopathologic data were collected from clinical records. The inclusion criteria for all the participants enrolled were: (i) primary OSCC (International Classification of Disease-10 diagnosis codes: C02.0, C02.1, C02.2, C02.3, C03.0, C03.1, C04.0, C04.1, C05.0, C06.0, C06.1, and C06.2), (ii) treatment-naïve patients from whom we had formalin-fixed paraffin-embedded (FFPE) primary biopsy tissue samples, and (iii) with a minimum follow-up of at least three years for alive patients. In addition, the exclusion criteria were: (i) OSCC with immediate postoperative death, (ii) recurrent disease, (iii) neoadjuvant chemo- or radiotherapy, and (iv) missing survival data.

All patients underwent surgery of the primary tumor with curative intention as well as neck dissection. None of the patients received any treatment before surgery, but complementary radiotherapy and/or chemotherapy were administered in 75 (60%) and 14 (11%) cases, respectively. During the follow-up period (6 to 230 months), 19 (15%) patients suffered from a second primary tumor in the oral cavity, and 51 (42%) patients died of OSCC. The clinical endpoint of this study was disease-specific survival (DSS), calculated as the period of time between the initial treatment and the death caused by the tumor or the presence of a non-treatable recurrence.

Samples and data from donors included in this study were provided by the Principado de Asturias BioBank (PT20/0161), integrated in the Spanish National Biobanks and Biomodels Network financed with European funds and they were processed following standard operating procedures with the appropriate approval of the Ethical and Scientific Committees.

2.2. Immunohistochemistry (IHC)

Tissue microarrays (TMAs) were constructed by collecting 1 mm tissue cores from the most morphologically representative areas of formalin-fixed, paraffin-embedded (FFPE) tissue blocks. Three individual cores were taken per tumor block. Then, 3 μm tissue sections dried on Flex IHC microscope slides (DakoCytomation, Glostrup, Denmark) were heated with high-pH Envision Flex Target Retrieval solution (Dako) and stained on an automatic staining workstation (Dako Autostainer Plus, Dako) using monoclonal antibodies against β-catenin (BD Biosciences, #610153, 1:200 dilution), CD8 (Dako, clone C8/144B, prediluted), FoxP3 (Cell Signaling Technology, Danvers, MA, USA, clone D6O8R, 1:100 dilution), PD-L1 antibody (PD-L1 IHC 22C3 pharmDx, Dako SK006, clone 22C3, 1:200 dilution), CD20 (Dako, clone L26, #M0755; 1:200 dilution), CD4 (Dako, clone 4B12, 1:80 dilution), CD68 (Agilent-Dako, Santa Clara, CA, USA, clone KP1, prediluted), and CD163 (Biocare Medical, Pacheco, CA, USA, clone 10D6, 1:100 dilution). The antibody-antigen complexes were visualized using the Dako EnVision Flex + Visualization System (Dako) and diaminobenzidine chromogen as a substrate.

Each TMA also contained three cores of morphologically normal oral mucosa from non-oncological patients undergoing oral surgery, used as internal controls. No staining was observed when the primary antibody was omitted. Negative control was included by replacing the primary antibody with serum and positive controls using appropriate positive control tissue slides.

The IHC results were independently evaluated by two observers (VB-L and FJS-S), blinded to clinical information. The number of CD20[+], CD68[+], CD163[+], CD4[+], CD8[+], and FOXP3[+] cells in both the tumor nests and the tumor stroma was counted in each 1 mm^2 area from three independent HPFs at 400x, as we previously reported [20–22]. The median value was used as a cut-off point to separate patients based on the tumoral and stromal CD8[+] lymphocyte densities into two groups, above (high density) and below (low density) the median number of positive stained cells for the total patient population. Additionally, TIL intensity in the tumor nests and the surrounding stroma was subdivided into three groups: negative, mild-moderate, or intense. PD-L1 expression in more than 10% of the tumor cells was previously found to significantly associate with poorer survival [23], and therefore established as a cut-off point for subsequent analyses. Since β-catenin staining intensity within the tumor showed a homogeneous pattern, a semiquantitative scoring system based on staining positivity was applied for IHC evaluation into three categories: negative (0), membrane staining, and (1) nuclear staining (2). The type of tumor immune microenvironment (TIME) based on the presence of TILs and PD-L1 expression was determined according to the classification reported by Teng et al. [24].

2.3. Statistical Analysis

Statistical analyses were carried out using SPSS software version 27 (IBM Co., Armonk, NY, USA). Continuous variables were expressed as the mean $\pm$ standard deviation (SD), and absolute and relative frequencies were calculated for categorical variables. Associations between the numbers of immune cells infiltrating the tumors and β-catenin expression were assessed by using the Kruskall–Wallis test. Fisher's exact test was used to evaluate the relationship between β-catenin expression, the type of TIME, and TIL intensity. Potential associations of β-catenin and the type of TIME with the different clinicopathological variables were assessed using the Chi-square or Fisher's exact test.

Disease-specific survival (DSS) was estimated using the Kaplan–Meier method, and the log-rank test was used for comparisons between survival rates. Hazard ratios (HR) with their 95% confidence intervals (CI) were calculated using univariable and multivariable Cox regression models. All tests were two-sided, and p-values less than 0.05 were considered statistically significant.

3. Results

3.1. Immunohistochemical Analysis of β-Catenin, PD-L1, and CD8$^+$ TIL Density in OSCC Patient Samples

β-catenin immunostaining was evaluated in 125 OSCC samples. Membrane β-catenin expression was detected in 102 (81.6%) tumors, nuclear β-catenin was only found in 2 cases (1.6%), and no expression of β-catenin neither in the membrane nor the nucleus was observed in 21 cases (16.8%). Representative images of positive β-catenin staining (membranous and nuclear) are shown in Figure 1. A total of 104 cases (85%) exhibited positive PD-L1 immunostaining in tumor cells. The mean CD8$^+$ TILs in the tumor nests was 47.88 ± 57.16 cells per mm^2 (range: 0.00 to 288.33), and the mean stromal CD8$^+$ TILs 178.45 ± 203.21 (range: 0.33 to 1202.67) cells per mm^2.

3.2. Associations between the Expression of β-Catenin, PD-L1, and CD8$^+$ TIL Density in OSCC

There was an association between the infiltration of CD8$^+$ T cells in both stroma and tumor nests, and β-catenin immunoexpression (Table 2). The mean number of stromal CD8$^+$ T cells was higher in tumors harboring negative β-catenin expression, and the lowest CD8$^+$ TIL density was found in those tumors with positive nuclear β-catenin (Kruskal–Wallis test, $p = 0.02$). Similarly, the mean number of tumoral CD8$^+$ T cells was higher in tumors with negative β-catenin expression, and also the lowest CD8$^+$ TIL density was seen in the subset of tumors with positive nuclear β-catenin (Kruskal–Wallis test, $p = 0.029$). Nevertheless, since only two cases exhibited nuclear β-catenin, these figures should be cautiously taken into account. Moreover, there was a tendency of association between PD-L1 positivity and membranous β-catenin expression, although this relationship did not reach statistical significance (Fisher's exact test, $p = 0.25$).

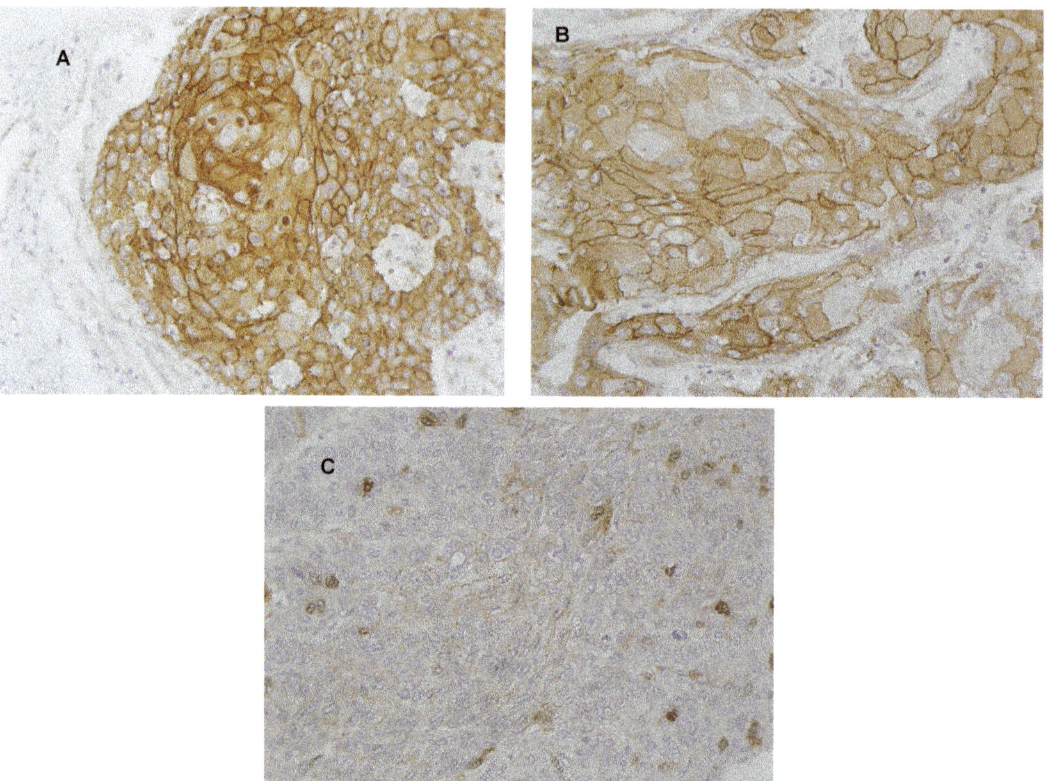

Figure 1. Immunohistochemical analysis of β-catenin in OSCC tissue specimens. Representative images of tumors showing (**A**) membrane and nuclear staining, (**B**) membrane staining, and (**C**) nuclear staining. Magnification 40×.

Table 2. Associations between stromal and tumoral CD8$^+$ TIL density and β-catenin expression in OSCC patients.

CD8$^+$ Compartment	β-Catenin Expression	No. Cases	Mean CD8$^+$ TIL Density (SD)	p
Stromal	Negative	21	280.15 (279.15)	
	Membrane	102	160.43 (179.83)	0.020
	Nucleus	2	30.00 (41.48)	
Tumoral	Negative	21	67.41 (54.74)	
	Membrane	102	44.63 (57.40)	0.029
	Nucleus	2	6.83 (3.50)	

SD: standard deviation. p-values were calculated using the Kruskal–Wallis test.

We also assessed the association between β-catenin expression and the type of TIME, which had a statistically significant relationship (Fisher's exact test, $p = 0.012$). Tumors harboring nuclear β-catenin were associated with type II TIME (i.e., immunological ignorance), whereas membranous β-catenin was predominantly associated with type IV (i.e., immune tolerance) (Table 3).

Table 3. Associations between β-catenin expression and the type of immune tumor microenvironment (TME).

Type of Immune TME	β-Catenin Expression			*p*
	Negative	Membrane	Nucleus	
Type I (PD-L1+/CD8⁺ high)	3 (16%)	10 (10%)	0 (0%)	
Type II (PD-L1−/CD8⁺ low)	1 (5%)	37 (37%)	2 (100%)	0.012
Type III (PD-L1+/CD8⁺ low)	2 (11%)	3 (3%)	0 (0%)	
Type IV (PD-L1−/CD8⁺ high)	13 (68%)	51 (50%)	0 (0%)	

p-value calculated using Fisher's exact test.

We next assessed the correlation between different immune cell subtypes and the density of stromal and tumoral TIL infiltration, and we found significant associations between intense TIL infiltration and tumoral CD68$^+$ ($p < 0.0001$), CD163$^+$ ($p = 0.006$), and tumoral CD8$^+$ ($p = 0.02$) (Table 4).

Table 4. Correlations between the mean numbers of CD68$^+$ and CD163$^+$ macrophages and CD8$^+$, CD20$^+$, CD4$^+$, and FOXP3$^+$ infiltrating TILs in the tumor nests and surrounding stroma, according to the intensity of TIL infiltration.

Mean (SD)	TILs			*p*
	Negative	Mild-Moderate	Intense	
Stromal CD68$^+$	110.55 (73.53)	140.56 (94.67)	118.72 (54.29)	0.20
Tumoral CD68$^+$	37.86 (36.65)	61.39 (47.64)	115.66 (36.22)	<0.0001
Stromal CD163$^+$	165.35 (90.93)	172.52 (109.10)	149.66 (81.73)	0.92
Tumoral CD163$^+$	25.79 (27.58)	36.19 (28.62)	54.33 (32.04)	0.006
Stromal CD8$^+$	168.86 (203.92)	202.73 (211.62)	83.66 (100.49)	0.11
Tumoral CD8$^+$	34.65 (46.44)	62.46 (64.67)	76.27 (70.72)	0.02
Stromal CD20$^+$	36.15 (73.23)	54.48 (88.84)	17.50 (35.96)	0.06
Tumoral CD20$^+$	1.21 (2.98)	2.34 (3.87)	0.50 (0.91)	0.26
Stromal CD4$^+$	52.81 (78.38)	61.48 (56.03)	19.27 (8.58)	0.08
Tumoral CD4$^+$	4.66 (10.40)	8.43 (14.83)	1.38 (1.28)	0.05
Stromal FOXP3$^+$	12.62 (18.38)	19.65 (31.70)	17.16 (15.30)	0.51
Tumoral FOXP3$^+$	2.10 (3.66)	4.30 (9.07)	4.83 (4.54)	0.20

The Kruskal–Wallis *p* values are shown.

3.3. Associations with Clinicopathological Features and Patient Survival

Absence/loss of β-catenin expression or nuclear β-catenin expression were more frequently observed in larger tumor sizes (T3–T4) (21% vs. 17% in T1-T2), tumors with neck lymph node metastasis (pN+) (25% vs. 15% in pN0 cases), and recurrent tumors (22% vs. 16% in non-recurrent tumors); however, these differences did not reach statistical significance. Regarding histological degree of differentiation, absence or nuclear β-catenin expression was more frequently detected in moderately and poorly differentiated tumors, (22% vs. 16% in well-differentiated tumors), almost reaching significance ($p = 0.06$).

Stromal/tumoral CD8$^+$ TILs were divided into two categories (high vs. low density) according to their respective median values. Only stromal CD8$^+$ TILs, but not tumoral CD8$^+$ TILs, were significantly associated with DSS (0.8 and 0.73 patient survival in high vs. low TIL infiltration, $p = 0.045$) (Figure 2A) as previously reported [20]. Regarding β-catenin expression, membrane staining was associated with better survival than nuclear β-catenin or no expression, but differences were not statistically significant (Figure 2B). However, when combining the stromal CD8$^+$ TILs with membranous β-catenin expression in the tumor, we found that patients harboring high CD8$^+$ TIL density (above the median value) and positive membranous β-catenin showed the highest DSS ($p = 0.012$) (Figure 2C).

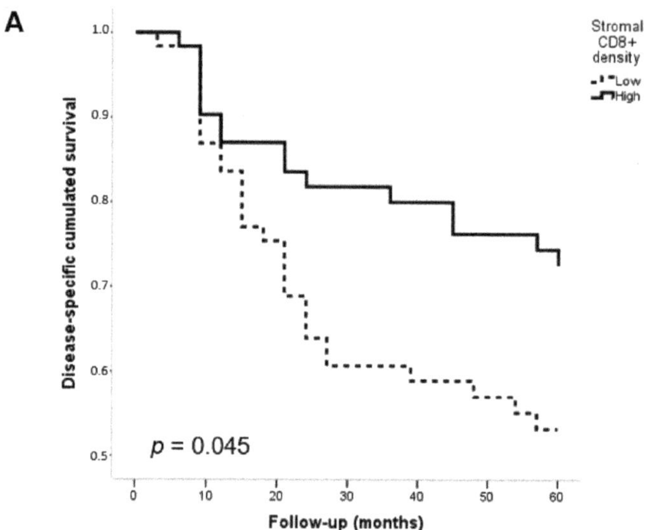

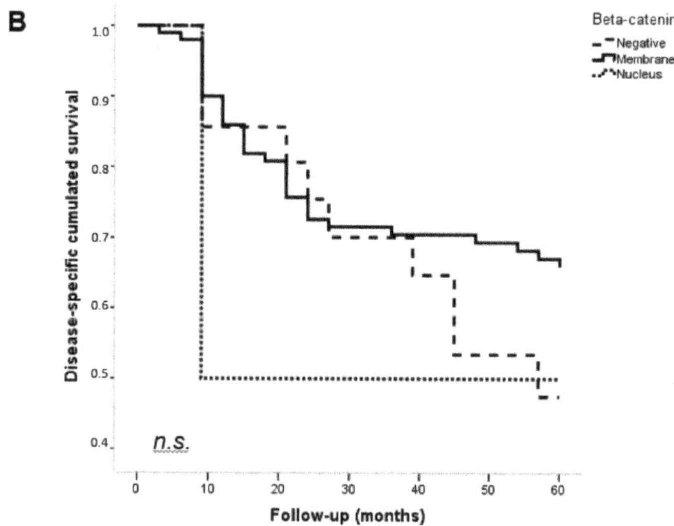

Figure 2. *Cont.*

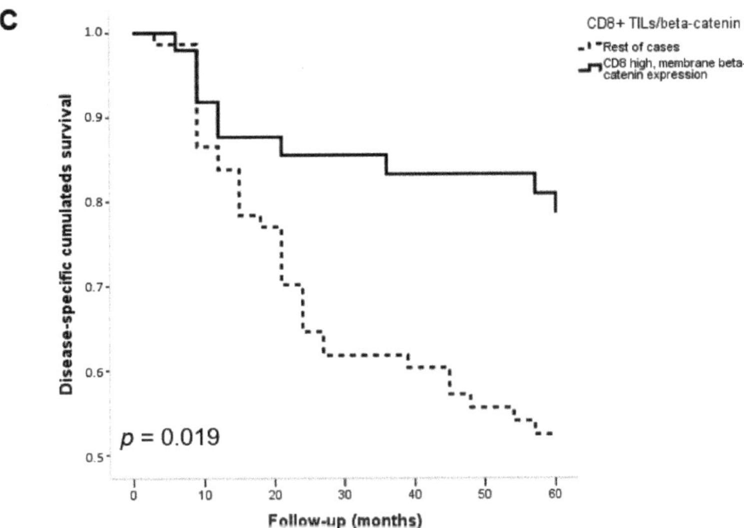

Figure 2. Kaplan–Meier disease-specific survival and overall survival curves in the cohort of 125 OSCC patients categorized by (**A**) High versus low stromal CD8$^+$ density (p = 0.045), (**B**) β-catenin immunostaining (membrane, nucleus, negative) (non-significant), and (**C**) High CD8$^+$ TILs stromal infiltration and membrane β-catenin immunostaining versus rest of cases (p = 0.019). p values estimated by log-rank test.

Multivariable Cox regression analysis, including T classification (T1-T2 vs. T3-T4), N classification (N0 vs. N+), differentiation grade (well-moderately vs. poorly differentiated), and stromal CD8$^+$ TIL infiltration combined with tumoral β-catenin expression (high CD8$^+$ TIL density and positive membranous β-catenin vs. all the rest of the cases) showed that the parameters independently associated with worse DSS were T3-T4 classification (HR = 2.61, 95% CI = 1.45–4.68, p = 0.001), and neck node metastasis (HR = 1.84, 95% CI = 1.06–3.18, p = 0.02). Conversely, high CD8$^+$ TIL density combined with membranous β-catenin was a significant independent predictor of better DSS (HR = 0.48, 95% CI = 0.26–0.88, p = 0.019). Finally, the degree of differentiation was not significantly associated with better DSS (HR = 0.56, 95% CI = 0.31–1.03, p = 0.06). If we included in the multivariable analysis the stromal CD8$^+$ TIL infiltration instead of this variable combined with tumoral β-catenin expression, T and N parameters retained their significant association with a worse DSS; however, CD8$^+$ TILs density did not reach a significant independent association with survival (HR = 0.60, 95% CI = 0.34–1.06, p = 0.08).

4. Discussion

β-catenin is an 88 kDa multifunctional and evolutionary-conserved protein, a member of the armadillo family of proteins, and involved in two independent processes, which are cell-cell adhesion and signal transduction [25]. β-catenin is considered as the 'gatekeeper' of the canonical Wnt signaling [26]. In the absence of Wnt ligands, this protein is involved in cell adhesion, thereby acting as a bridge between E-cadherin and cytoskeleton-associated actin to form adherent junctions between adjacent cells in the stratified squamous epithelium of oral mucosa [27]. In this series, expression of membranous β-catenin was detected in 81.6% tumors, nuclear β-catenin was only found in two cases (1.6%), and β-catenin expression was absent in 16.8%. Akinyamoju et al. [28] reported positive β-catenin expression in 70.7% OSCC samples. Laxmidevi et al. [29] reported 56.6% of β-catenin positivity and Zaid [30] found 67.1% in OSCC.

Herein, we did not find any significant relationship between β-catenin expression and the clinicopathological characteristics of our sample. Nevertheless, abnormal β-catenin expression (defined as absence/loss of expression or nuclear expression) was more frequent in T3-T4 tumors, with the presence of nodal metastasis as well as recurrent OSCC tumors. Similar to our results, Al-Rawi et al. [31] found no significant associations with either clinicopathological features or prognosis. In a study with 30 OSCC samples, β-catenin expression loss evaluated by immunohistochemistry was an unreliable marker of nodal metastasis; however, the loss of this protein was associated with a lower degree of differentiation [32]. Conversely, Tanaka et al. [33] reported a significant reduction in the expression level of β-catenin in those tumors with lymph node metastasis compared with non-metastatic cases. Interestingly, we found a trend between β-catenin expression loss and poor tumor differentiation. Concordantly, Zaid [30] and Zhao et al. [34] showed a significant reduction in β-catenin expression in poor histopathological grades in OSCC and in esophageal squamous cell carcinoma, respectively. Since only two cases showed nuclear β-catenin staining in our OSCC cohort, consequently, these results should be cautiously interpreted. Some studies reported that reduced nuclear β-catenin is related with a more aggressive tumor behavior in non-small cell lung cancer [35], whereas Pukkila et al. [36] did not find such an association in squamous cell carcinomas of the oropharynx and hypopharynx. Furthermore, nuclear β-catenin expression was associated with shorter survival rates. Moreover, in a meta-analysis that included 2746 OSCC patients, Ramos-García and Gonzalez-Moles [17] showed that aberrant β-catenin expression was significantly associated with poor overall survival, disease-free survival, higher tumor size, neck lymph node metastasis, and moderately-poorly differentiated tumors.

Among the reasons that may explain the differences and inconsistencies from the available data are small sample sizes in most studies, methodological differences in immunohistochemical evaluation and scoring, often subjective, and different antibodies/epitopes used.

In this study, we showed that β-catenin expression in the nucleus of oral cancer cells was associated with the lowest density of CD8$^+$ TILs in the TME, both in the tumor as well as in the surrounding stroma, and with a tendency to a low tumoral PD-L1 expression. The interaction between PD-1 and PD-L1 inhibits the activation and effector functions of CD8$^+$ T cells and induces suppressive Treg cells, ultimately leading to cancer immune evasion [8]. Oncogenic pathways may contribute to the protumoral effects of cancer-intrinsic PD-L1. Specifically, the Wnt/β-catenin signaling pathway could exert a relevant role in facilitating PD-L1-mediated cell proliferation, migration, and invasion. PD-L1 expression induces ERK phosphorylation and activates the Wnt/β-catenin signaling pathway, which upregulates downstream target genes, as reported in colorectal carcinoma and in non-small cell lung cancer [37,38]. In turn, activation of β-catenin increases PD-L1 transcription and promotes immune evasion [8]. According to these data, PD-L1 and β-catenin are reciprocally implicated in a bidirectional positive feedback loop. Therefore, blocking Wnt/β-catenin signaling could increase anti-PD-L1 treatment efficacy [37].

Tumors are closely related to the TME, and there is a continuous interplay between tumor cells and stromal cells. It has been recently reported that most tumors are inversely related to a T cell-inflamed gene expression signature, potentially linked to Wnt/β-catenin pathway activation [9]. Here, we found a relationship between tumoral and stromal CD8$^+$ TIL density and β-catenin expression in OSCC. Furthermore, tumors harboring nuclear β-catenin were associated with a type II TIME (i.e., immunological ignorance), as reported in other HNSCC subsites [39]. Notably, a T cell-inflamed phenotype is associated with the efficacy of an immune checkpoint blockade, whereas non-T-cell inflamed tumors rarely benefit from anti-PD-1 therapies [2,39]. We also found an association between the nuclear β-catenin and poor survival, in accordance with previous reports [40]. The molecular mechanism by which tumoral β-catenin regulates PD-L1-mediated immunosuppression remains unknown [41]. However, it has been recently shown that β-catenin can be activated by both Wnt ligand and epidermal growth factor receptor (EGFR) activation (frequent in

OSCC), which leads to the binding of the β-catenin/TCF/LEF complex to the promoter region of the *CD274* gene to induce PD-L1 expression [8].

The Wnt/β-catenin pathway has been identified as one of the most important oncogenic signaling pathways associated with tumor immune evasion [42], and its mutation leads to a non-T cell inflammatory tumor phenotype. The Wnt/β-catenin signaling pathway also regulates the differentiation of CD4[+] helper T cells [2] and limits the immunosuppressive activity of Treg cells by modulating FOXP3 transcriptional activity, which has been associated with a lack of T-cell infiltration in the TME of metastatic melanoma and other cancer types [43]. Additionally, the Wnt/β-catenin pathway has also been implicated in the regulation of innate immunity. Colon cancer cells induce IL-β in macrophages via Snail, which is a product of Wnt target genes [44]. Moreover, macrophages usually express higher PD-L1 levels compared with tumor cells [45].

Furthermore, we also found a significant relationship between the density of TILs within the tumor nests and stroma, with the density of CD68[+] and CD163[+] macrophages infiltrating the tumor, and also with the tumor-infiltrating CD8[+] T cells. In a recent review, Li et al. [2] summarized the effects of Wnt/β-catenin signaling on cancer immunosurveillance. Briefly, Wnt ligands released by cancer cells induce the canonical Wnt signaling pathway, whose hallmark is the accumulation of β-catenin protein into the nucleus, thereby leading to the inhibition of CD8[+] T-cell infiltration within the TME and production of Treg cells with subsequent inhibition of cytotoxic T-cell activity. Glycogen synthase kinase 3 (GSK3) inhibition in cancer cells promotes β-catenin activation and subsequently PD-L1 stabilization, driving in turn cytotoxic T-cell activity exhaustion. On the other hand, non-canonical Wnt signaling is initiated by Wnt5a-type ligands [11]. Wnt1 ligands related with the canonical Wnt pathway promote invasion and inhibit apoptosis in OSCC [46], whereas Wnt5a, Wnt5b, Wnt7a, and Wnt7b ligands, all of them related to the non-canonical Wnt pathway, enhance cell migration and invasion in OSCC [47–50]. The relationship between β-catenin and a non-T-cell inflamed microenvironment has been previously reported by Luke et al. [9]. In an integrative TCGA analysis by separating tumors according to their T-cell inflamed status, they found that 33.9% of tumors were non-T-cell-inflamed, which can be regulated by oncogenic events [9,51]. Furthermore, it was also revealed that over 90% of tumor types across the TCGA showed an inverse correlation between Wnt/β-catenin pathway activation and a T-cell-inflamed gene expression signature.

As far as we know, this is the first study to analyze the relationship between β-catenin expression and the immune infiltrate TME in OSCC, mainly focused on CD8[+] T cells.

5. Conclusions

In conclusion, our findings support that the Wnt/β-catenin pathway could be a molecular target of paramount relevance in an effort to elicit an infiltration of CD8[+] T cells in the OSCC TME and potentially increase the efficacy of immunotherapy in the clinical setting. Hence, β-catenin emerges as a valuable molecular target to improve the efficacy of immunotherapy in OSCC by increasing TIL density in the TME. Notably, a combination of stromal CD8[+] T-cell infiltration and membranous β-catenin expression in the tumor was found to be an independent predictor of better DSS for OSCC patients.

Author Contributions: Conceptualization, P.L.-F., J.M.G.-P. and J.C.D.V.; methodology, P.L.-F., J.M.G.-P., J.C.D.V., J.P.R. and T.R.-S.; software, P.L.-F., T.R.-S., V.B.-L., E.G.-G., H.E.T.-R. and F.J.S.-S.; validation, P.L.-F. and F.J.S.-S.; formal analysis, J.C.D.V., P.L.-F., E.G.-G., H.E.T.-R., V.B.-L. and F.J.S.-S.; investigation, P.L.-F., J.C.D.V. and J.M.G.-P.; resources, J.P.R., F.J.S.-S., H.E.T.-R. and V.B.-L.; data curation, P.L.-F., J.C.D.V., J.P.R., J.M.G.-P. and E.G.-G.; writing—original draft preparation, J.C.D.V. and J.M.G.-P.; writing—review and editing, J.C.D.V., J.M.G.-P. and P.L.-F.; visualization, H.E.T.-R., V.B.-L., P.L.-F. and E.G.-G.; supervision, P.L.-F., J.C.D.V., J.M.G.-P. and J.P.R.; project administration, P.L.-F., J.C.D.V. and J.M.G.-P.; funding acquisition, P.L.-F., J.C.D.V.; J.M.G.-P. and J.P.R. received funding for the study. All authors have read and agreed to the published version of the manuscript.

Funding: This research was funded by the Instituto de Salud Carlos III through the projects PI19/01255, PI19/00560 and PI22/00167, and co-funded by the European Union, and also the Instituto de Investigación Sanitaria del Principado de Asturias (ISPA), Fundación Bancaria Caja de Ahorros de Asturias-IUOPA, and Universidad de Oviedo. Also through the grant "Ayudas para Grupos de Investigación de Organismos del Principado de Asturias 2021–2023" (IDI/2021/000079), funded by Principado de Asturias through FICYT and the FEDER Funding Program from the European Union.

Institutional Review Board Statement: This study was conducted in accordance with the Declaration of Helsinki and approved by the Regional Ethics Committee from Principado de Asturias (date of approval 14 May 2019; approval number: 136/19).

Informed Consent Statement: Informed consent was obtained from all participants whose materials were analyzed in the study. Written informed consent has been obtained from the patient(s) to publish this paper.

Ethics Approval and Consent to Participate: All experimental procedures were conducted in accordance to the Declaration of Helsinki, and approved by Institutional Ethics Committee of the HUCA and by the Regional Ethics Committee from Principado de Asturias (date of approval 14 May 2019; approval number 136/19, for the project PI19/01255). Samples and data from donors included in this study were provided by the Principado de Asturias BioBank (PT20/0161), integrated in the Spanish National Biobanks and Biomodels Network financed with European funds and they were processed following standard operating procedures with the appropriate approval of the Ethical and Scientific Committees.

Data Availability Statement: Data supporting the present study are available from the corresponding author (JCV) upon reasonable request.

Acknowledgments: We want to particularly acknowledge for its collaboration the Principado de Asturias BioBank (PT17/0015/0023 and PT20/00161), financed jointly by Servicio de Salud del Principado de Asturias, Instituto de Salud Carlos III and Fundación Bancaria Cajastur, and part of the Spanish National Biobanks Network.

References

1. Sung, H.; Ferlay, J.; Siegel, R.L.; Laversanne, M.; Soerjomataram, I.; Jemal, A.; Freddie, B. Global Cancer Statistics 2020: GLOBOCAN estimates of incidence and mortality worldwide for 36 cancers in 185 countries. *CA. Cancer J. Clin.* **2021**, *71*, 209–249. [CrossRef] [PubMed]
2. Li, X.; Xiang, Y.; Li, F.; Yin, C.; Li, B.; Ke, V. WNT/β-Catenin signaling pathway regulating T cell-inflammation in the tumor microenvironment. *Front. Immunol.* **2019**, *10*, 2293. [CrossRef] [PubMed]
3. Chen, J.; Jiang, C.C.; Jin, L.; Zhang, X.D. Regulation of PD-L1: A novel role of pro-survival signalling in cancer. *Ann. Oncol.* **2016**, *27*, 409–416. [CrossRef] [PubMed]
4. Chen, C.; Luo, L.; Xu, C.; Yang, X.; Liu, T.; Luo, J.; Shi, W.; Yang, L.; Zheng, Y.; Yang, J. Tumor specificity of WNT ligands and receptors reveals universal squamous cell carcinoma oncogenes. *BMC Cancer* **2022**, *22*, 790. [CrossRef] [PubMed]
5. Thommen, D.D.; Schumacher, T.N. T cell dysfunction in cancer. *Cancer Cell* **2018**, *33*, 547–562. [CrossRef] [PubMed]
6. Arneth, B. Tumor Microenvironment. *Medicina* **2019**, *56*, 15. [CrossRef]
7. Naito, Y.; Saito, K.; Shiiba, K.; Ohuchi, A.; Saigenji, K.; Nagura, H.; Ohtani, H. CD8+ T cells infiltrated within cancer cell nests as a prognostic factor in human colorectal cancer. *Cancer Res.* **1998**, *58*, 3491–3494.
8. Du, L.; Lee, J.H.; Jiang, H.; Wang, C.; Wang, V.; Zheng, Z.; Shao, F.; Xu, D.; Xia, V.; Li, J.; et al. β-Catenin induces transcriptional expression of PD-L1 to promote glioblastoma immune evasion. *J. Exp. Med.* **2020**, *217*, e20191115. [CrossRef]
9. Luke, J.J.; Bao, R.; Sweis, R.F.; Spranger, S.; Gajewski, T.F. WNT/β-catenin pathway activation correlates with immune exclusion across human cancers. *Clin. Cancer Res.* **2019**, *25*, 3074–3083. [CrossRef]
10. MacDonald, B.T.; Tamai, K.; He, X. Wnt/beta-catenin signaling: Components, mechanisms, and diseases. *Dev. Cell.* **2009**, *17*, 9–26. [CrossRef]
11. Xie, J.; Huang, L.; Lu, Y.G.; Zheng, D.L. Roles of the Wnt signaling pathway in head and neck squamous cell carcinoma. *Front. Mol. Biosci.* **2021**, *7*, 590912. [CrossRef] [PubMed]
12. Patni, A.P.; Harishankar, M.K.; Joseph, J.P.; Sreeshma, B.; Jayaraj, R.; Devi, A. Comprehending the crosstalk between Notch, Wnt and Hedgehog signaling pathways in oral squamous cell carcinoma—Clinical implications. *Cell Oncol.* **2021**, *44*, 473–494. [CrossRef] [PubMed]
13. Cadigan, K.M.; Waterman, M.L. TCF/LEFs and Wnt signaling in the nucleus. *Cold Spring Harb. Perspect. Biol.* **2012**, *4*, a007906. [CrossRef]
14. Wang, Y.; Zheng, C.; Liu, F.; Ou, Y. Clinical significance of activated Wnt/β-catenin signaling in apoptosis inhibition of oral cancer. *Open Life Sci.* **2021**, *16*, 1045–1052. [CrossRef]

15. Iwai, S.; Yonekawa, A.; Harada, C.; Hamada, M.; Katagiri, W.; Nakazawa, M.; Yura, Y. Involvement of the Wnt-β-catenin pathway in invasion and migration of oral squamous carcinoma cells. *Int. J. Oncol.* **2010**, *37*, 1095–1103. [CrossRef] [PubMed]

16. Noguti, J.D.E.; Moura, C.F.; Hossaka, T.A.; Franco, M.; Oshima, C.T.; Dedivitis, R.A.; Ribeiro, D.A. The role of canonical WNT signaling pathway in oral carcinogenesis: A comprehensive review. *Anticancer Res.* **2012**, *32*, 873–878.

17. Ramos-García, P.; González-Moles, M.Á. Prognostic and clinicopathological significance of the aberrant expression of β-catenin in oral squamous cell carcinoma: A systematic review and meta-analysis. *Cancers* **2022**, *14*, 479. [CrossRef]

18. Ridge, J.A.; Lydiatt, W.M.; Patel, S.G.; Glastonbury, C.M.; Brandwein-Gensler, M.; Gossein, R.A.; Shah, J.P. Oral Cavity. In *AJCC Cancer Staging Manual*, 8th ed.; Editor Amin, M.B., Ed.; Springer: Chicago, IL, USA, 2017; pp. 79–94.

19. Müller, S. Update from the 4th edition of the World Health Organization of Head and Neck Tumours: Tumours of the oral cavity and mobile tongue. *Head Neck Pathol.* **2017**, *11*, 33–40. [CrossRef]

20. Lequerica-Fernández, P.; Suárez-Canto, J.; Rodríguez-Santamarta, T.; Rodrigo, J.P.; Suárez-Sánchez, F.J.; Blanco-Lorenzo, V.; Domínguez-Iglesias, F.; García-Pedrero, J.M.; de Vicente, J.C. Prognostic relevance of CD4$^+$, CD8$^+$ and FOXP3$^+$ TILs in oral squamous cell carcinoma and correlations with PD-L1 and cancer stem cell markers. *Biomedicines* **2021**, *9*, 653. [CrossRef]

21. Suárez-Sánchez, F.J.; Lequerica-Fernández, P.; Suárez-Canto, J.; Rodrigo, J.P.; Rodríguez-Santamarta, T.; Domínguez-Iglesias, F.; García-Pedrero, J.; de Vicente, J.C. Macrophages in oral carcinomas: Relationship with cancer stem cell markers and PD-L1 expression. *Cancers* **2020**, *12*, 1764. [CrossRef]

22. Suárez-Sánchez, F.J.; Lequerica-Fernández, P.; Rodrigo, J.P.; Hermida-Prado, F.; Suárez-Canto, J.; Rodríguez-Santamarta, T.; Domínguez-Iglesias, F.; García-Pedrero, J.; de Vicente, J.C. Tumor-infiltrating CD20$^+$ B lymphocytes: Significance and prognostic implications in oral cancer microenvironment. *Cancers* **2021**, *13*, 395. [CrossRef] [PubMed]

23. de Vicente, J.C.; Rodríguez-Santamarta, T.; Rodrigo, J.P.; Blanco-Lorenzo, V.; Allonca, E.; García-Pedrero, J.M. PD-L1 expression in tumor cells is an independent unfavorable prognostic factor in oral squamous cell carcinoma. *Cancer Epidemiol. Biomark. Prev.* **2019**, *28*, 546–554. [CrossRef] [PubMed]

24. Teng, M.W.; Ngiow, S.F.; Ribas, A.; Smyth, M.J. Classifying cancers based on T-cell infiltration and PD-L1. *Cancer Res.* **2015**, *75*, 2139–2145. [CrossRef] [PubMed]

25. Steinhart, Z.; Angers, S. Wnt signaling in development and tissue homeostasis. *Development* **2018**, *145*, dev146589. [CrossRef]

26. Shang, S.; Hua, F.; Hu, Z.W. The regulation of beta-catenin activity and function in cancer: Therapeutic opportunities. *Oncotarget* **2017**, *8*, 33972–33989. [CrossRef]

27. Huber, A.H.; Weis, W.I. The structure of the beta-catenin/E-cadherin complex and the molecular basis of diverse ligand recognition by beta-catenin. *Cell* **2001**, *105*, 391–402. [CrossRef]

28. Akinyamoju, A.O.; Lawal, A.O.; Adisa, A.O.; Adeyemi, B.F.; Kolude, B.J. Immunohistochemical expression of E-cadherin and β-catenin in oral squamous cell carcinoma. *West Afr. Coll. Surg.* **2023**, *13*, 43–47. [CrossRef]

29. Laxmidevi, L.B.; Angadi, P.V.; Pillai, R.K.; Chandreshekar, C. Aberrant β-catenin expression in the histologic differentiation of oral squamous cell carcinoma and verrucous carcinoma: An immunohistochemical study. *J. Oral Sci.* **2010**, *52*, 633–640. [CrossRef]

30. Zaid, K.W. Immunohistochemical assessment of E-cadherin and β-catenin in the histological differentiations of oral squamous cell carcinoma. *Asian Pac. J. Cancer Prev.* **2014**, *15*, 8847–8853. [CrossRef]

31. Al-Rawi, N.; Al Ani, M.; Quadri, A.; Hamdoon, Z.; Awwad, A.; Al Kawas, S.; Al Nuaimi, A. Prognostic significance of E-Cadherin, β-Catenin and cyclin D1 in oral squamous cell carcinoma: A tissue microarray study. *Histol. Histopathol.* **2021**, *36*, 1073–1083. [CrossRef]

32. Mahomed, F.; Altini, M.; Meer, S. Altered E-cadherin/beta-catenin expression in oral squamous carcinoma with and without nodal metastasis. *Oral Dis.* **2007**, *13*, 386–392. [CrossRef]

33. Tanaka, N.; Odajima, T.; Ogi, K.; Ikeda, T.; Satoh, M. Expression of E-cadherin, alpha-catenin, and beta-catenin in the process of lymph node metastasis in oral squamous cell carcinoma. *Br. J. Cancer* **2003**, *89*, 557–563. [CrossRef]

34. Zhao, X.J.; Li, H.; Chen, H.; Liu, Y.X.; Zhang, L.H.; Liu, S.X.; Feng, Q.L. Expression of e-cadherin and beta-catenin in human esophageal squamous cell carcinoma: Relationships with prognosis. *World J. Gastroenterol.* **2003**, *9*, 225–232. [CrossRef]

35. Pirinen, R.T.; Hirvikoski, P.; Johansson, R.T.; Hollmén, S.; Kosma, V.M. Reduced expression of alpha-catenin, beta-catenin, and gamma-catenin is associated with high cell proliferative activity and poor differentiation in non-small cell lung cancer. *J. Clin. Pathol.* **2001**, *54*, 391–395. [CrossRef] [PubMed]

36. Pukkila, M.J.; Virtaniemi, J.A.; Kumpulainen, E.J.; Pirinen, R.T.; Johansson, R.T.; Valtonen, H.J.; Juhola, M.T.; Kosma, V.M. Nuclear beta catenin expression is related to unfavourable outcome in oropharyngeal and hypopharyngeal squamous cell carcinoma. *J. Clin. Pathol.* **2001**, *54*, 42–47. [CrossRef] [PubMed]

37. Ma, Y.; Marinkova, R.; Nenkov, M.; Jin, L.; Huber, O.; Sonnemann, J.; Peca, N.; Gabler, N.; Chen, Y. Tumor-intrinsic PD-L1 exerts an oncogenic function through the activation of the Wnt/β-catenin pathway in human non-small cell lung cancer. *Int. J. Mol. Sci.* **2022**, *23*, 11031. [CrossRef]

38. Fu, L.; Fan, J.; Maity, S.; McFadden, G.; Shi, Y.; Kong, W. PD-L1 interacts with Frizzled 6 to activate β-catenin and form a positive feedback loop to promote cancer stem cell expansion. *Oncogene* **2022**, *41*, 1100–1113. [CrossRef] [PubMed]

39. Sánchez-Canteli, M.; Juesas, L.; Garmendia, V.; Otero-Rosales, M.; Calvo, A.; Alvarez-Fernández, M.; Astudillo, A.; Montuenga, V.; García-Pedrero, J.M.; Rodrigo, J.P. Tumor-intrinsic nuclear β-catenin associates with an immune ignorance phenotype and a poorer prognosis in head and neck squamous cell carcinomas. *Int. J. Mol. Sci.* **2022**, *23*, 11559. [CrossRef]

40. Anastas, J.N.; Moon, R.T. WNT signalling pathways as therapeutic targets in cancer. *Nat. Rev. Cancer* **2013**, *13*, 11–26. [CrossRef]

41. Han, C.; Fu, Y.X. β-Catenin regulates tumor-derived PD-L1. *J. Exp. Med.* **2020**, *217*, e20200684. [CrossRef]
42. Fu, C.; Liang, X.; Cui, W.; Ober-Blöbaum, J.L.; Vazzana, J.; Shrikant, P.A.; Lee, K.P.; Clausen, B.E.; Mellman, I.; Jiang, A. β-catenin in dendritic cells exerts opposite functions in cross-priming and maintenance of CD8+ T cells through regulation of IL-10. *Proc. Natl. Acad. Sci. USA* **2015**, *112*, 2823–2828. [CrossRef] [PubMed]
43. van Loosdregt, J.; Fleskens, V.; Tiemessen, M.M.; Mokry, M.; van Boxtel, R.; Meerding, J.; Pals, C.E.G.M.; Kurek, D.; Baert, M.R.M.; Delemarre, E.M.; et al. Canonical Wnt signaling negatively modulates regulatory T cell function. *Immunity* **2013**, *39*, 298–310. [CrossRef] [PubMed]
44. Kaler, P.; Augenlicht, V.; Klampfer, L. Activating mutations in β-catenin in colon cancer cells alter their interaction with macrophages; the role of snail. *PLoS ONE* **2012**, *7*, e45462. [CrossRef] [PubMed]
45. Tang, H.; Liang, Y.; Anders, R.A.; Taube, J.M.; Qiu, X.; Mulgaonkar, A.; Liu, X.; Harrington, S.M.; Guo, J.; Xin, Y.; et al. PD-L1 on host cells is essential for PD-L1 blockade-mediated tumor regression. *J. Clin. Investig.* **2018**, *128*, 580–588. [CrossRef]
46. Rhee, C.S.; Sen, M.; Lu, D.; Wu, C.; Leoni, L.; Rubin, J.; Corr, M.; Carson, D.A. Wnt and frizzled receptors as potential targets for immunotherapy in head and neck squamous cell carcinomas. *Oncogene* **2002**, *21*, 6598–6605. [CrossRef] [PubMed]
47. Prgomet, V.; Andersson, V.; Lindberg, P. Higher expression of WNT5A protein in oral squamous cell carcinoma compared with dysplasia and oral mucosa with a normal appearance. *Eur. J. Oral Sci.* **2017**, *125*, 237–246. [CrossRef] [PubMed]
48. Zhang, W.; Yan, Y.; Gu, M.; Wang, X.; Zhu, H.; Zhang, S.; Wang, W. High expression levels of Wnt5a and Ror2 in laryngeal squamous cell carcinoma are associated with poor prognosis. *Oncol. Lett.* **2017**, *14*, 2232–2238. [CrossRef]
49. Sakamoto, T.; Kawano, S.; Matsubara, R.; Goto, Y.; Jinno, T.; Maruse, Y.; Kaneko, N.; Hashiguchi, Y.; Hattori, T.; Tanaka, S.; et al. Critical roles of Wnt5a-Ror2 signaling in aggressiveness of tongue squamous cell carcinoma and production of matrix metalloproteinase-2 via ΔNp63β-mediated epithelial-mesenchymal transition. *Oral Oncol.* **2017**, *69*, 15–25. [CrossRef]
50. Xie, H.; Ma, Y.; Li, J.; Chen, H.; Xie, V.; Chen, M.; Zhao, X.; Tang, S.; Zhao, S.; Zhang, Y.; et al. WNT7A Promotes EGF-induced migration of oral squamous cell carcinoma cells by activating β-catenin/MMP9-mediated signaling. *Front. Pharmacol.* **2020**, *11*, 98. [CrossRef]
51. Trujillo, J.A.; Sweis, R.F.; Bao, R.; Luke, J.J. T cell-inflamed versus non-T cell-inflamed tumors: A conceptual framework for cancer immunotherapy drug development and combination therapy selection. *Cancer Immunol. Res.* **2018**, *6*, 990–1000. [CrossRef]

 *biomedicines*

Article

Mometasone Furoate Inhibits the Progression of Head and Neck Squamous Cell Carcinoma via Regulating Protein Tyrosine Phosphatase Non-Receptor Type 11

Lin Qiu [1,†], Qian Gao [1,†], Anqi Tao [1], Jiuhui Jiang [2,*] and Cuiying Li [1,*]

[1]   Central Laboratory, Peking University School and Hospital of Stomatology & National Center for Stomatology & National Clinical Research Center for Oral Diseases & National Engineering Research Center of Oral Biomaterials and Digital Medical Devices, Beijing 100081, China; 2111110525@stu.pku.edu.cn (L.Q.); gaoqian1995@bjmu.edu.cn (Q.G.); 2011110515@stu.pku.edu.cn (A.T.)

[2]   Department of Orthodontics, Peking University School and Hospital of Stomatology & National Center for Stomatology & National Clinical Research Center for Oral Diseases & National Engineering Research Center of Oral Biomaterials and Digital Medical Devices, Beijing 100081, China

*   Correspondence: drjiangw@163.com (J.J.); kqlicuiying@bjmu.edu.cn (C.L.)

†   These authors contributed equally to this work.

Abstract: Mometasone furoate (MF) is a kind of glucocorticoid with extensive pharmacological actions, including inhibiting tumor progression; however, the role of MF in head and neck squamous cell carcinoma (HNSCC) is still unclear. This study aimed to evaluate the inhibitory effect of MF against HNSCC and investigate its underlying mechanisms. Cell viability, colony formation, cell cycle and cell apoptosis were analyzed to explore the effect of MF on HNSCC cells. A xenograft study model was used to investigate the effect of MF on HNSCC in vivo. The core targets of MF for HNSCC were identified using network pharmacology analysis, TCGA database analysis and real-time PCR. Molecular docking was performed to determine the binding energy. Protein tyrosine phosphatase non-receptor type 11 (PTPN11)-overexpressing cells were constructed, and then, the cell viability and the expression levels of proliferation- and apoptosis-related proteins were detected after treatment with MF to explore the role of PTPN11 in the inhibitory effect of MF against HNSCC. After cells were treated with MF, cell viability and the number of colonies were decreased, the cell cycle was arrested and cell apoptosis was increased. The xenograft study results showed that MF could inhibit cell proliferation via promoting cell apoptosis in vivo. PTPN11 was shown to be the core target of MF against HNSCC via network pharmacology analysis, TCGA database analysis and real-time PCR. The molecular docking results revealed that PTPN11 exhibited the strongest ability to bind to MF. Finally, MF could attenuate the effects of increased cell viability and decreased cell apoptosis caused by PTPN11 overexpression, suggesting that MF can inhibit the progression of HNSCC by regulating PTPN11. MF targeted PTPN11, promoting cell cycle arrest and cell apoptosis, and consequently exerting effective anti-tumor activity.

Keywords: head and neck squamous cell carcinoma; mometasone furoate; proliferation; apoptosis; network pharmacology; PTPN11

Citation: Qiu, L.; Gao, Q.; Tao, A.; Jiang, J.; Li, C. Mometasone Furoate Inhibits the Progression of Head and Neck Squamous Cell Carcinoma via Regulating Protein Tyrosine Phosphatase Non-Receptor Type 11. *Biomedicines* **2023**, *11*, 2597. https://doi.org/10.3390/biomedicines11102597

Academic Editors: Vui King Vincent-Chong and Giuseppe Minervini

Received: 20 July 2023
Revised: 18 September 2023
Accepted: 19 September 2023
Published: 22 September 2023

1. Introduction

Head and neck squamous cell carcinoma (HNSCC) is the sixth most common form of malignant tumor in the world. The incidence rate is increasing and is expected to increase by 30% by 2030 [1]. The traditional treatment for HNSCC includes surgery followed by chemoradiotherapy. Despite the multimodal treatment strategy, over half of patients experience relapse or metastasis. For patients with recurrent/metastatic HNSCC, combined chemotherapy with platinum and paclitaxel or 5-fluorouracil plus the EGFR monoclonal antibody cetuximab is the standard first-line therapeutic regimen, but consequential drug resistance is common and finally leads to the limited efficacy of treatment [2]. The FDA-approved immune checkpoint inhibitor pembrolizumab is an antibody to PD1 and can effectively improve the survival rate of recurrent/metastatic HNSCC patients, although only the patients who express PD-L1 can benefit from immune checkpoint inhibitor therapy. Meanwhile, the serious adverse reactions caused by immune checkpoint inhibitor therapies need to be taken into consideration for administration. To date, the median survival of recurrent/metastatic HNSCC patients is only 11.6 months [3]. Therefore, drug treatment methods that are more effective and present fewer side effects are still urgently required.

Glucocorticoids are a class of steroidal hormones that bind to the glucocorticoid receptor to become involved in the regulation of multiple key biological processes, including inflammation and glucose metabolism [4]. While classically used as an anti-inflammation drug, recently accumulating evidence demonstrates that glucocorticoids can also treat malignant tumors. It is well established that glucocorticoids are the cornerstone of lymphatic cancer treatment due to their verified functions of arresting cell growth and promoting apoptosis [5,6]. Most importantly, the activation of cell cycle arrest and apoptosis are common ways for drugs to perform anti-cancer activity in multiple cancers, including HNSCC [7,8]. Mometasone furoate (MF) is a kind of glucocorticoid with extensive pharmacological action. Previous studies regarding MF have mostly focused on its use as a treatment for inflammation. In a recent study, MF inhibited the growth and induced the apoptosis of acute leukemia cells by regulating the PI3K signaling pathway [9]. For the above-mentioned reasons, MF was identified as a promising anti-cancer drug for HNSCC. Nevertheless, the inhibitory role and underlying mechanisms of MF against HNSCC remain to be determined.

Network pharmacology is a comparatively comprehensive and systematic way to predict the potential targets of clinical drugs [10,11]. Molecular docking is an important method to predict the binding between targets and drugs. The combination of these two methods provides a better reference for the application of clinical drugs and the repurposing of precise and effective therapeutic drugs. Therefore, it is an effective adjuvant method to screen the targets and underlying mechanisms of MF against HNSCC.

Protein tyrosine phosphatase non-receptor type 11 (PTPN11) is a member of the protein tyrosine phosphatase (PTP) family and is the first proto-oncogene receptor tyrosine phosphatase. PTPs work in coordination with protein tyrosine kinases (PTKs) to balance the phosphorylation status of tyrosine in signaling proteins, which determines multiple cellular processes through the transduction of signaling cascades. PTPN11 is required by most receptor tyrosine kinases (RTKs) to activate the downstream signaling pathways. As a result, PTPN11 plays a central role in the activation of oncogenic signaling pathways, such as PI3K/AKT [12], RAS/Raf/MAPK [13] and Jak/STAT [14]. It is widely acknowledged that PTPN11 is highly expressed in many tumors [15], and its aberrant expression is closely related to a poorer prognosis in a range of tumors [16,17]. In HNSCC, PTPN11 is overexpressed and participates in the invasion and metastasis of cells via activating the ERK1/2-Snail/Twist1 pathway [18]. Overexpressed PTPN11 could contribute to antigen-processing-machinery-component downregulation, thus leading to cytotoxic T lymphocyte evasion [19]. Accordingly, PTPN11 could be regarded as an effective target for tumor therapy in HNSCC [20].

2. Materials and Methods

2.1. Cell Culture

Cells from two human HNSCC cell lines, WSU-HN6 and CAL-27, were purchased from the American Type Culture Collection. Both cells were cultured in high-glucose DMEM (Gibco, New York, NY, USA) with 10% FBS (Gibco, New York, NY, USA) and 1% penicillin/streptomycin solution (Solarbio, Beijing, China) and maintained in 5% CO_2 at 37 °C.

2.2. Cell Counting Kit-8 Assay

The cell viability was measured using CCK-8 (Beyotime, Shanghai, China) according to the manufacturer's protocol. Cells were seeded into 96-well plates at a density of

4000 cells/well. After 24, 48 and 72 h, the supernatant was replaced by serum-free DMEM and CCK-8 solution (10:1) and incubated for 2 h at 37 °C. The optical density (OD) at 450 nm was measured using an automatic microplate reader (BioTek ELX808, Biotek Instruments, Vermont, VT, USA).

Cell viability was calculated using the formula: cell viability = [(experimental wells' OD − blank wells' OD)/(control wells' OD − blank wells' OD)] × 100%); OD450 = experimental wells' OD − blank wells' OD.

2.3. Colony Formation Assay

Cells were plated into 6-well plates at a density of 400 cells/well and cultured for 7 days. The culture medium was changed every two days. When the cell clone was visible to the naked eye, it was washed with phosphate-buffered saline (PBS), fixed with formaldehyde for 15 min and stained with 0.5% crystal violet at room temperature for 20 min. Finally, the cell clones were rinsed with PBS and the photographs were captured via a scanner (HP Scanjet G4050, China Hewlett-Packard Co., Ltd., Beijing, China).

2.4. Flow Cytometric Analysis

Cells were seeded into 6-well plates at 3×10^5 cells/well and cultured for 48 h. For the analysis of the cell cycle, the cells were fixed in 70% ethanol overnight at 4 °C, rinsed twice with PBS and stained with 500 µL of buffer, 25 µL of PI and 10 µL of RNase A at 37 °C for 30 min using the Cell Cycle and Apoptosis Analysis Kit (Beyotime, Shanghai, China) according to the manufacturer's instructions. For the analysis of apoptosis, the cells were rinsed twice with PBS and were stained with 5 µL of Annexin V-FITC and 5 µL of PI for 15 min using the Annexin V-FITC Apoptosis Detection Kit (Solarbio, Beijing, China) according to the manufacturer's instructions. Then, the DNA content and apoptosis rate of cells were examined via flow cytometry (Beckman, Brea, CA, USA).

2.5. Western Blotting

The Western blotting protocol was based on our earlier publication [21]. Briefly, collected tissues and cells were lysed using RIPA buffer (Solarbio, Beijing, China) complemented with 100 mM PMSF on ice. The BCA Protein Assay Kit (Beyotime, Shanghai, China) was employed to detect the concentrations of proteins. A total of 20 µg of protein from each sample was added to each lane of a 4–12% SDS polyacrylamide gel to run SDS-PAGE, and then the proteins were transferred to PVDF membranes and blocked with 5% BSA for 1 h at room temperature. The membranes were incubated with primary antibodies overnight at 4 °C and the corresponding secondary antibodies for 1 h at room temperature. The antibodies used included ki67 (1:1000, Beyotime, Shanghai, China), PCNA (1:1000, Beyotime, Shanghai, China), cleaved caspase-3 (1:1000, CST, Danvers, MA, USA), Bax (1:1000, Wanleibio, Shenyang,, China), Bcl-2 (1:1000, Wanleibio, Shenyang, China), GAPDH (1:10,000, Proteintech, Wuhan, China), PTPN11 (1:1000, Wanleibio, Shenyang, China) and HRP-labeled goat anti-rabbit IgG (1:1000, Beyotime, Shanghai, China).

2.6. Real-Time PCR Analysis

Total RNA was extracted using Trizol (Invitrogen, Carlsbad, CA, USA), and then the equivalent RNA of each group was reverse-transcribed into cDNA using a Prime ScriptTM RT reagent Kit (TaKaRa, Gunma, Japan). Real-time PCR reactions were conducted in 10 µL of mixture including 1 µL of each cDNA sample, 0.5 µL of specific forward primers (10 µM), 0.5 µL of specific reverse primers (10 µM), 5 µL of 2X Universal SYBR Green Fast qPCR Mix (ABclonal, Wuhan, China) and 3 µL of double-distilled water. The conditions were 95 °C for 5 min followed by 40 cycles of 95 °C for 15 s, 60 °C for 30 s and 72 °C for 30 s, and finally 72 °C for 10 min. All data were normalized to GAPDH. The primer sequences are listed in Table 1.

Table 1. The real-time PCR primers.

Gene	Primer Sequence
EGFR	F: GGTGAGTGGCTTGTCTGGAA
EGFR	R: CCTTACGCCCTTCACTGTGT
GBR2	F: AAGCTACTGCAGACGACGAG
GBR2	R: CTTGGCTCTGGGGATTTTGC
IGF1R	F: AGGCTGGGGCTCTTGTTTAC
IGF1R	R: CCTCTCTCGAGTTCGCCTG
SRC	F: TTCTGCTGTTGACTGGCTGT
SRC	R: TGAGGATGGTCAGGTTGTGC
PTPN11	F: CGTCATGCGTGTTAGGAACG
PTPN11	R: TCTCTCCGTATTCCCCTGGA
MAPK1	F: TCCTTTGAGCCGTTTGGAGG
MAPK1	R: AGTACATACTGCCGCAGGTC

2.7. Xenograft Study Models

All animal experiments were approved by the Institutional Animal Care and Use Committee of the Peking University Health Science Center (No. LA2022229), following the Committee of Peking University Health Science Center's Animals Usage Guidelines and performed using the approved protocols of the Animal Ethical and Welfare Committee. Healthy male BALB/cA-nu mice (4–5 weeks) were injected with 5×10^6 cells in 100 µL of PBS subcutaneously (two groups were injected with WSU-HN6 cells, and the other two groups were injected with CAL-27 cells, four mice/group). When the tumor volume was about 100 mm^3, 15 mg/kg MF was administered to mice twice per week orally by gavage and an equal volume of DMSO alone (0 mg/kg MF) was given to the control group. Tumor size was measured every 3 days and estimated using the formula: $V = \text{length} \times \text{width}^2/2$. After 4 weeks, the mice were euthanized and the tumors were removed entirely to conduct HE and Western blotting.

2.8. Hematoxylin–Eosin Staining

Hematoxylin–eosin (HE) staining was performed to determine whether organ toxicity occurred due to the MF treatment in vivo. The corresponding tissues from mice were collected and fixed in 4% paraformaldehyde, embedded in paraffin wax and 5 µm sections were cut and mounted onto slides. The slides were deparaffinized, rehydrated and stained with hematoxylin and eosin. Then, the images were captured using a fluorescence microscope (Nikon, Minato-ku, Japan).

2.9. Bioinformatics Analysis

Pharmmapper (http://www.lilab-ecust.cn/pharmmapper/, accessed on 13 April 2023) was employed to screen for potential targets of MF. HNSCC-related genes were obtained from GeneCards (https://www.genecards.org/, accessed on 13 April 2023) and the Comparative Toxicogenomics Database (CTD, http://ctdbase.org/, accessed on 13 April 2023). The intersection of potential MF targets and HNSCC-related genes was set as the potential targets of MF in HNSCC; these genes were imported into STRING (https://string-db.org/, accessed on 14 April 2023) to construct a PPI network, and then degree centrality (DC), closeness centrality (CC), eigenvector centrality (EC), betweenness centrality (BC), local average connectivity (LAC) and network centrality (NC) were used to identify the core targets of MF against HNSCC via Cytoscape 3.7.1.

To further investigate the underlying mechanisms of MF against HNSCC, the potential targets were used to perform Gene Ontology (GO) and Kyoto Encyclopedia of Genes and Genomes (KEGG) pathway analysis using the Database for Annotation Visualization and Integrated Discovery (DAVID, https://david.ncifcrf.gov/, accessed on 15 April 2023).

To valid the results of the network pharmacology analysis, the RNA sequencing (FPKM) and clinical information about HNSCC were accessed before August 2022 from The Cancer Genome Atlas (TCGA) database (https://portal.gdc.cancer.gov/, accessed on

29 August 2023). Then, the expression levels of MF core targets were detected and survival analysis was performed.

2.10. Molecular Docking

The structure of MF was downloaded from PubChem (https://pubchem.ncbi.nlm.nih.gov/, Compound CID: 441336, accessed on 13 April 2023) and is shown in Figure 1. The protein structures were obtained from PDB (http://www.rcsb.org/, PDB ID: EGFR: 1m14, GRB2: 1bm2, IGF1R: 1igr, SRC: 1a07, PTPN11: 2shp, MAPK1: 1pme, accessed on 15 April 2023). PyMOL software (version 4.6.0) was used to remove the water and small-molecule ligands of the protein. AutoDockTools (version 1.5.6) was employed to hydrotreat the protein molecules, obtain PDBQT files and determine active pockets. We ran the Vina 1.1.2 program to calculate the binding energy. Finally, the optimal combination model was visualized via PyMOL. A smaller binding energy indicated a stronger binding force between MF and the target proteins. A binding energy ≤ -5.0 kcal/mol indicated they could be combined, and a binding energy ≤ -7.0 kcal/mol indicated that they exhibited excellent binding strength.

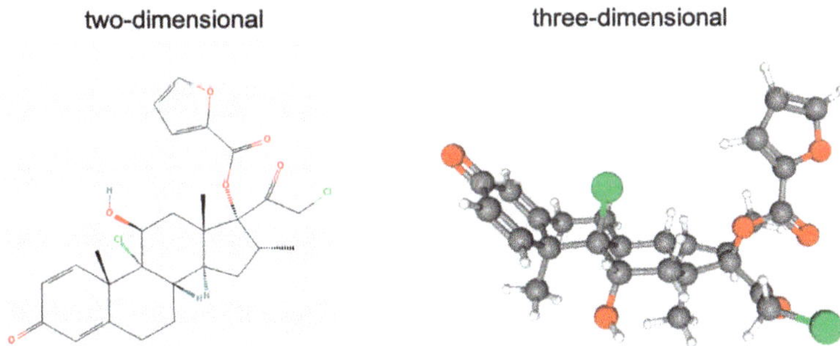

Figure 1. The structure of MF (image downloaded from PubChemd, Compound CID: 441336).

2.11. Construction of PTPN11-Overexpression Plasmid and Cell Lines

To explore whether PTPN11 acts as a downstream target of MF, we constructed stable PTPN11-overexpressing WSU-HN6 and CAL-27 cells. The PCDH plasmid was employed as the control group (scramble cells). Human full-length PTPN11 cDNA was amplified and cloned into the PCDH plasmid using the ClonExpress II One Step Cloning Kit (Vazyme, Nanjing, China) according to the manufacturer's protocol. The primer is shown in Table 2. The plasmid with the correct sequence was transfected into 293T cells with VSVG and PAX8 plasmids. Lentivirus supernatants were harvested at 48 h and were utilized to infect WSU-HN6 and CAL-27 cells at 80% confluency. Puromycin was added 48 h after infection to obtain the positive PTPN11-overexpressed WSU-HN6 and CAL-27 cells. Real-time PCR and Western blotting were performed to detect the overexpression efficiency at the RNA level and protein level, respectively, according to the methods described detailed in the corresponding sections. Additionally, the stable PTPN11-overexpressing cells were treated with MF, and then cell viability and Western blotting were conducted to detect the role of PTPN11 in the anti-tumor efficacy of MF.

Table 2. Sequences of oligonucleotides used for PTPN11 overexpression.

Primer	Oligonucleotides Sequence
PTPN11-OE-F	GGGGGAGGAGGGGGATCCGGAATGACATCGCGGAGATGGT
PTPN11-OE-R	GATCCTTCGCGGCCGCGATCCTCATCTGAAACTTTTCTGC

2.12. Statistical Analysis

GraphPad Prism 8.4.3 software (San Diego, CA, USA) was employed for the statistical analyses. All experiments were repeated three times to ensure the validity of the data. All data are expressed as the mean ± standard deviation (SD, $n = 3$). Data for more than two groups were tested for homogeneity, followed by one-way ANOVA analysis. $p < 0.05$ was considered statistically significant.

3. Results

3.1. MF Inhibited the Proliferation of HNSCC Cells

To determine the cell cytotoxicity of MF, we first detected the half-maximal inhibitory concentration (IC_{50}) of cells treated with MF, which decreased as time increased in CAL-27 cells (Figure 2B). The IC_{50} of WSU-HN6 cells treated with MF was decreased after 48 h of culture compared with 24 h of culture (24 h: 50.57 µM; 48 h: 25.07 µM). Meanwhile, after 72 h of culture, the IC_{50} of WSU-HN6 (72 h: 32.07 µM) was slightly higher than that of 48 h of culture (Figure 2A). According to the IC_{50}, we chose 0, 10, 20 and 50 µM for further experiments. Then, cell viability was detected after cells were treated with different doses of MF and incubated for different periods of time (24, 48 and 72 h). Cell viability gradually decreased as the dose and time increased (Figure 2C,D), indicating that MF could inhibit cell proliferation in a time- and dose-dependent manner. Similar results were observed in the colony formation assay. The number of colonies in the MF-treated groups decreased with increasing concentrations of MF (Figure 2E,F). Accordingly, these data recapitulated the inhibitory effect of MF on the proliferation of HNSCC cells.

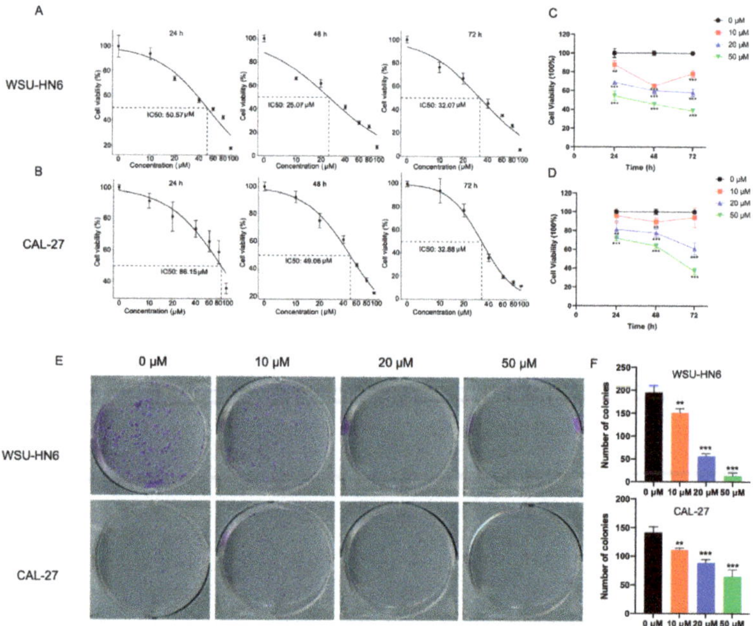

Figure 2. The effect of MF on the cell proliferation of HNSCC cells in vitro. (**A,B**) The IC_{50} of WSU-HN6 cells (**A**) and CAL-27 cells (**B**) were detected after cells were treated with 0, 10, 20, 40, 60, 80 and 100 µM MF for 24, 48 and 72 h; (**C,D**) the percentage of viability WSU-HN6 cells (**C**) and CAL-27 cells (**D**) after cells were treated with 0, 10, 20 and 50 µM MF for 24, 48 and 72 h; (**E**) the colony formation of WSU-HN6 and CAL-27 cells after cells were treated with 0, 10, 20 and 50 µM MF for 7 days; (**F**) the quantitative analysis results of (**E**). ** $p < 0.01$, *** $p < 0.001$, $n = 3$. Asterisks represent differences between cells treated with 10, 20 and 50 µM MF and cells treated with 0 µM MF.

3.2. MF Regulated the Cell Cycle and Induced Apoptosis In Vitro

Since glucocorticoids suppress lymphoid progression via regulating the cell cycle and apoptosis, we studied the cell cycle and apoptosis of cells treated with MF further to obtain insight into the proliferation inhibition effect of MF. Compared with the control group, cells treated with MF performed S-phase arrest (Figure 3A–D). Meanwhile, the cell apoptosis rate increased significantly elevated with increasing concentrations of MF (Figure 3E–H). This notion was further supported by the observation that the protein expression levels of the cell proliferation markers ki67, PCNA and anti-apoptotic protein Bcl-2 decreased and the apoptosis-related proteins Bax and cleaved caspase-3 increased with increasing concentrations of MF (Figure 3I,J). Collectively, these results demonstrated that MF inhibited cell proliferation via inducing cell cycle arrest and cell apoptosis.

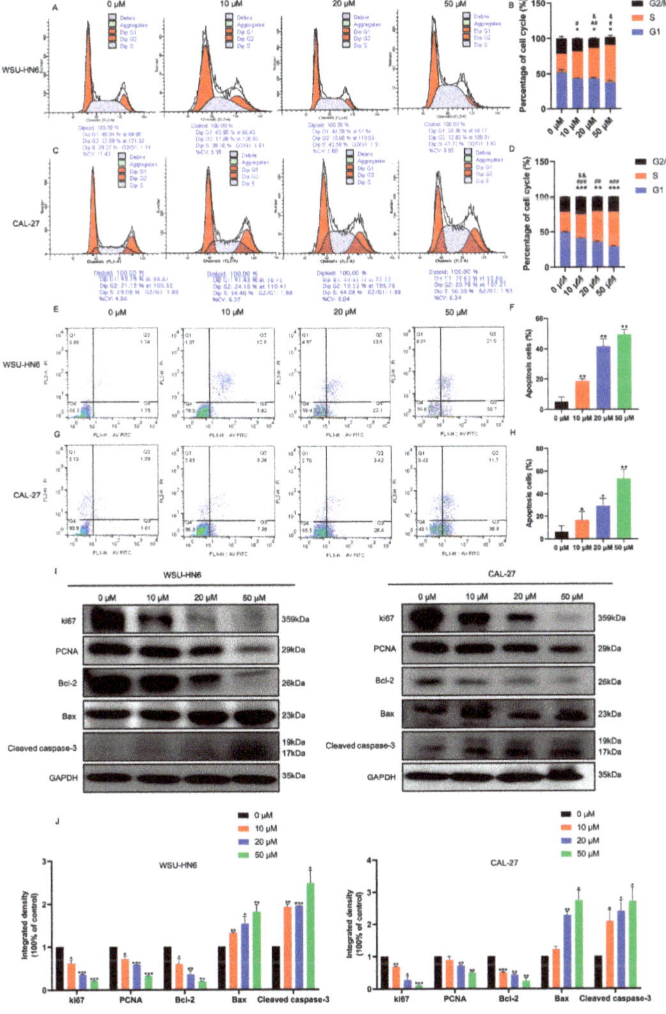

Figure 3. MF induced cell cycle arrest and cell apoptosis of HNSCC cells in vitro. (**A–D**) Distribution (**A,C**) and statistical analysis (**B,D**) of cell cycles of WSU-HN6 and CAL-27 cells, respectively, treated with 0, 10, 20 and 50 µM MF for 24 h; (**E–H**) representative flow cytometry plots (**E,G**) and statistical analysis of the cell apoptosis rate (**F,H**) of WSU-HN6 and CAL-27 cells, respectively, treated with 0,

10, 20 and 50 μM MF for 24 h; (**I**) the expression levels of proliferation- and apoptosis-related proteins in WSU-HN6 and CAL-27 cells treated with 0, 10, 20 and 50 μM MF; (**J**) the quantitative analysis results of (**I**). *, #, & $p < 0.05$; **, ##, && $p < 0.01$; ***, ### $p < 0.001$; $n = 3$. Asterisks represents the difference between G1 phase, apoptosis or the corresponding protein expression levels of cells treated with 10, 20 and 50 μM MF (MF-treated groups) and cells treated with 0 μM MF (control group), pounds represent the difference between S phase in MF-treated groups and control group; ampersands represents the difference between G2/M phase in MF-treated groups and control group.

3.3. MF Suppressed Tumor Growth In Vivo

We then examined whether MF could inhibit cell proliferation in vivo. Nude mice were injected with HNSCC cells and then orally administered MF at a concentration of 15 mg/kg when the tumor volume was about 100 mm³ (the tenth day after cell injection). Mice given an equal volume of DMSO (0 mg/kg MF) were employed as the control group. Compared with the control group, the tumor volume in the MF-treated group was remarkably decreased after 19 days of tumor inoculation, indicating that MF suppressed HNSCC cell growth in vivo (Figure 4A,B). There was no organ toxicity in the MF-treated group relative to the control group (Figure 4C). Western blotting revealed the downregulated protein expression levels of ki67, PCNA and Bcl-2 and the upregulated protein expression levels of Bax and Cleaved caspase-3 in tumors of mice treated with MF (Figure 4D,E).

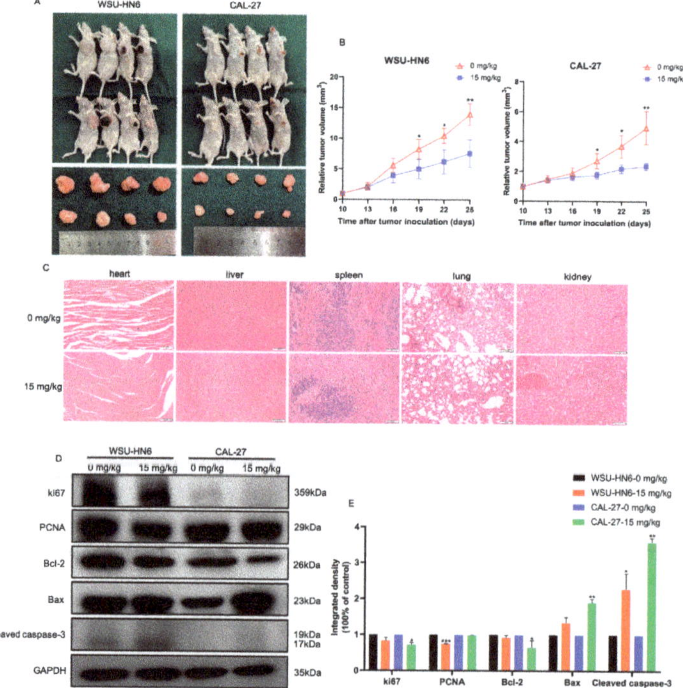

Figure 4. MF inhibited HNSCC progression in vivo. (**A**) Representative images of tumor-bearing mice and tumor samples; (**B**) tumor volume of the mice treated with 15 mg/kg MF compared with the mice treated with 0 mg/kg MF; (**C**) HE staining images of heart, liver, spleen, lung and kidneys of the mice treated with 0 mg/kg and 15 mg/kg MF, 20×; (**D**) the expression levels of proliferation- and apoptosis-related proteins of the mice treated with 0 mg/kg and 15 mg/kg MF; (**E**) the quantitative analysis results of (**D**). * $p < 0.05$, ** $p < 0.01$, *** $p < 0.001$, $n = 3$. Asterisks represents the difference between mice treated with 15 mg/kg MF and mice treated with 0 mg/kg MF.

3.4. PTPN11 Was the Core Target of MF against HNSCC

We then performed network pharmacology analysis to identify the potential targets of MF against HNSCC. A total of 168 genes were set as the potential targets of MF against HNSCC which were obtained from the intersection of 225 genes of MF targets from PharmMapper, and 4748 genes and 16,986 genes associated with HNSCC from GeneCards and CTD, respectively (Figure 5A). The GO analysis showed that the 168 genes were mainly associated with the response to steroid hormone and nuclear receptor activity (Figure 5B). KEGG analysis revealed that proteoglycans in cancer was the most significant enrichment signaling pathway for the potential targets of MF against HNSCC (Figure 5C, Table 3). Overall, 168 genes were imported into the STRING database and then a primary PPI network with 394 edges and 134 nodes (namely targets) was obtained; the 134 targets were examined with a two-step topology analysis and finally 10 targets were identified (Figure 5D). To further narrow the potential functional targets, we analyzed the intersection of the 10 targets and 26 genes enriched with proteoglycans in cancer. As a result, eight genes, *EGFR, ESR1, GRB2, IGF1, IGF1R, MAPK1, PTPN11* and *SRC*, were identified (Figure 5E). *ESR1* and *IGF1* were ruled out due to their non-significant and downregulated mRNA expression levels in HNSCC compared with normal tissues after analyzing the TCGA database (Figure 6A). We also analyzed the relationship between the expression levels of these potential targets and patient survival, and found that only the expression levels of *EGFR, IGF1R, PTPN11* and *ESR1* were significantly associated with patient prognosis (Supplementary Figure S1). We then detected the mRNA expression levels of six genes after cells were treated with MF. Interestingly, only the expression level of *PTPN11* notably decreased in a MF dose-dependent manner (Figure 6B). The molecular docking analysis revealed that PTPN11 showed the strongest combination with MF and the binding energy was −8.3 kcal/mol (Figure 6C). Taken together, these data strongly indicated that MF was involved in multiple tumor-associated signaling pathways, and, most importantly, that MF may exert its anti-tumor activity by targeting PTPN11.

Table 3. The top ten KEGG pathways.

ID	Description	Gene Ratio	*p* Value	p. Adjust	q Value
hsa05205	Proteoglycans in cancer	26/150	4.1534×10^{-15}	1.109×10^{-12}	5.2901×10^{-13}
hsa01522	Endocrine resistance	18/150	1.4301×10^{-13}	1.9092×10^{-11}	9.1077×10^{-12}
hsa05215	Prostate cancer	17/150	1.5834×10^{-12}	1.4092×10^{-10}	6.7226×10^{-11}
hsa01521	EGFR tyrosine kinase inhibitor resistance	15/150	1.0373×10^{-11}	6.9241×10^{-10}	3.3031×10^{-10}
hsa04068	FoxO signaling pathway	18/150	2.4517×10^{-11}	1.3092×10^{-9}	6.2455×10^{-10}
hsa04659	Th17 cell differentiation	16/150	9.1379×10^{-11}	4.0664×10^{-9}	1.9398×10^{-9}
hsa04917	Prolactin signaling pathway	13/150	3.5466×10^{-10}	1.3528×10^{-8}	6.4532×10^{-9}
hsa04926	Relaxin signaling pathway	16/150	1.5842×10^{-9}	5.2872×10^{-8}	2.5222×10^{-8}
hsa05145	Toxoplasmosis	15/150	1.7878×10^{-9}	5.3037×10^{-8}	2.5301×10^{-8}
hsa04933	AGE-RAGE signaling pathway in diabetic complications	14/150	3.5417×10^{-9}	9.3914×10^{-8}	4.48×10^{-8}

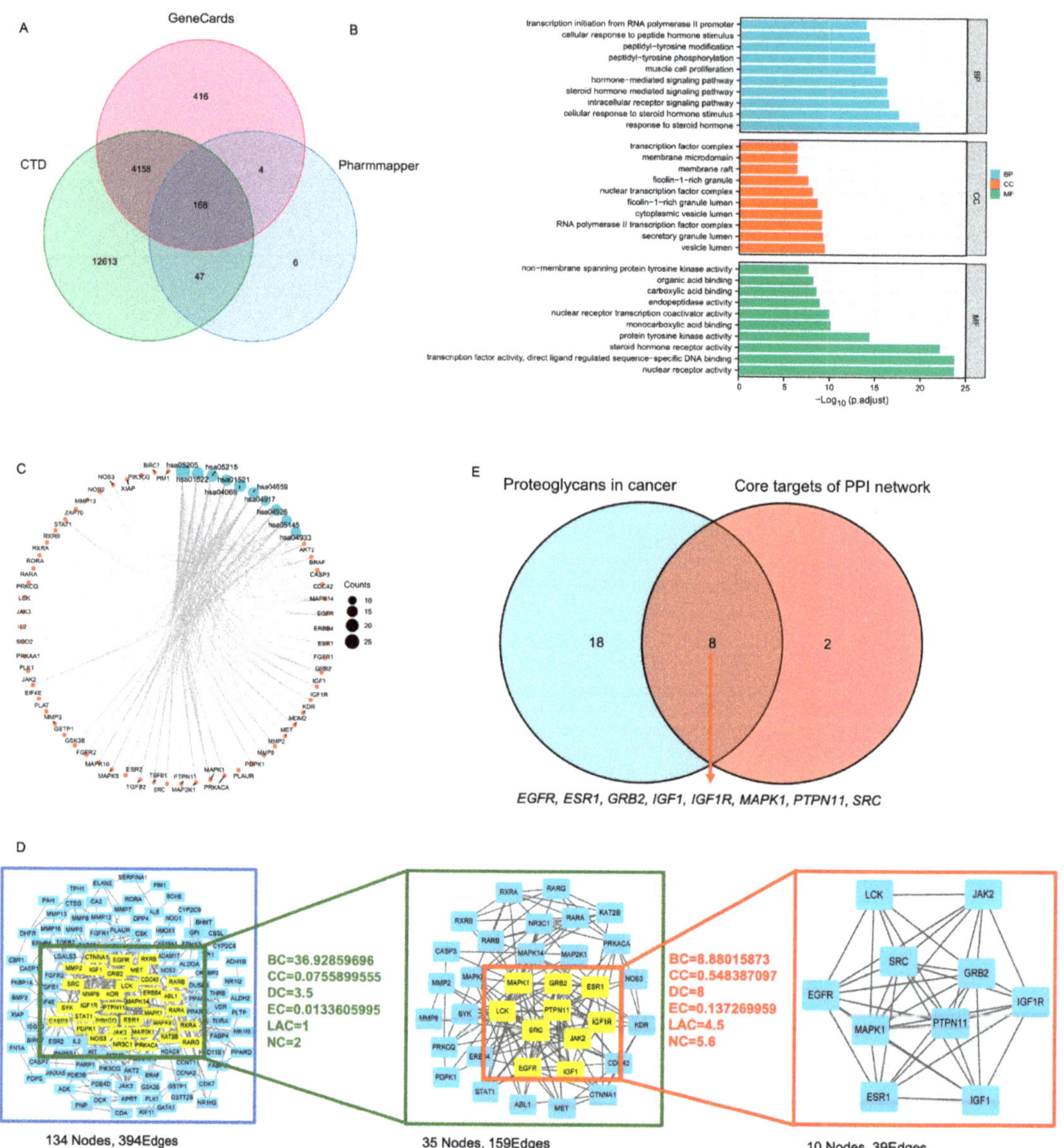

Figure 5. Network pharmacology analysis of the MF targets against HNSCC. (**A**) Venn diagram of MF targets and HNSCC-related genes; (**B**) chart of the top ten GO enrichments; (**C**) chart of the top ten KEGG pathways; (**D**) the PPI network constructed by Cytoscape via two-step topology analysis. The parameters of the first step are shown in green text and the parameters of the second step are shown in red text; (**E**) Venn diagram of 10 targets of MF against HNSCC and 26 genes enriched with proteoglycans in cancer.

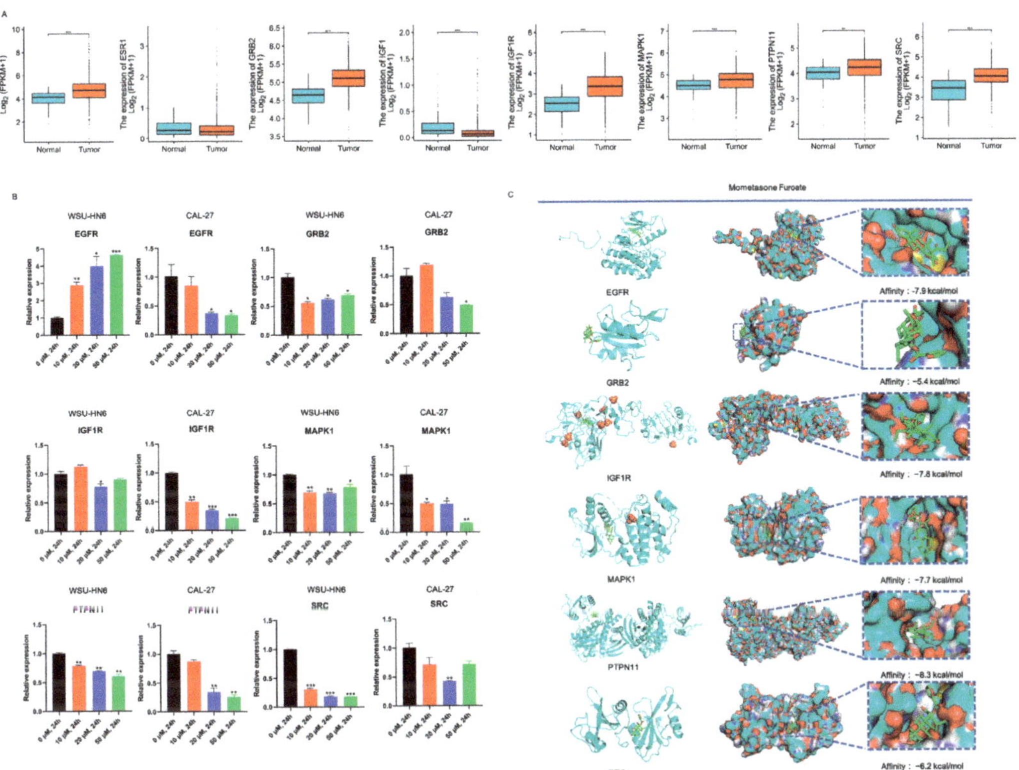

Figure 6. PTPN11 was the core target of MF against HNSCC. (**A**) The expression of the eight targets in TCGA database. The blue boxes represent normal tissue ($n = 44$) and the red boxes represent HNSCC tumor tissue ($n = 504$); (**B**) the mRNA expression levels of the six targets of MF against HNSCC after cells were treated with 0, 10, 20 and 50 μM MF for 24 h; (**C**) molecular docking of the six targets of MF against HNSCC. The affinity represents the binding energy. * $p < 0.05$, ** $p < 0.01$, *** $p < 0.001$, $n = 3$. Asterisks represents the differences between HNSCC tumor tissues and normal tissue or cells treated with 10, 20 and 50 μM MF and cells treated with 0 μM MF.

3.5. MF Exerted Anti-Tumor Activity by Targeting PTPN11

To further investigate whether the inhibition of PTPN11 was required for the anti-tumor efficacy of MF, we constructed stable PTPN11-overexpressing cells. The efficiency of overexpression was confirmed at the mRNA and protein levels (Figure 7A–C). Interestingly, we observed that cells treated with MF revealed a significantly decreased expression level for PTPN11. However, there was still relatively higher expression of PTPN11 compared to the control group cells (Figure 7A–C), indicating that MF could not completely inhibit the overexpression of PTPN11. As shown in Figure 7D, the cell viability of PTPN11-overexpressing cells was significantly higher than that of scramble cells, indicating that PTPN11 could promote the proliferation of HNSCC cells, which was consistent with a previous study [22]. After cells were treated with MF, significantly decreased cell viability was observed in scramble cells and PTPN11-overexpressing cells, and the cell viability of the scramble cells was slightly lower than that of PTPN11-overexpressing cells, suggesting that PTPN11 played a central role in the inhibitory effect of MF on cell proliferation. Similar results were observed using Western blotting (Figure 7E,F): after MF treatment, the protein expression levels of ki67, PCNA and Bcl-2 in scramble cells were slightly lower than those of PTPN11-overexpressing cells. Taken together, it was shown that MF could target PTPN11 to exert anti-tumor activity.

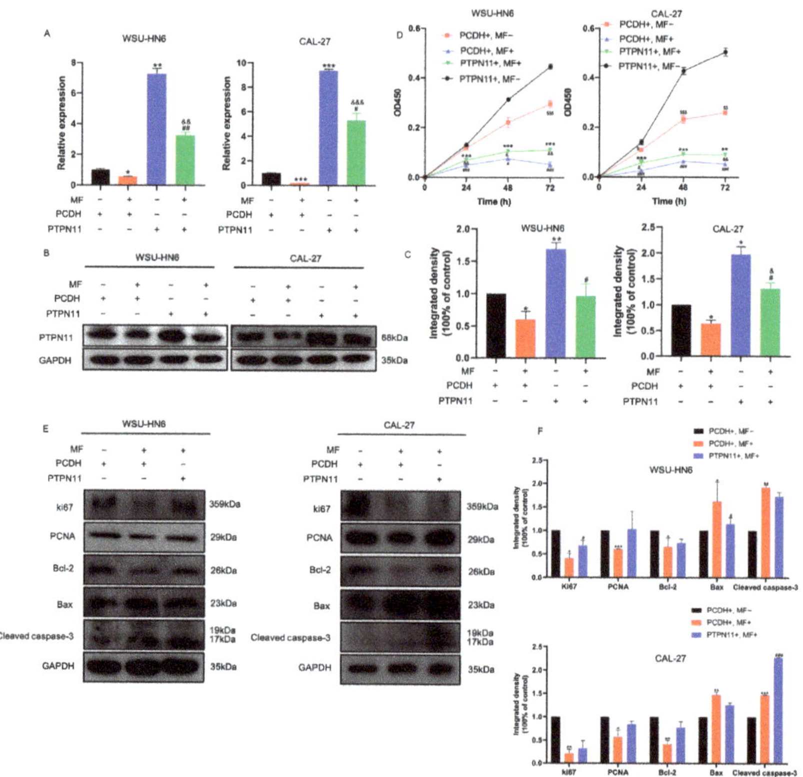

Figure 7. MF inhibited the HNSCC progression induced by PTPN11 overexpression. (**A,B**) The mRNA (**A**) and protein (**B**) expression levels of PTPN11 of the PTPN11-overexpressing cells treated with MF; (**C**) the quantitative analysis results of (**B**); (**D**) the percentage of viable PTPN11-overexpressing cells treated with and without MF; (**E**) the expression levels of proliferation- and apoptosis-related proteins of the PTPN11-overexpressing cells treated with MF; (**F**) the quantitative analysis results of (**E**). *, #, &, \$ $p < 0.05$; **, ##, &&, \$\$ $p < 0.01$; ***, &&&, ###, \$\$\$ $p < 0.001$; $n = 3$. Asterisks represent the difference between the PCDH+, MF+ group or the PTPN11+, MF− group and the PCDH+, MF− group (**A,C**). Pounds represent the difference between the PTPN11+, MF+ group and the PTPN11+, MF− group (**A,C**). Ampersands represent the difference between the PTPN11+, MF+ group and the PCDH+, MF+ group (**A,C**). Asterisks represent the difference between the PTPN11+, MF+ group cand the PTPN11+, MF− group (**D,F**). Pounds represent the difference between the PCDH+, MF− group and the PCDH+, MF+ group (**D,F**). Ampersands represents the difference between the PCDH+, MF+ group and the PTPN11+, MF+ group (**D,F**). Dollars represent the difference between the PTPN11+, MF− group and the PCDH+, MF− group (**D,F**).

4. Discussion

While they are classically used as anti-inflammation drugs, accumulating evidence has demonstrated that glucocorticoids can also treat malignant tumors such as multiple myeloma [23], prostate cancer [24], colorectal cancer [25] and breast cancer [26]. Inflammation is closely related to the development of most types of cancers. Cancer-extrinsic inflammation, triggered by elements such as viral infections and autoimmune disease, has been reported as an induced factor for tumor initiation and progression. Cancer-intrinsic inflammation is involved in the recruitment and activation of inflammatory cells, inducing an immunosuppressive tumor microenvironment and eventually accelerating malignant progression. Therefore, anti-inflammatory drugs targeting the inflammatory tumor microenvironment have been identified as the pivotal determinant for conventional chemotherapy and immunotherapy efficacy [27]. In addition, it has been reported that the overexpression of inflammation-related factors is related to drug resistance. Increased circulation of the cytokine interleukin-6 was associated with acquired resistance to dasatinib [28], which was further confirmed by a phase II clinical trial in HNSCC patients [29]. Furthermore, anti-inflammatory drugs were shown to attenuate the toxicity of chemotherapeutic agents [30]. For instance, when combined with docetaxel and celecoxib to treat patients with metastatic prostate cancer, they were shown to reduce hematologic toxicity [31]. Emerging studies have reported that local application of MF remarkably reduced acute radiation dermatitis after radiotherapy for HNSCC [32] and breast cancer [33]. Based on the aforementioned intimate relationship between inflammation and tumors, anti-inflammatory drugs seem to be a powerful method to enhance therapeutic efficiency in HNSCC.

MF is a traditional glucocorticoid with an adequate anti-inflammatory capacity. However, previous studies regarding MF have mainly focused on its use as a treatment for inflammation, and researchers have only recently begun to delineate the role of MF in cancer. Since glucocorticoids can promote cell apoptosis via binding with glucocorticoid receptors in lymphoid cells, the most extensive application of glucocorticoids in cancer is to treat lymphoid malignancies [34]. Consistent with this, it was reported that MF could inhibit the growth of acute leukemia cells through promoting cell apoptosis [9], but there was still no exploration of the pharmacological effect of MF on solid tumors. Here, we found that MF could inhibit HNSCC cell proliferation both in vitro and in vivo. Mechanistically, the protein expression level of the anti-apoptotic molecule Bcl-2 was significantly decreased and that of the pro-apoptotic Bax was increased after cells were treated with MF, which was consistent with a previous study that showed that the downregulation of Bcl-2 was essential for glucocorticoid-induced cell apoptosis [34]. Meanwhile, the decreased Bcl-2 expression subsequently released cytochrome C into the cytoplasm and then promoted the activation of cleaved caspase-3, which also is a sign of cell apoptosis. Our results also revealed that the protein expression level of cleaved caspase-3 was remarkably increased after MF treatment. These results supported the notion that MF could exert its anti-tumor effect via promoting cell apoptosis.

To better understand the inhibitory effect of MF in HNSCC, we then conducted network pharmacology and molecular docking analyses, which were effective methods to provide a more comprehensive perspective of MF's mechanism against HNSCC. KEGG analysis demonstrated that the potential target genes of MF against HNSCC were mainly associated with proteoglycans in cancer signaling pathways, which was related to various biological processes in cancer [35]. According to a series of bioinformatics analyses, PTPN11 seemed to be the most important target of MF against HNSCC. Our experimental results confirmed that MF decreased the mRNA expression level of PTPN11 in a dose-dependent manner. It is widely known that sustained activation of PTPN11 is responsible for the occurrence, development and prognosis of multiple cancers. Our pan-cancer analysis of PTPN11 through the TCGA database also confirmed abnormal expression of PTPN11 in multiple tumors and a significant correlation with patient prognosis (Supplemental Figure S2). Allosteric inhibition of PTPN11 via SHP099 suppressed the RTK-driven human cancer cells by inhibiting the RAS-ERK signaling pathway [36]. RAS was overexpressed in HNSCC cells and was responsible for tumor growth [37]. Therefore, we assumed that MF exerted its excellent anti-tumor capacity via targeting PTPN11. To verify this, we constructed PTPN11-overexpressing cells to explore whether MF performs its anti-tumor effect by suppressing PTPN11. More interestingly, we observed that MF could decrease the mRNA and protein expression levels of PTPN11, as well as inhibit the cell proliferation and increased cell apoptosis after PTPN11 is overexpressed. However, MF could not eliminate the exogenous overexpression of PTPN11 and the promotion of cell growth caused by

PTPN11 overexpression. We predicted that there were eight genes targeted by MF against HNSCC via the network pharmacology analysis. Thus, MF also has the potential to bind with other targets, resulting in a competitive combination between the targets and MF, which may explain why MF could not inhibit PTPN11 completely. Further studies are needed to clarify this issue. In addition, among the eight targets, EGFR was upregulated in 80–90% of HNSCC patients and the overexpression of EGFR was associated with a poor prognosis [38]. The FDA-approved EGFR monoclonal antibody cetuximab has been combined with chemotherapy or radiotherapy as a first-line therapy for HNSCC patients, but the acquisition of EGRF resistance was a frequent occurrence, eventually leading to therapy failure. Furthermore, PTPN11 inhibitors could overcome EGFR resistance in NSCLC [39]. Thus, we reasonably assumed that a single usage of MF could target PTPN11 to produce a marked anti-tumor effect. The combination of MF and PTPN11 inhibitors could overcome drug resistance to achieve a better therapeutic effect on HNSCC.

Overall, we proved that MF could exert an excellent anti-tumor effect through regulating PTPN11, which provides a theoretical basis for its clinical application. However, the detailed molecular mechanism still needs further exploration for clinical drug use.

5. Conclusions

We preliminarily revealed that MF plays a vital role in suppressing the proliferation of HNSCC via regulating the expression level of PTPN11. This provided a new perspective on the treatment of HNSCC with MF.

Supplementary Materials: The following supporting information can be downloaded at: https://www.mdpi.com/article/10.3390/biomedicines11102597/s1, Figure S1: Survival curve of patients with HNSCC in TCGA database; Figure S2: Bioinformatic analysis of PTPN11 in multiple cancers in the TCGA database. (A) The expression characteristics of PTPN11 in multiple cancers. ACC: adrenocortical carcinoma, BLCA: bladder urothelial carcinoma, BRCA: breast invasive carcinoma, CESC: cervical endocervical adenocarcinoma and squamous cell carcinoma, CHOL: cholangiocarcinoma, COAD: colon adenocarcinoma, DLBC: lymphoid neoplasm diffuse large B-cell lymphoma, ESCA: esophageal carcinoma, GBM: glioblastoma multiforme, HNSC: head and neck squamous cell carcinoma, KICH: kidney chromophobe, KIRC: kidney renal clear cell carcinoma, KIRP: kidney renal papillary cell carcinoma, LAML: acute myeloid leukemia, LGG: brain lower grade glioma, LIHC: liver hepatocellular carcinoma, LUAD: lung adenocarcinoma, LUSC: lung squamous cell carcinoma, MESO: mesothelioma, OV: ovarian serous cystadenocarcinoma, PAAD: pancreatic adenocarcinoma, PCPG: pheochromocytoma and paraganglioma, PRAD: prostate adenocarcinoma, READ: rectum adenocarcinoma, SARC: sarcoma, SKCM: skin cutaneous melanoma, STAD: stomach adenocarcinoma, TGCT: testicular germ cell tumors, THCA: thyroid carcinoma, THYM: thymoma, UCEC: uterine corpus endometrial carcinoma, UCS: uterine carcinosarcoma, UVM: uveal melanoma. ** $p < 0.01$, *** $p < 0.001$, Asterisks were normal compared with tumor. (B) The survival curve of PTPN11 in multiple cancers.

Author Contributions: L.Q. and Q.G. contributed equally to this work and share first authorship. Conceptualization, L.Q. and Q.G.; methodology, L.Q., Q.G. and A.T.; data curation, L.Q. and Q.G.; writing—original draft preparation, L.Q. and Q.G.; writing—review and editing, L.Q., Q.G., J.J. and C.L.; funding acquisition, C.L. and J.J. All authors have read and agreed to the published version of the manuscript.

Funding: This research was funded by the National Natural Science Foundation of China, grant number 81072214 and 30371547.

Institutional Review Board Statement: The ethics approval statements of all animal experiments were approved by the Institutional Animal Care and Use Committee of the Peking University Health Science Center and followed the Committee of Peking University Health Science Center's Animals Usage Guideline. The approval number was "LA2022229".

Data Availability Statement: The data supporting this study are available from the corresponding author upon reasonable request.

Conflicts of Interest: The authors declare no conflict of interest.

References

1. Sung, H.; Ferlay, J.; Siegel, R.L.; Laversanne, M.; Soerjomataram, I.; Jemal, A.; Bray, F. Global Cancer Statistics 2020: GLOBOCAN Estimates of Incidence and Mortality Worldwide for 36 Cancers in 185 Countries. *CA Cancer J. Clin.* **2021**, *71*, 209–249. [CrossRef] [PubMed]
2. Ortiz-Cuaran, S.; Bouaoud, J.; Karabajakian, A.; Fayette, J.; Saintigny, P. Precision Medicine Approaches to Overcome Resistance to Therapy in Head and Neck Cancers. *Front. Oncol.* **2021**, *11*, 614332. [CrossRef] [PubMed]
3. Tahara, M.; Muro, K.; Hasegawa, Y.; Chung, H.C.; Lin, C.C.; Keam, B.; Takahashi, K.; Cheng, J.D.; Bang, Y.-J. Pembrolizumab in Asia-Pacific patients with advanced head and neck squamous cell carcinoma: Analyses from KEYNOTE-012. *Cancer Sci.* **2018**, *109*, 771–776. [CrossRef]
4. Mayayo-Peralta, I.; Zwart, W.; Prekovic, S. Duality of glucocorticoid action in cancer: Tumor-suppressor or oncogene? *Endocr. Relat. Cancer* **2021**, *28*, R157–R171. [CrossRef] [PubMed]
5. Huang, Y.; Cai, G.Q.; Peng, J.P.; Shen, C. Glucocorticoids induce apoptosis and matrix metalloproteinase-13 expression in chondrocytes through the NOX4/ROS/p38 MAPK pathway. *J. Steroid. Biochem. Mol. Biol.* **2018**, *181*, 52–62. [CrossRef]
6. Cari, L.; De Rosa, F.; Nocentini, G.; Riccardi, C. Context-Dependent Effect of Glucocorticoids on the Proliferation, Differentiation, and Apoptosis of Regulatory T Cells: A Review of the Empirical Evidence and Clinical Applications. *Int. J. Mol. Sci.* **2019**, *20*, 1142. [CrossRef]
7. Liu, X.; Suo, H.; Zhou, S.; Hou, Z.; Bu, M.; Liu, X.; Xu, W. Afatinib induces pro-survival autophagy and increases sensitivity to apoptosis in stem-like HNSCC cells. *Cell Death Dis.* **2021**, *12*, 728. [CrossRef]
8. Raudenska, M.; Balvan, J.; Masarik, M. Cell death in head and neck cancer pathogenesis and treatment. *Cell Death Dis.* **2021**, *12*, 192. [CrossRef]
9. Wang, X.; Shi, J.; Gong, D. Mometasone furoate inhibits growth of acute leukemia cells in childhood by regulating PI3K signaling pathway. *Hematology* **2018**, *23*, 478–485. [CrossRef]
10. Xu, K.; Qin, X.S.; Zhang, Y.; Yang, M.Y.; Zheng, H.S.; Li, L.; Yang, X.; Xu, Q.; Li, Y.; Xu, P.; et al. Lycium ruthenicum Murr. anthocyanins inhibit hyperproliferation of synovial fibroblasts from rheumatoid patients and the mechanism study powered by network pharmacology. *Phytomedicine* **2023**, *118*, 154949. [CrossRef]
11. Tang, B.H.; Dong, Y. Network pharmacology and bioinformatics analysis on the underlying mechanisms of baicalein against oral squamous cell carcinoma. *J. Gene Med.* **2023**, *25*, e3490. [CrossRef] [PubMed]
12. Liu, M.; Gao, S.; Elhassan, R.M.; Hou, X.B.; Fang, H. Strategies to overcome drug resistance using SHP2 inhibitors. *Acta Pharm. Sin. B* **2021**, *11*, 3908–3924. [CrossRef] [PubMed]
13. Song, Y.H.; Zhao, M.; Zhang, H.Q.; Yu, B. Double-edged roles of protein tyrosine phosphatase SHP2 in cancer and its inhibitors in clinical trials. *Pharmacol. Ther.* **2022**, *230*, 107966. [CrossRef] [PubMed]
14. Ruess, D.A.; Heynen, G.J.; Ciecielski, K.J.; Ai, J.Y.; Berninger, A.; Kabacaoglu, D.; Görgülü, K.; Dantes, Z.; Wörmann, S.M.; Diakopoulos, K.N.; et al. Mutant KRAS-driven cancers depend on PTPN11/SHP2 phosphatase. *Nat. Med.* **2018**, *24*, 954–960. [CrossRef]
15. Li, S.; Wang, X.; Li, Q.; Li, C. Role of SHP2/PTPN11 in the occurrence and prognosis of cancer: A systematic review and meta-analysis. *Oncol. Lett.* **2023**, *25*, 19. [CrossRef] [PubMed]
16. Richards, C.E.; Elamin, Y.Y.; Carr, A.; Gately, K.; Rafee, S.; Cremona, M.; Hanrahan, E.; Smyth, R.; Ryan, D.; Morgan, R.K.; et al. Protein Tyrosine Phosphatase Non-Receptor 11 (PTPN11/Shp2) as a Driver Oncogene and a Novel Therapeutic Target in Non-Small Cell Lung Cancer (NSCLC). *Int. J. Mol. Sci.* **2023**, *24*, 10545. [CrossRef]
17. Hoffmann, L.; Coras, R.; Kobow, K.; López-Rivera, J.A.; Lal, D.; Leu, C.; Najm, I.; Nürnberg, P.; Herms, J.; Harter, P.N.; et al. Ganglioglioma with adverse clinical outcome and atypical histopathological features were defined by alterations in PTPN11/KRAS/NF1 and other RAS-/MAP-Kinase pathway genes. *Acta Neuropathol.* **2023**, *145*, 815–827. [CrossRef]
18. Wang, H.C.; Chiang, W.F.; Huang, H.H.; Shen, Y.Y.; Chiang, H.C. Src-homology 2 domain-containing tyrosine phosphatase 2 promotes oral cancer invasion and metastasis. *BMC Cancer* **2014**, *14*, 442. [CrossRef]
19. Leibowitz, M.S.; Srivastava, R.M.; Andrade Filho, P.A.; Egloff, A.M.; Wang, L.; Seethala, R.R.; Ferrone, S.; Ferris, R.L. SHP2 is overexpressed and inhibits pSTAT1-mediated APM component expression, T-cell attracting chemokine secretion, and CTL recognition in head and neck cancer cells. *Clin. Cancer Res.* **2013**, *19*, 798–808. [CrossRef]
20. Yuan, X.; Bu, H.; Zhou, J.; Yang, C.Y.; Zhang, H. Recent Advances of SHP2 Inhibitors in Cancer Therapy: Current Development and Clinical Application. *J. Med. Chem.* **2020**, *63*, 11368–11396. [CrossRef]
21. Qiu, L.; Liu, H.; Wang, S.; Dai, X.H.; Shang, J.W.; Lian, X.L.; Wang, G.H.; Zhang, J. FKBP11 promotes cell proliferation and tumorigenesis via p53-related pathways in oral squamous cell carcinoma. *Biochem. Biophys. Res. Commun.* **2021**, *559*, 183–190. [CrossRef] [PubMed]
22. Xie, H.; Huang, S.; Li, W.; Zhao, H.; Zhang, T.; Zhang, D. Upregulation of Src homology phosphotyrosyl phosphatase 2 (Shp2) expression in oral cancer and knockdown of Shp2 expression inhibit tumor cell viability and invasion in vitro. *Oral. Surg. Oral. Med. Oral. Pathol. Oral. Radiol.* **2014**, *117*, 234–242. [CrossRef] [PubMed]
23. Kaiser, M.F.; Hall, A.; Walker, K.; Sherborne, A.; Tute, R.M.D.; Newnham, N.; Roberts, S.; Ingleson, E.; Bowles, K.; Garg, M.; et al. Daratumumab, Cyclophosphamide, Bortezomib, Lenalidomide, and Dexamethasone as Induction and Extended Consolidation Improves Outcome in Ultra-High-Risk Multiple Myeloma. *J. Clin. Oncol.* **2023**, *41*, 3945–3955. [CrossRef] [PubMed]

24. Chi, K.N.; Fleshner, N.; Chiuri, V.E.; Bruwaene, S.V.; Hafron, J.; McNeel, D.G.; Porre, P.D.; Maul, R.S.; Daksh, M.; Zhong, X.; et al. Niraparib with Abiraterone Acetate and Prednisone for Metastatic Castration-Resistant Prostate Cancer: Phase II QUEST Study Results. *Oncologist* **2023**, *28*, e309–e312. [CrossRef]

25. Ahmed, A.; Reinhold, C.; Breunig, E.; Phan, T.S.; Dieteich, L.; Kostadinova, F.; Urwyler, C.; Merk, V.M.; Noti, M.; da Silva, I.T.; et al. Immune escape of colorectal tumours via local LRH-1/Cyp11b1-mediated synthesis of immunosuppressive glucocorticoids. *Mol. Oncol.* **2023**, *17*, 1545–1566. [CrossRef]

26. Cairat, M.; Rahmoun, M.A.; Gunter, M.J.; Heudel, P.E.; Severi, G.; Dossus, L.; Fournier, A. Use of systemic glucocorticoids and risk of breast cancer in a prospective cohort of postmenopausal women. *BMC Med.* **2021**, *19*, 186. [CrossRef]

27. Diakos, C.I.; Charles, K.A.; McMillan, D.C.; Clarke, S.J. Cancer-related inflammation and treatment effectiveness. *Lancet Oncol.* **2014**, *15*, e493–e503. [CrossRef]

28. Sen, B.; Saigal, B.; Parikh, N.; Gallick, G.; Johnson, F.M. Sustained Src inhibition results in signal transducer and activator of transcription 3 (STAT3) activation and cancer cell survival via altered Janus-activated kinase-STAT3 binding. *Cancer Res.* **2009**, *69*, 1958–1965. [CrossRef]

29. Stabile, L.P.; Egloff, A.M.; Gibson, M.K.; Gooding, W.E.; Ohr, J.; Zhou, P.; Rothenberger, N.J.; Wang, J.; Geiger, J.L.; Flaherty, J.T.; et al. IL6 is associated with response to dasatinib and cetuximab: Phase II clinical trial with mechanistic correlatives in cetuximab-resistant head and neck cancer. *Oral. Oncol.* **2017**, *69*, 38–45. [CrossRef]

30. Wu, Q.J.; Li, W.; Zhao, J.; Sun, W.; Yang, Q.Q.; Chen, C.; Xia, P.; Zhu, J.; Huang, G.; Yong, C.; et al. Apigenin ameliorates doxorubicin-induced renal injury via inhibition of oxidative stress and inflammation. *Biomed. Pharmacother.* **2021**, *137*, 111308. [CrossRef]

31. Albouy, B.; Tourani, J.M.; Allain, P.; Rolland, F.; Staerman, F.; Eschwege, P.; Pfister, C. Preliminary results of the Prostacox phase II trial in hormonal refractory prostate cancer. *BJU Int.* **2007**, *100*, 770–774. [CrossRef] [PubMed]

32. Liao, Y.; Feng, G.; Dai, T.; Long, F.; Tang, J.; Pu, Y.; Zheng, X.; Cao, S.; Xu, S.; Du, X. Randomized, self-controlled, prospective assessment of the efficacy of mometasone furoate local application in reducing acute radiation dermatitis in patients with head and neck squamous cell carcinomas. *Medicine* **2019**, *98*, e18230. [CrossRef] [PubMed]

33. Hindley, A.; Zain, Z.; Wood, L.; Whitehead, A.; Sanneh, A.; Barber, D.; Hornsby, R. Mometasone furoate cream reduces acute radiation dermatitis in patients receiving breast radiation therapy: Results of a randomized trial. *Int. J. Radiat. Oncol. Biol. Phys.* **2014**, *90*, 748–755. [CrossRef] [PubMed]

34. Jing, D.; Bhadri, V.A.; Beck, D.; Thoms, J.A.; Yakob, N.A.; Wong, J.W.; Knezevic, K.; Pimanda, J.E.; Lock, R.B. Opposing regulation of BIM and BCL2 controls glucocorticoid-induced apoptosis of pediatric acute lymphoblastic leukemia cells. *Blood* **2015**, *125*, 273–283. [CrossRef]

35. Iozzo, R.V.; Sanderson, R.D. Proteoglycans in cancer biology, tumour microenvironment and angiogenesis. *J. Cell. Mol. Med.* **2011**, *15*, 1013–1031. [CrossRef]

36. Chen, Y.N.; LaMarche, M.J.; Chan, H.M.; Fekkes, P.; Garcia-Fortanet, J.; Acker, M.G.; Antonakos, B.; Chen, C.H.-T.; Chen, Z.; Cooke, V.G.; et al. Allosteric inhibition of SHP2 phosphatase inhibits cancers driven by receptor tyrosine kinases. *Nature* **2016**, *535*, 148–152. [CrossRef]

37. Tateishi, K.; Tsubaki, M.; Takeda, T.; Yamatomo, Y.; Imano, M.; Satou, T.; Nishida, S. FTI-277 and GGTI-289 induce apoptosis via inhibition of the Ras/ERK and Ras/mTOR pathway in head and neck carcinoma HEp-2 and HSC-3 cells. *J. BUON* **2021**, *26*, 606–612.

38. Solomon, L.W.; Frustino, J.L.; Loree, T.R.; Brecher, M.L.; Alberico, R.A.; Sullivan, M. Ewing sarcoma of the mandibular condyle: Multidisciplinary management optimizes outcome. *Head Neck* **2008**, *30*, 405–410. [CrossRef]

39. Sun, Y.; Meyers, B.A.; Czako, B.; Leonard, P.; Mseeh, F.; Harris, A.L.; Wu, Q.; Johnson, S.; Parker, C.A.; Cross, J.B.; et al. Allosteric SHP2 Inhibitor, IACS-13909, Overcomes EGFR-Dependent and EGFR-Independent Resistance Mechanisms toward Osimertinib. *Cancer Res.* **2020**, *80*, 4840–4853. [CrossRef]

Article

Development of a Novel Anti-CD44 Variant 7/8 Monoclonal Antibody, C$_{44}$Mab-34, for Multiple Applications against Oral Carcinomas

Hiroyuki Suzuki *,†, Kazuki Ozawa †, Tomohiro Tanaka, Mika K. Kaneko and Yukinari Kato *

Department of Antibody Drug Development, Tohoku University Graduate School of Medicine, 2-1 Seiryo-machi, Aoba-ku, Sendai 980-8575, Miyagi, Japan
* Correspondence: hiroyuki.suzuki.b4@tohoku.ac.jp (H.S.); yukinari.kato.e6@tohoku.ac.jp (Y.K);
  Tel.: +81-29-853-3944 (H.S. & Y.K.)
† These authors contributed equally to this work.

Abstract: Cluster of differentiation 44 (CD44) has been investigated as a cancer stem cell (CSC) marker as it plays critical roles in tumor malignant progression. The splicing variants are overexpressed in many carcinomas, especially squamous cell carcinomas, and play critical roles in the promotion of tumor metastasis, the acquisition of CSC properties, and resistance to treatments. Therefore, each CD44 variant (CD44v) function and distribution in carcinomas should be clarified for the establishment of novel tumor diagnosis and therapy. In this study, we immunized mouse with a CD44 variant (CD44v3–10) ectodomain and established various anti-CD44 monoclonal antibodies (mAbs). One of the established clones (C$_{44}$Mab-34; IgG$_1$, kappa) recognized a peptide that covers both variant 7- and variant 8-encoded regions, indicating that C$_{44}$Mab-34 is a specific mAb for CD44v7/8. Moreover, C$_{44}$Mab-34 reacted with CD44v3–10-overexpressed Chinese hamster ovary-K1 (CHO) cells or the oral squamous cell carcinoma (OSCC) cell line (HSC-3) by flow cytometry. The apparent K_D of C$_{44}$Mab-34 for CHO/CD44v3–10 and HSC-3 was 1.4×10^{-9} and 3.2×10^{-9} M, respectively. C$_{44}$Mab-34 could detect CD44v3–10 in Western blotting and stained the formalin-fixed paraffin-embedded OSCC in immunohistochemistry. These results indicate that C$_{44}$Mab-34 is useful for detecting CD44v7/8 in various applications and is expected to be useful in the application of OSCC diagnosis and therapy.

Keywords: CD44; CD44 variant 7/8; monoclonal antibody; flow cytometry; immunohistochemistry

Citation: Suzuki, H.; Ozawa, K.; Tanaka, T.; Kaneko, M.K.; Kato, Y. Development of a Novel Anti-CD44 Variant 7/8 Monoclonal Antibody, C$_{44}$Mab-34, for Multiple Applications against Oral Carcinomas. *Biomedicines* **2023**, *11*, 1099. https://doi.org/10.3390/biomedicines11041099

Academic Editor: Vui King Vincent-Chong

Received: 27 February 2023
Revised: 29 March 2023
Accepted: 3 April 2023
Published: 5 April 2023

1. Introduction

Head and neck cancers mainly arise from the oral cavity, pharynx, larynx, and nasal cavity. These tumors exhibit strong associations with smoking tobacco products, alcohol, and infection with human papillomavirus (HPV) types 16 and 18 [1]. The estimated number of new cases in the oral cavity and pharynx in the United States increased from 35,310 in 2008 to 54,540 in 2023 due to rising HPV-positive cases [2–4]. Mortality rates continue to increase for the oral cavity cancers associated with HPV infection (cancers of the tongue, tonsil, and oropharynx) by about 2% per year in men and 1% per year in women [2].

Although many different histologies exist in head and neck cancers, head and neck squamous cell carcinoma (HNSCC) is the common type. The treatment options for HNSCC include surgery, chemo-radiation, molecular targeted therapy, immunotherapy, or a combination of these modalities [5]. Despite the development in cancer treatment, metastasis and drug resistance remain the main causes of death [6]. Although survival can be improved, the impairment due to surgery and the toxicities of treatments deteriorate the patient's quality of life. Thus, the 5-year survival rate remains stagnant at approximately 50% [1].

Cancer stem cells (CSCs) play critical roles in tumor development through their important properties, including self-renewal, resistance to therapy, and tumor metastasis [7–9].

Studies have reported the importance of CSCs in HNSCC development [10] and regulation by both intrinsic and extrinsic mechanisms in the tumor microenvironment [11]. Several cell surface receptors and intracellular proteins have been reported as applicable CSC markers in HNSCC [12,13]. Among them, cluster of differentiation 44 (CD44) is one of the important CSC markers in solid tumors, and it was first applied to study HNSCC-derived CSCs [14]. Notably, CD44-high CSCs from HNSCC tumors exhibited the properties of epithelial to mesenchymal transition, including elevated migration, invasiveness, and stemness [15]. Furthermore, CD44-high cells could form lung metastases in immunodeficient mice, in contrast to CD44-low cells, which failed to exhibit a similar metastatic proliferation of cancer cells [16]. Therefore, specific monoclonal antibodies (mAbs) against CD44 are required for the isolation of CD44-high CSCs and the analysis of their properties in detail.

CD44 is a multifunctional transmembrane protein that binds to the extracellular matrix, including hyaluronic acid (HA) [17]. Human CD44 has 19 exons, 10 of which are constant or present in all variants and make up the standard form of CD44 (CD44s). The CD44 variants (CD44v) are produced by alternative splicing and consist of the 10 constant exons in any combination with the remaining nine variant exons [18]. The CD44 isoforms have both overlapping and unique roles. Both CD44s and CD44v (pan-CD44) possess HA-binding motifs that promote interaction with the microenvironment and facilitate the activation of various signaling pathways [19].

Overexpression of CD44v has been observed in many types of carcinomas and is considered a promising target for tumor diagnosis and therapy [20,21]. There is growing evidence that CD44v plays important roles in the promotion of tumor metastasis, the acquisition of CSC properties [22], and resistance to chemotherapy and radiotherapy [23,24]. Several variant exon-encoded regions have been reported to promote tumorigenesis through their interacting proteins. The v3-encoded region is modified by heparan sulfate, which promotes the recruitment of heparin-binding growth factors such as fibroblast growth factors. Thus, the v3-encoded region functions as a co-receptor of receptor tyrosine kinases [25]. Furthermore, the v6-encoded region has been reported to be essential for the activation of c-MET through the formation of ternary complexes with HGF [26]. Moreover, the v8–10-encoded region mediates oxidative stress resistance through the regulation of intracellular redox states. [27]. Therefore, CD44v-specific mAbs are required not only for the understanding of each variant function but also for CD44v-targeting tumor diagnosis and therapy. However, the function and distribution of the variant-encoded region in tumors have not been fully understood.

Our group has developed the Cell-Based Immunization and Screening (CBIS) method and established a novel anti-pan-CD44 mAb, C_{44}Mab-5 (IgG$_1$, kappa) [28]. We also established another anti-pan-CD44 mAb, C_{44}Mab-46 (IgG$_1$, kappa) [29], using the immunization of CD44v3–10 ectodomain (CD44ec). We determined the epitopes of C_{44}Mab-5 and C_{44}Mab-46 in the standard exons (1 to 5)-encoding sequences [30–32]. We further showed that both C_{44}Mab-5 and C_{44}Mab-46 are available for flow cytometry, Western blot, and immunohistochemistry in oral SCC (OSCC) [28] and esophageal SCC [29]. Furthermore, we have also investigated the antitumor effects using recombinant C_{44}Mab-5 in mouse xenograft models of oral OSCC [33]. We converted the mouse IgG$_1$ subclass antibody (C_{44}Mab-5) into an IgG$_{2a}$ subclass antibody (5-mG$_{2a}$) and further produced a defucosylated version (5-mG$_{2a}$-f) using FUT8-deficient ExpiCHO-S (BINDS-09) cells. The 5-mG$_{2a}$-f showed moderate in vitro ADCC and CDC activities against HSC-2 and SAS OSCC cell lines. Furthermore, the 5-mG$_{2a}$-f significantly suppressed the xenograft growth of HSC-2 and SAS compared to control mouse IgG [33]. Here, we have developed a novel anti-CD44v7/8 mAb, C_{44}Mab-34 (IgG$_1$, kappa), and examined its applications to flow cytometry, Western blotting, and immunohistochemical analyses.

2. Materials and Methods

2.1. Cell Lines

Chinese hamster ovary (CHO)-K1, a human glioblastoma cell line (LN229), and mouse multiple myeloma P3X63Ag8U.1 (P3U1) cell lines were obtained from the American Type

Culture Collection (ATCC, Manassas, VA, USA). The human OSCC cell line, HSC-3, was obtained from the Japanese Collection of Research Bioresources (Osaka, Japan). CHO-K1 and P3U1 were cultured in Roswell Park Memorial Institute (RPMI)-1640 medium (Nacalai Tesque, Inc., Kyoto, Japan), supplemented with 100 U/mL penicillin, 100 μg/mL streptomycin, 0.25 μg/mL amphotericin B (Nacalai Tesque, Inc.), and 10% heat-inactivated fetal bovine serum (FBS; Thermo Fisher Scientific, Inc., Waltham, MA, USA).

LN229 and HSC-3 were cultured in Dulbecco's modified Eagle medium (DMEM) (Nacalai Tesque, Inc.), supplemented with 10% (v/v) FBS, 100 U/mL of penicillin (Nacalai Tesque, Inc.), 100 μg/mL streptomycin (Nacalai Tesque, Inc.), and 0.25 μg/mL amphotericin B (Nacalai Tesque, Inc.). LN229/CD44ec was cultured in the presence of 0.5 mg/mL of G418 (Nacalai Tesque, Inc.).

All the cells were grown in a humidified incubator at 37 °C with 5% CO_2.

2.2. Plasmid Construction and Establishment of Stable Transfectants

Human CD44v3–10 open reading frame (ORF) was obtained from the RIKEN BRC through the National Bio-Resource Project of the MEXT, Japan. CD44s cDNA was amplified using a HotStar HiFidelity Polymerase Kit (Qiagen Inc., Hilden, Germany) using LN229 cDNA as a template. The CD44s and CD44v3–10 ORFs were subcloned into a pCAG-Ble-ssPA16 vector possessing signal sequence and N-terminal PA16 tag (GLEGGVAMP-GAEDDVV) [28,34 37], which is detected by NZ-1, which was originally developed as an anti-human podoplanin mAb [38–53].

CHO/CD44s and CHO/CD44v3–10 were established by transfecting the plasmids into CHO K1 cells using a Neon transfection system (Thermo Fisher Scientific, Inc.). CD44ec-secreting LN229 (LN229/CD44ec) was established by transfecting pCAG-Neo/PA-CD44ec-RAP-MAP into LN229 cells using the Neon transfection system. The amino acid sequences of the tag system in this study were as follows: PA tag [43,47,51], 12 amino acids (GVAMP-GAEDDVV); RAP tag [54,55], 12 amino acids (DMVNPGLEDRIE); and MAP tag [56,57], 12 amino acids (GDGMVPPGIEDK).

2.3. Purification of CD44ec

The purification of CD44ec from the culture supernatant of LN229/CD44ec was performed using an anti-RAP tag mAb (clone PMab-2) and a RAP peptide (GDDMVNPGLEDRIE) [54,55]. The culture supernatant (5 L) was passed through a 2 mL bed volume of PMab-2-sepharose, and the process was repeated three times. After washing the beads with 100 mL of phosphate-buffered saline (PBS, Nacalai Tesque, Inc.), CD44ec was eluted with 0.1 mg/mL of a RAP peptide in a step-wise manner (2 mL × 10). The purity of CD44ec was determined by Coomassie Brilliant Blue (CBB) staining using the Bio-Safe CBB G-250 Stain (Bio-Rad Laboratories, Inc., Berkeley, CA, USA) (Supplemental Figure S1A).

2.4. Hybridomas

Female BALB/c mouse was purchased from CLEA Japan (Tokyo, Japan). The Animal Care and Use Committee of Tohoku University approved the animal experiments (permit number: 2019NiA-001). The immunization of CD44ec was performed as described previously [29].

The splenic cells were fused with P3U1 cells using polyethylene glycol 1500 (PEG1500; Roche Diagnostics, Indianapolis, IN, USA). The culture supernatants of hybridomas were screened using an enzyme-linked immunosorbent assay (ELISA) against CD44ec. The supernatants were further screened using CHO/CD44v3–10 and parental CHO-K1 cells by flow cytometry using SA3800 Cell Analyzers (Sony Corp. Tokyo, Japan).

C_{44}Mab-34 was purified from the cultured supernatants of C_{44}Mab-34-producing hybridomas using Ab-Capcher ExTra (ProteNova Co., Ltd., Kagawa, Japan). The purity of C_{44}Mab-34 was determined by CBB staining (Supplemental Figure S1B).

2.5. ELISA

Fifty-eight synthesized peptides, which cover the CD44v3–10 extracellular domain [30], were synthesized by Sigma-Aldrich Corp (St. Louis, MO, USA). The peptides (1 µg/mL) or CD44ec were immobilized on Nunc Maxisorp 96-well immunoplates (Thermo Fisher Scientific Inc) for 30 min at 37 °C. The immunoplate washing was performed with PBS containing 0.05% (v/v) Tween 20 (PBST; Nacalai Tesque, Inc.). After the blocking with 1% (w/v) bovine serum albumin (BSA) in PBST, C_{44}Mab-34 (10 µg/mL) was added to each well. Then, the wells were further incubated with anti-mouse immunoglobulins peroxidase-conjugate (1:2000 diluted; Agilent Technologies Inc., Santa Clara, CA, USA). One-Step Ultra TMB (Thermo Fisher Scientific Inc.) was used for enzymatic reactions. An iMark microplate reader (Bio-Rad Laboratories, Inc.) was used to measure the optical density at 655 nm.

2.6. Flow Cytometry

CHO/CD44v3–10, CHO-K1, and HSC-3 were harvested using 0.25% trypsin and 1 mM ethylenediamine tetraacetic acid (EDTA; Nacalai Tesque, Inc.). The cells were treated with C_{44}Mab-34, C_{44}Mab-46, or blocking buffer (control) (0.1% BSA in PBS) for 30 min at 4 °C. Then, the cells were treated with anti-mouse IgG conjugated with Alexa Fluor 488 (1:2000; Cell Signaling Technology, Inc, Danvers, MA, USA) for 30 min at 4 °C. The SA3800 Cell Analyzer and SA3800 software ver. 2.05 (Sony Corporation) were used for fluorescence data collection and analysis, respectively.

2.7. Dissociation Constant (K_D) Determination by Flow Cytometry

Serially diluted C_{44}Mab-34 was treated with CHO/CD44v3–10 and HSC-3 cells. Then, the cells were treated with anti-mouse IgG conjugated with Alexa Fluor 488 (1:200). BD FACSLyric and BD FACSuite software version 1.3 (BD Biosciences) were used for fluorescence data collection and analysis, respectively. The GeoMean of each histogram, including primary mAb (C_{44}Mab-34) + secondary Ab (Alexa Fluor 488-conjugated anti-mouse IgG) and only secondary Ab (for background), was determined. We further withdrew the background from each data and determined the dissociation constant (K_D) by GraphPad Prism 8 (the fitting binding isotherms to built-in one-site binding models; GraphPad Software, Inc., La Jolla, CA, USA).

2.8. Western Blot Analysis

The total cell lysates (10 µg of protein) were denatured by sodium dodecyl sulfate (SDS) sample buffer (Nacalai Tesque, Inc.) in the presence of 2-mercaptoethanol and separated on 7.5% or 5–20% polyacrylamide gels (FUJIFILM Wako Pure Chemical Corporation, Osaka, Japan) and transferred onto polyvinylidene difluoride (PVDF) membranes (Merck KGaA, Darmstadt, Germany). After blocking with 4% skim milk (Nacalai Tesque, Inc.) in PBST, the membranes were incubated with 10 µg/mL of C_{44}Mab-34, 10 µg/mL of C_{44}Mab-46, 1 µg/mL of NZ-1, or 1 µg/mL of an anti-β-actin mAb (clone AC-15; Sigma-Aldrich Corp.) and then incubated with peroxidase-conjugated anti-mouse immunoglobulins (diluted 1:1000; Agilent Technologies, Inc.) for C_{44}Mab-34, C_{44}Mab-46, and anti-β-actin. The chemiluminescence signals were obtained with ImmunoStar LD (FUJIFILM Wako Pure Chemical Corporation) and detected using a Sayaca-Imager (DRC Co. Ltd., Tokyo, Japan).

2.9. Immunohistochemical Analysis

Formalin-fixed paraffin-embedded (FFPE) sections of the OSCC tissue array (OR601c) were purchased from US Biomax Inc. (Rockville, MD, USA). The OSCC tissue array was autoclaved in EnVision FLEX Target Retrieval Solution High pH (Agilent Technologies, Inc.) for 20 min. After blocking with SuperBlock T20 (Thermo Fisher Scientific, Inc.), the sections were incubated with C_{44}Mab-34 (10 µg/mL) and C_{44}Mab-46 (1 µg/mL) for 1 h at room temperature and then treated with the EnVision+ Kit for mouse (Agilent Technologies Inc.) for 30 min. The chromogenic reaction was conducted using 3,3'-diaminobenzidine tetrahydrochloride (DAB; Agilent Technologies Inc.). The counterstaining was performed using hematoxylin (FUJIFILM Wako Pure Chemical Corporation). To examine the sections and obtain images, we used a Leica DMD108 (Leica Microsystems GmbH, Wetzlar, Germany).

3. Results

3.1. Development of an Anti-CD44v7/8 mAb, C_{44}Mab-34

In this study, we purified human CD44ec as an immunogen (Figure 1). One mouse was immunized with CD44ec, and hybridomas were seeded into 96-well plates. The supernatants were first screened by the reactivity to CD44ec by ELISA. Subsequently, the supernatants, which were positive for CHO/CD44v3–10 cells and negative for CHO-K1 cells, were further selected using flow cytometry. Finally, anti-CD44 mAb-producing clones were established by limiting dilution. Among them, C_{44}Mab-34 (IgG$_1$, kappa) was shown to recognize CD44p421–440 (GHQAGRRMDMDSSHSTTLQP), which corresponds to the variant 7- and variant 8-encoded sequence (Supplementary Table S1). In contrast, C_{44}Mab-34 never recognized other CD44v3–10 extracellular regions. These results indicate that C_{44}Mab-34 specifically recognizes the border region between variants 7 and 8.

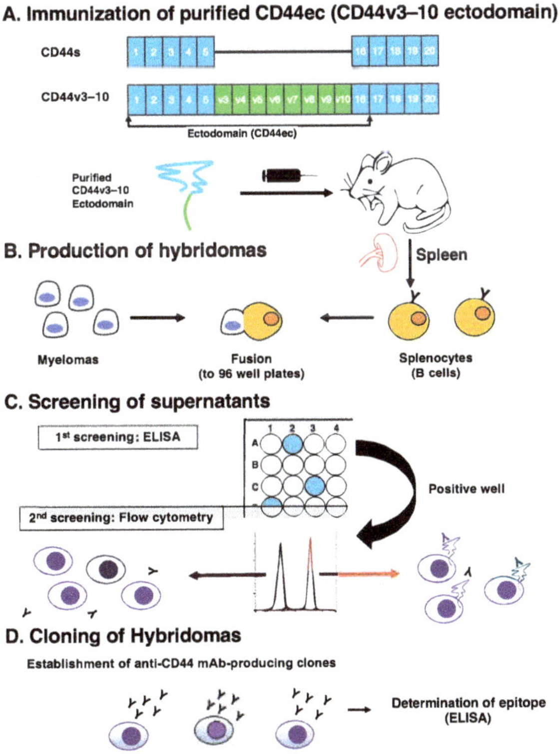

Figure 1. A schematic illustration of anti-human CD44 mAb production. (**A**) Purified CD44v3–10 ectodomain was intraperitoneally injected into BALB/c mouse. (**B**) Hybridomas were produced by fusion of the splenocytes and P3U1 cells. (**C**) The screening was performed by enzyme-linked immunosorbent assay (ELISA) and flow cytometry using parental CHO-K1 and CHO/CD44v3–10cells. (**D**) After cloning and additional screening, a clone C_{44}Mab-34 (IgG$_1$, kappa) was established. Furthermore, the binding epitopes were determined by ELISA using peptides, which cover the extracellular domain of CD44v3–10.

3.2. Flow Cytometric Analysis of C$_{44}$Mab-34 to CD44-Expressing Cells

We next investigated the reactivity of C$_{44}$Mab-34 against CHO/CD44v3–10 and CHO/CD44s cells by flow cytometry. C$_{44}$Mab-34 recognized CHO/CD44v3–10cells in a dose-dependent manner (Figure 2A) but not CHO/CD44s (Figure 2B) or CHO-K1 (Figure 2C) cells. The CHO/CD44 cells were recognized by an anti-pan-CD44 mAb, C$_{44}$Mab-46 [29] (Supplemental Figure S2). C$_{44}$Mab-34 also recognized the OSCC cell line HSC-3 (Figure 2D) in a dose-dependent manner.

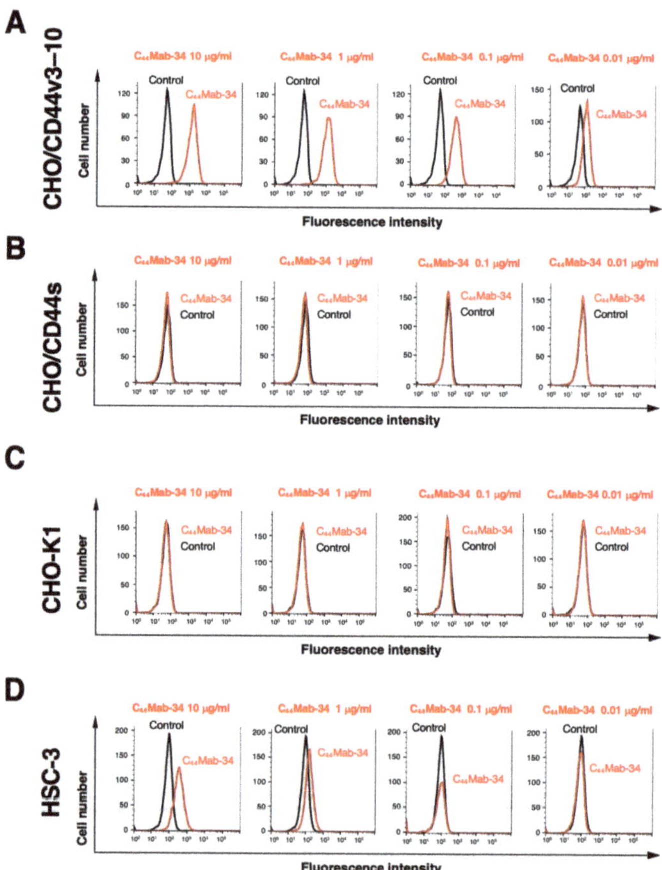

Figure 2. Flow cytometry using C$_{44}$Mab-34 against CD44-expressing cells. CHO/CD44v3–10 (**A**), CHO/CD44s (**B**), CHO-K1 (**C**), and HSC-3 (**D**) cells were treated with C$_{44}$Mab-34, followed by treatment with anti-mouse IgG conjugated with Alexa Fluor 488 (red line). The black line represents the negative control (blocking buffer).

We next determined the binding affinity of C$_{44}$Mab-34 with CHO/CD44v3–10 and HSC-3 using flow cytometry. The K_D of CHO/CD44v3–10 and HSC-3 was 1.4×10^{-9} and 3.2×10^{-9} M, respectively. These results indicate that C$_{44}$Mab-34 possesses a moderate affinity for CD44v3–10-expressing cells (Figure 3).

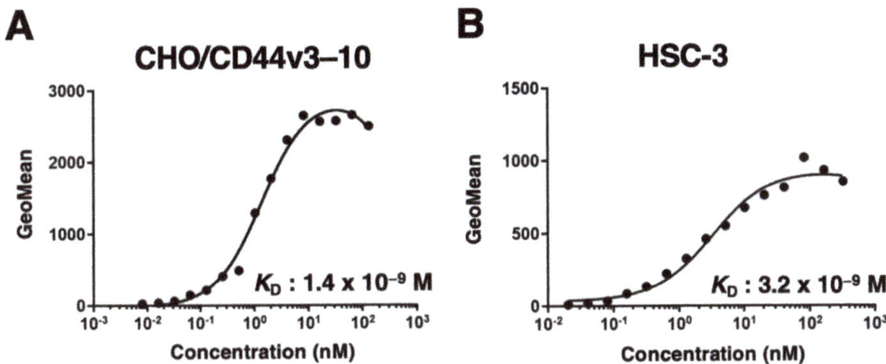

Figure 3. The binding affinity of C_{44}Mab-34 to CD44-expressing cells. CHO/CD44v3–10 (**A**) and HSC-3 (**B**) cells were suspended in serially diluted C_{44}Mab-34 at indicated concentrations. Then, cells were treated with anti-mouse IgG conjugated with Alexa Fluor 488. Fluorescence data were collected, followed by the calculation of the dissociation constant (K_D) by GraphPad PRISM 8.

3.3. Western Blot Analysis

We next performed Western blot analysis to assess the sensitivity of C_{44}Mab-34. As shown in Figure 4A, an anti-pan-CD44 mAb, C_{44}Mab-46, recognized the lysates from both CHO/CD44s (75~100 kDa) and CHO/CD44v3–10 (>180 kDa). C_{44}Mab-34 mainly detected CD44v3–10 as more than 180-kDa bands. However, C_{44}Mab-34 did not detect any bands from the lysates of CHO-K1 and CHO/CD44s cells (Figure 4B). These results indicate that C_{44}Mab-34 can detect CD44v3–10.

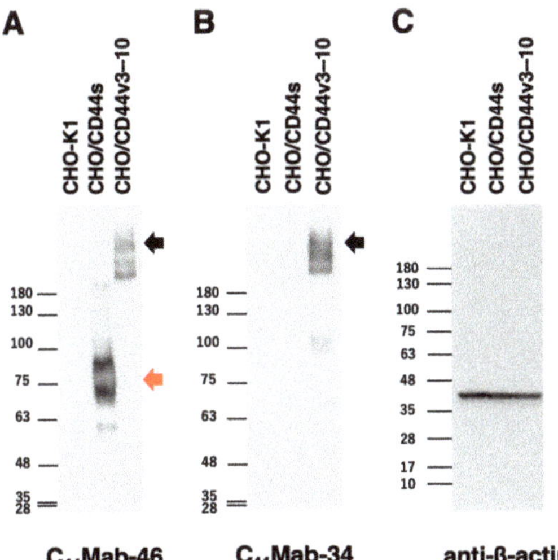

Figure 4. Western blot analysis using C_{44}Mab-34. The cell lysates from CHO-K1, CHO/CD44s, and CHO/CD44v3–10 (10 µg) were electrophoresed and transferred onto polyvinylidene fluoride (PVDF) membranes. The membranes were incubated with 10 µg/mL of C_{44}Mab-46 (**A**), 10 µg/mL of C_{44}Mab-34 (**B**), and 1 µg/mL of an anti-β-actin mAb (**C**). Then, the membranes were incubated with anti-mouse immunoglobulins conjugated with peroxidase for C_{44}Mab-46, C_{44}Mab-34, and an anti-β-actin mAb. The red arrows indicate the CD44s (75~100 kDa). The black arrows indicate the CD44v3–10.

3.4. Immunohistochemical Analysis against Tumor Tissues Using C_{44}Mab-34

We next examined whether C_{44}Mab-34 could be used for immunohistochemical analyses using FFPE sections. We used sequential sections of an OSCC tissue microarray. In a well-differentiated OSCC section, the clear membranous staining in OSCC was observed by C_{44}Mab-34 and C_{44}Mab-46 (Figure 5A,B). In an OSCC section with the stromal-invaded phenotype, C_{44}Mab-34 strongly stained stromal-invaded OSCC and could clearly distinguish tumor cells from stromal tissues (Figure 5C). In contrast, C_{44}Mab-46 stained both invaded tumor cells and surrounding stroma cells (Figure 5D). In Figure 5E,F, C_{44}Mab-34 and C_{44}Mab-46 never stained tumor tissue, but clear stromal staining was observed by C_{44}Mab-46 (Figure 5F). We have summarized the data of the immunohistochemical analysis of CD44 expression in tumor cells in Table 1; C_{44}Mab-34 stained 42 out of 49 (86%) cases of OSCC. These results indicate that C_{44}Mab-34 is useful for the immunohistochemical analysis of FFPE tumor sections.

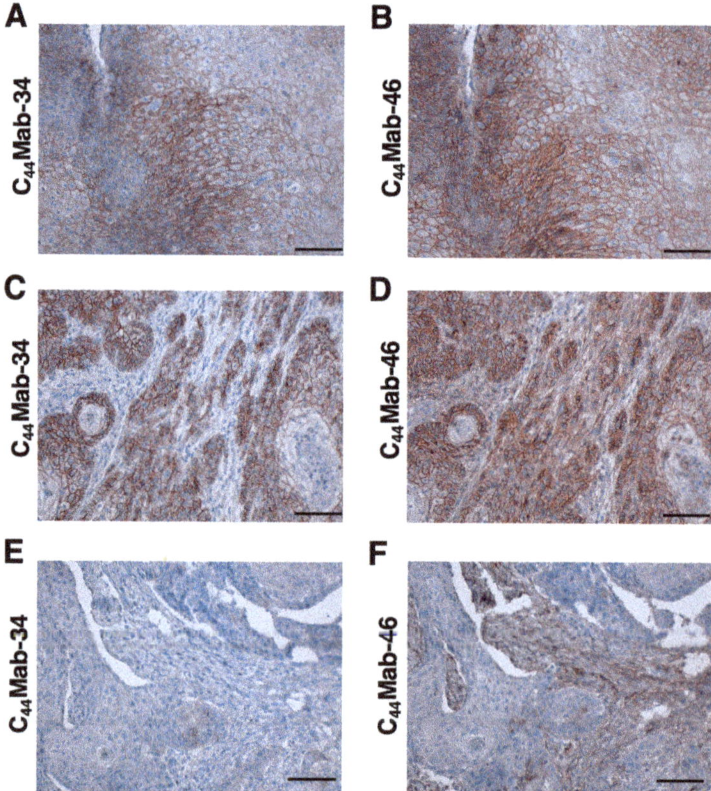

Figure 5. Immunohistochemical analysis using C_{44}Mab-34 and C_{44}Mab-46 against OSCC tissues. After antigen retrieval, serial sections of OSCC tissue array (Catalog number: OR601c) were incubated with 10 µg/mL of C_{44}Mab-34 (**A, C, E**) or 1 µg/mL of C_{44}Mab-46 (**B, D, F**), followed by treatment with the Envision+ kit. The chromogenic reaction was conducted using 3,3′-diaminobenzidine tetrahydrochloride (DAB). The counterstaining was performed using hematoxylin. Scale bar = 100 µm.

Table 1. Immunohistochemical analysis using C_{44}Mab-34 and C_{44}Mab-46 against OSCC.

No	Age	Sex	Organ/Anatomic Site	Pathology Diagnosis	TNM	C_{44}Mab-34	C_{44}Mab-46
1	78	M	Tongue	Squamous cell carcinoma of tongue	T2N0M0	+	+
2	40	M	Tongue	Squamous cell carcinoma of tongue	T2N0M0	+	++
3	35	F	Tongue	Squamous cell carcinoma of tongue	T2N0M0	+++	++
4	61	M	Tongue	Squamous cell carcinoma of tongue	T2N0M0	+++	+++
5	41	F	Tongue	Squamous cell carcinoma of tongue	T2N0M0	+	+
6	64	M	Tongue	Squamous cell carcinoma of right side of tongue	T2N2M0	+	++
7	76	M	Tongue	Squamous cell carcinoma of tongue	T1N0M0	+	++
8	50	F	Tongue	Squamous cell carcinoma of tongue	T2N0M0	+++	++
9	44	M	Tongue	Squamous cell carcinoma of tongue	T2N1M0	+++	++
10	53	F	Tongue	Squamous cell carcinoma of tongue	T1N0M0	+	++
11	46	F	Tongue	Squamous cell carcinoma of tongue	T2N0M0	-	+
12	50	M	Tongue	Squamous cell carcinoma of root of tongue	T3N1M0	+++	+
13	36	F	Tongue	Squamous cell carcinoma of tongue	T1N0M0	+++	+++
14	63	F	Tongue	Squamous cell carcinoma of tongue	T1N0M0	+	+
15	46	M	Tongue	Squamous cell carcinoma of tongue	T2N0M0	++	-
16	58	M	Tongue	Squamous cell carcinoma of tongue	T2N0M0	+	+
17	61	M	Lip	Squamous cell carcinoma of lower lip	T1N0M0	+++	+++
18	57	M	Lip	Squamous cell carcinoma of lower lip	T2N0M0	++	+++
19	61	M	Lip	Squamous cell carcinoma of lower lip	T1N0M0	+++	++
20	60	M	Gum	Squamous cell carcinoma of gum	T3N0M0	+	+
21	60	M	Gum	Squamous cell carcinoma of gum	T1N0M0	+++	+++
22	69	M	Gum	Squamous cell carcinoma of upper gum	T3N0M0	+	++
23	53	M	Bucca cavioris	Squamous cell carcinoma of bucca cavioris	T2N0M0	++	+
24	55	M	Bucca cavioris	Squamous cell carcinoma of bucca cavioris	T1N0M0	++	+
25	58	M	Tongue	Squamous cell carcinoma of base of tongue	T1N0M0	+	++
26	63	M	Oral cavity	Squamous cell carcinoma	T1N0M0	++	++
27	48	F	Tongue	Squamous cell carcinoma of tongue	T1N0M0	++	+
28	80	M	Lip	Squamous cell carcinoma of lower lip	T1N0M0	+++	+++
29	77	M	Tongue	Squamous cell carcinoma of base of tongue	T2N0M0	+++	++
30	59	M	Tongue	Squamous cell carcinoma of tongue	T2N0M0	++	-
31	77	F	Tongue	Squamous cell carcinoma of tongue	T1N0M0	+	++
32	56	M	Tongue	Squamous cell carcinoma of root of tongue	T2N1M0	+	+
33	60	M	Tongue	Squamous cell carcinoma of tongue	T2N1M0	+	++
34	62	M	Tongue	Squamous cell carcinoma of tongue	T2N0M0	+++	++
35	67	F	Tongue	Squamous cell carcinoma of tongue	T2N0M0	+++	++
36	47	F	Tongue	Squamous cell carcinoma of tongue	T2N0M0	+++	+++
37	37	M	Tongue	Squamous cell carcinoma of tongue	T2N1M0	-	-
38	55	F	Tongue	Squamous cell carcinoma of tongue	T2N0M0	++	++
39	56	F	Bucca cavioris	Squamous cell carcinoma of bucca cavioris	T2N0M0	++	+
40	49	M	Bucca cavioris	Squamous cell carcinoma of bucca cavioris	T1N0M0	-	-
41	45	M	Bucca cavioris	Squamous cell carcinoma of bucca cavioris	T2N0M0	-	-
42	42	M	Bucca cavioris	Squamous cell carcinoma of bucca cavioris	T3N0M0	++	++
43	44	M	Jaw	Squamous cell carcinoma of right drop jaw	T1N0M0	+	+++
44	40	F	Tongue	Squamous cell carcinoma of base of tongue	T2N0M0	-	++
45	49	M	Bucca cavioris	Squamous cell carcinoma of bucca cavioris	T1N0M0	+++	+++
46	56	F	Tongue	Squamous cell carcinoma of base of tongue	T2N0M0	-	-
47	42	M	Bucca cavioris	Squamous cell carcinoma of bucca cavioris	T3N0M0	+++	+++
48	87	F	Face	Squamous cell carcinoma of left side of face	T2N0M0	+	+
49	50	M	Gum	Squamous cell carcinoma of gum	T2N0M0	-	-

4. Discussion

Head and neck cancer is the seventh most common type of cancer worldwide, and it exhibits aggressive development in clinical settings [58]. Head and neck cancer remains a complex disease with a profound impact on patients and their quality of life after surgical ablation and therapies. Knowledge of the disease has been accumulated with regard to tumor biology and prevention, and therapeutic options have been simultaneously developed [58]. HNSCC is the most common type of head and neck cancer, and it has been revealed as the second-highest CD44-expressing cancer type in the Pan-Cancer Atlas [59]. CD44 overexpression is associated with poor prognosis and resistance to therapy [60–62]. Reduced CD44 expression leads to the growth suppression of tumor cells [17,63]. Therefore, CD44 is considered an important target for mAb therapies. In this study, we developed a novel anti-CD44v7/8 mAb, C_{44}Mab-34, and showed multiple applications to flow cytometry (Figures 2 and 3), Western blotting (Figure 4), and the immunohistochemistry of OSCC (Figure 5).

An anti-CD44v7/8 mAb (clone VFF-17) was previously developed, and it has been mainly used for the immunohistochemistry of normal tissue and tumors [64,65]. The epitope of VFF-17 mAb was determined by binding studies with fusion proteins encoding v7 or v8 exons, either alone or in combination [66]. However, a detailed amino acid sequence of the epitope has not been determined. As shown in Supplementary Table S1, C_{44}Mab-34 recognizes CD44p421–440 [GHQAGRRMD (included in v7) + MDSSH-STTLQP (included in v8)]. In contrast, C_{44}Mab-34 has never recognized CD44p411–430 (FNPISHPMGRGHQAGRRMD (included in v7) + M (included in v8)) or CD44p431–450 (DSSHSTTLQPTANPNTGLVE (included in v8)). These results suggest that C_{44}Mab-34 recognizes the border sequence between v7 and v8. In addition, CD44 is known to be heavily glycosylated [67], and the glycosylation pattern is thought to depend on the host cells. Since the epitope of C_{44}Mab-34 contains predicted and confirmed *O*-glycan sites [67], further studies are needed on whether the recognition of C_{44}Mab-34 is affected by glycosylation.

Among many CD44v types, CD44v8–10, CD44v6–10, CD44v4–10, and CD44v3–10 were mainly detected in SCC cells by semi-quantitative RT-PCR analysis (manuscript submitted). Since C_{44}Mab-34 recognizes the border sequence between v7 and v8 (Supplemental Table S1), C_{44}Mab-34 can distinguish CD44v8–10 and the longer CD44v (v6-10, v4-10, and v3-10). Furthermore, the inclusion of these variants (from v8-10 to the longer variants) is promoted by EGF signaling [68,69]. If the expression of CD44v8-10 and the longer variants are differently regulated in normal and tumor cells, C_{44}Mab-34 could contribute to tumor diagnosis and therapy. We are now investigating the C_{44}Mab-34 reactivity against other tumor tissues together with the epitope analyses.

ΔNp63 is known as a marker of basal cells of stratified epithelium and SCC [70]. ΔNp63 mediates HA metabolism and signaling [71]. Specifically, ΔNp63 directly binds to the p63-binding sequence on the promoter/enhancer region of the CD44 gene [71]. In whole-exome sequencing data analysis from 74 HNSCC–normal pairs, the ΔNp63-encoded gene, *TP63*, was identified as a significantly mutated gene that results in the activation of the ΔNp63 pathway [72]. The relationship between ΔNp63 activation and CD44 transcription should be investigated in future studies. Furthermore, the mechanism of the variant 7/8 inclusion by alternative splicing remains to be determined.

An anti-pan CD44 mAb, RG7356, demonstrated some efficacy and an acceptable safety profile in the phase I study. However, the study was terminated due to no evidence of a clinical and dose–response relationship with RG7356 [73]. Furthermore, a variant 6-specific CD44 mAb-drug conjugate (bivatuzumab–mertansine) was also evaluated in clinical trials. However, lethal epidermal necrolysis halted further development. The efficient accumulation of mertansine in the skin was most likely responsible for the high toxicity [74,75]. Therefore, the therapeutic effects of CD44 mAbs have been disappointing until now.

Near-infrared photoimmunotherapy (NIR-PIT) is a novel tumor therapy that uses a targeted mAb–photoabsorber conjugate (APC) [76]. The mAb binds to the targeted cell surface antigen, and the photoactivatable dye IRDye700DX (IR700) induces the disruption of the cellular membrane after NIR-light exposure. Since NIR-light exposure can be performed at tumor sites locally, APC can exert antitumor effect selectivity while minimizing damage to the surrounding tissue [77,78]. Preclinical studies indicate that NIR-PIT induces tumor necrosis and immunogenic cell death through the induction of innate and adaptive immunity [79]. A first-in-human phase I and II trial of NIR-PIT with RM-1929 (an anti-epidermal growth factor receptor mAb, cetuximab–IR700 conjugate) in patients with inoperable HNSCC was conducted and exhibited the efficacy [80].

A preclinical study of anti-CD44 mAb-based NIR-PIT has been reported [81]. The study used an anti-mouse/human pan-CD44 mAb, IM7, conjugated with IR700 (CD44–IR700) in a syngeneic mouse model of OSCC. The CD44–IR700 can induce significant antitumor responses after a single injection of the conjugate and NIR-light exposure in CD44-expressing OSCC tumors [81]. As shown in Figure 5D,F, a pan-CD44 mAb, C_{44}Mab-46, recognized not only tumor cells but also stromal tissue and probably immune cells, which are important for antitumor immunity. Therefore, CD44v is a promising tumor antigen for NIR-PIT, which could be a new modality for OSCC with locoregional recurrence.

We have previously produced recombinant antibodies that are converted to a mouse IgG_{2a} subclass from mouse IgG_1. Furthermore, we produced defucosylated IgG_{2a} mAbs using fucosyltransferase 8-deficient CHO-K1 cells to potentiate antibody-dependent cellular cytotoxicity. The defucosylated mAbs showed potent antitumor activity in mouse xenograft models [33,82–88]. Therefore, a class-switched and defucosylated version of C_{44}Mab-34 is required to evaluate the antitumor activity in vivo.

Supplementary Materials: The following supporting information can be downloaded at: https://www.mdpi.com/article/10.3390/biomedicines11041099/s1, Table S1, The determination of the binding epitope of C_{44}Mab-34 by ELISA. Figure S1, CBB staining of CD44ec and C_{34}Mab-34. Figure S2, Recognition of CHO/CD44s and CHO/CD44v3–10 by C_{44}Mab-46 by flow cytometry.

Author Contributions: K.O., H.S. and T.T. performed the experiments. M.K.K. and Y.K. designed the experiments. H.S. and K.O. analyzed the data. K.O., H.S. and Y.K. wrote the manuscript. All authors have read and agreed to the manuscript.

Funding: This research was supported in part by the Japan Agency for Medical Research and Development (AMED) under grant numbers: JP22ama121008 (to Y.K.), JP22am0401013 (to Y.K.), JP22bm1004001 (to Y.K.), JP22ck0106730 (to Y.K.), and JP21am0101078 (to Y.K.), and by the Japan Society for the Promotion of Science (JSPS) Grants-in-Aid for Scientific Research (KAKENHI) grant nos. 21K20789 (to T.T.), 22K06995 (to H.S.), 21K07168 (to M.K.K.), and 22K07224 (to Y.K.).

Institutional Review Board Statement: The animal study protocol was approved by the Animal Care and Use Committee of Tohoku University (permit number: 2019NiA-001, approved on 1 April 2019) for studies involving animals.

Informed Consent Statement: Not applicable.

Data Availability Statement: The data presented in this study are available in the article and supplementary material.

Conflicts of Interest: The authors have no conflict of interest to declare.

References

1. Johnson, D.E.; Burtness, B.; Leemans, C.R.; Lui, V.W.Y.; Bauman, J.E.; Grandis, J.R. Head and neck squamous cell carcinoma. *Nat. Rev. Dis. Prim.* **2020**, *6*, 92. [CrossRef]
2. Siegel, R.L.; Miller, K.D.; Wagle, N.S.; Jemal, A. Cancer statistics, 2023. *CA Cancer J. Clin.* **2023**, *73*, 17–48. [CrossRef] [PubMed]
3. Kang, J.J.; Yu, Y.; Chen, L.; Zakeri, K.; Gelblum, D.Y.; McBride, S.M.; Riaz, N.; Tsai, C.J.; Kriplani, A.; Hung, T.K.W.; et al. Consensuses, controversies, and future directions in treatment deintensification for human papillomavirus-associated oropharyngeal cancer. *CA Cancer J. Clin.* **2023**, *73*, 164–197. [CrossRef]
4. Jemal, A.; Siegel, R.; Ward, E.; Hao, Y.; Xu, J.; Murray, T.; Thun, M.J. Cancer statistics, 2008. *CA Cancer J. Clin.* **2008**, *58*, 71–96. [CrossRef] [PubMed]

5. Xing, D.T.; Khor, R.; Gan, H.; Wada, M.; Ermongkonchai, T.; Ng, S.P. Recent research on combination of radiotherapy with targeted therapy or immunotherapy in head and neck squamous cell carcinoma: A review for radiation oncologists. *Cancers* **2021**, *13*, 5716. [CrossRef] [PubMed]

6. Muzaffar, J.; Bari, S.; Kirtane, K.; Chung, C.H. Recent advances and future directions in clinical management of head and neck squamous cell carcinoma. *Cancers* **2021**, *13*, 338. [CrossRef]

7. Maitland, N.J.; Collins, A.T. Cancer stem cells—A therapeutic target? *Curr. Opin. Mol. Ther.* **2010**, *12*, 662–673.

8. Prince, M.E.; Ailles, L.E. Cancer stem cells in head and neck squamous cell cancer. *J. Clin. Oncol.* **2008**, *26*, 2871–2875. [CrossRef]

9. Ailles, L.E.; Weissman, I.L. Cancer stem cells in solid tumors. *Curr. Opin. Biotechnol.* **2007**, *18*, 460–466. [CrossRef]

10. Keysar, S.B.; Le, P.N.; Miller, B.; Jackson, B.C.; Eagles, J.R.; Nieto, C.; Kim, J.; Tang, B.; Glogowska, M.J.; Morton, J.J.; et al. Regulation of head and neck squamous cancer stem cells by PI3K and SOX2. *J. Natl. Cancer Inst.* **2017**, *109*, djw189. [CrossRef]

11. de Miranda, M.C.; Melo, M.I.A.; Cunha, P.D.S.; Gentilini, J.J.; Faria, J.; Rodrigues, M.A.; Gomes, D.A. Roles of mesenchymal stromal cells in the head and neck cancer microenvironment. *Biomed. Pharmacother.* **2021**, *144*, 112269. [CrossRef]

12. Yu, S.S.; Cirillo, N. The molecular markers of cancer stem cells in head and neck tumors. *J. Cell Physiol.* **2020**, *235*, 65–73. [CrossRef]

13. Tahmasebi, E.; Alikhani, M.; Yazdanian, A.; Yazdanian, M.; Tebyanian, H.; Seifalian, A. The current markers of cancer stem cell in oral cancers. *Life Sci.* **2020**, *249*, 117483. [CrossRef]

14. Prince, M.E.; Sivanandan, R.; Kaczorowski, A.; Wolf, G.T.; Kaplan, M.J.; Dalerba, P.; Weissman, I.L.; Clarke, M.F.; Ailles, L.E. Identification of a subpopulation of cells with cancer stem cell properties in head and neck squamous cell carcinoma. *Proc. Natl. Acad. Sci. USA* **2007**, *104*, 973–978. [CrossRef]

15. Lee, Y.; Shin, J.H.; Longmire, M.; Wang, H.; Kohrt, H.E.; Chang, H.Y.; Sunwoo, J.B. CD44+ cells in head and neck squamous cell carcinoma suppress T-cell-mediated immunity by selective constitutive and inducible expression of PD-L1. *Clin. Cancer Res.* **2016**, *22*, 3571–3581. [CrossRef]

16. Davis, S.J.; Divi, V.; Owen, J.H.; Bradford, C.R.; Carey, T.E.; Papagerakis, S.; Prince, M.E. Metastatic potential of cancer stem cells in head and neck squamous cell carcinoma. *Arch. Otolaryngol. Head Neck Surg.* **2010**, *136*, 1260–1266. [CrossRef]

17. Ponta, H.; Sherman, L.; Herrlich, P.A. CD44: From adhesion molecules to signalling regulators. *Nat. Rev. Mol. Cell Biol.* **2003**, *4*, 33–45. [CrossRef]

18. Yan, Y.; Zuo, X.; Wei, D. Concise review: Emerging role of CD44 in cancer stem cells: A promising biomarker and therapeutic target. *Stem Cells Transl. Med.* **2015**, *4*, 1033–1043. [CrossRef] [PubMed]

19. Slevin, M.; Krupinski, J.; Gaffney, J.; Matou, S.; West, D.; Delisser, H.; Savani, R.C.; Kumar, S. Hyaluronan-mediated angiogenesis in vascular disease: Uncovering RHAMM and CD44 receptor signaling pathways. *Matrix Biol.* **2007**, *26*, 58–68. [CrossRef] [PubMed]

20. Naor, D.; Wallach-Dayan, S.B.; Zahalka, M.A.; Sionov, R.V. Involvement of CD44, a molecule with a thousand faces, in cancer dissemination. *Semin. Cancer Biol.* **2008**, *18*, 260–267. [CrossRef] [PubMed]

21. Günthert, U.; Hofmann, M.; Rudy, W.; Reber, S.; Zöller, M.; Haussmann, I.; Matzku, S.; Wenzel, A.; Ponta, H.; Herrlich, P. A new variant of glycoprotein CD44 confers metastatic potential to rat carcinoma cells. *Cell* **1991**, *65*, 13–24. [CrossRef]

22. Guo, Q.; Yang, C.; Gao, F. The state of CD44 activation in cancer progression and therapeutic targeting. *Febs J.* **2021**. [CrossRef]

23. Mesrati, M.H.; Syafruddin, S.E.; Mohtar, M.A.; Syahir, A. CD44: A multifunctional mediator of cancer progression. *Biomolecules* **2021**, *11*, 1850. [CrossRef]

24. Morath, I.; Hartmann, T.N.; Orian-Rousseau, V. CD44: More than a mere stem cell marker. *Int. J. Biochem. Cell Biol.* **2016**, *81*, 166–173. [CrossRef]

25. Bennett, K.L.; Jackson, D.G.; Simon, J.C.; Tanczos, E.; Peach, R.; Modrell, B.; Stamenkovic, I.; Plowman, G.; Aruffo, A. CD44 isoforms containing exon V3 are responsible for the presentation of heparin-binding growth factor. *J. Cell Biol.* **1995**, *128*, 687–698. [CrossRef]

26. Orian-Rousseau, V.; Chen, L.; Sleeman, J.P.; Herrlich, P.; Ponta, H. CD44 is required for two consecutive steps in HGF/c-Met signaling. *Genes Dev.* **2002**, *16*, 3074–3086. [CrossRef]

27. Ishimoto, T.; Nagano, O.; Yae, T.; Tamada, M.; Motohara, T.; Oshima, H.; Oshima, M.; Ikeda, T.; Asaba, R.; Yagi, H.; et al. CD44 variant regulates redox status in cancer cells by stabilizing the xCT subunit of system xc(-) and thereby promotes tumor growth. *Cancer Cell* **2011**, *19*, 387–400. [CrossRef]

28. Yamada, S.; Itai, S.; Nakamura, T.; Yanaka, M.; Kaneko, M.K.; Kato, Y. Detection of high CD44 expression in oral cancers using the novel monoclonal antibody, C(44)Mab-5. *Biochem. Biophys. Rep.* **2018**, *14*, 64–68. [CrossRef]

29. Goto, N.; Suzuki, H.; Tanaka, T.; Asano, T.; Kaneko, M.K.; Kato, Y. Development of a novel anti-CD44 monoclonal antibody for multiple applications against esophageal squamous cell carcinomas. *Int. J. Mol. Sci.* **2022**, *23*, 5535. [CrossRef]

30. Takei, J.; Asano, T.; Suzuki, H.; Kaneko, M.K.; Kato, Y. Epitope mapping of the anti-CD44 monoclonal antibody (C44Mab-46) using alanine-scanning mutagenesis and surface plasmon resonance. *Monoclon. Antib. Immunodiagn. Immunother.* **2021**, *40*, 219–226. [CrossRef]

31. Asano, T.; Kaneko, M.K.; Takei, J.; Tateyama, N.; Kato, Y. Epitope mapping of the anti-CD44 monoclonal antibody (C44Mab-46) using the REMAP method. *Monoclon. Antib. Immunodiagn. Immunother.* **2021**, *40*, 156–161. [CrossRef] [PubMed]

32. Asano, T.; Kaneko, M.K.; Kato, Y. Development of a novel epitope mapping system: RIEDL insertion for epitope mapping method. *Monoclon. Antib. Immunodiagn. Immunother.* **2021**, *40*, 162–167. [CrossRef] [PubMed]

33. Takei, J.; Kaneko, M.K.; Ohishi, T.; Hosono, H.; Nakamura, T.; Yanaka, M.; Sano, M.; Asano, T.; Sayama, Y.; Kawada, M.; et al. A defucosylated antiCD44 monoclonal antibody 5mG2af exerts antitumor effects in mouse xenograft models of oral squamous cell carcinoma. *Oncol. Rep.* **2020**, *44*, 1949–1960. [CrossRef] [PubMed]
34. Kato, Y.; Yamada, S.; Furusawa, Y.; Itai, S.; Nakamura, T.; Yanaka, M.; Sano, M.; Harada, H.; Fukui, M.; Kaneko, M.K. PMab-213: A monoclonal antibody for immunohistochemical analysis against pig podoplanin. *Monoclon. Antib. Immunodiagn. Immunother.* **2019**, *38*, 18–24. [CrossRef] [PubMed]
35. Furusawa, Y.; Yamada, S.; Itai, S.; Sano, M.; Nakamura, T.; Yanaka, M.; Fukui, M.; Harada, H.; Mizuno, T.; Sakai, Y.; et al. PMab-210: A monoclonal antibody against pig podoplanin. *Monoclon. Antib. Immunodiagn. Immunother.* **2019**, *38*, 30–36. [CrossRef]
36. Furusawa, Y.; Yamada, S.; Itai, S.; Nakamura, T.; Yanaka, M.; Sano, M.; Harada, H.; Fukui, M.; Kaneko, M.K.; Kato, Y. PMab-219: A monoclonal antibody for the immunohistochemical analysis of horse podoplanin. *Biochem. Biophys. Rep.* **2019**, *18*, 100616. [CrossRef]
37. Furusawa, Y.; Yamada, S.; Itai, S.; Nakamura, T.; Takei, J.; Sano, M.; Harada, H.; Fukui, M.; Kaneko, M.K.; Kato, Y. Establishment of a monoclonal antibody PMab-233 for immunohistochemical analysis against Tasmanian devil podoplanin. *Biochem. Biophys. Rep.* **2019**, *18*, 100631. [CrossRef]
38. Kato, Y.; Kaneko, M.K.; Kuno, A.; Uchiyama, N.; Amano, K.; Chiba, Y.; Hasegawa, Y.; Hirabayashi, J.; Narimatsu, H.; Mishima, K.; et al. Inhibition of tumor cell-induced platelet aggregation using a novel anti-podoplanin antibody reacting with its platelet-aggregation-stimulating domain. *Biochem. Biophys. Res. Commun.* **2006**, *349*, 1301–1307. [CrossRef]
39. Chalise, L.; Kato, A.; Ohno, M.; Maeda, S.; Yamamichi, A.; Kuramitsu, S.; Shiina, S.; Takahashi, H.; Ozone, S.; Yamaguchi, J.; et al. Efficacy of cancer-specific anti-podoplanin CAR-T cells and oncolytic herpes virus G47Delta combination therapy against glioblastoma. *Mol. Ther. Oncolytics* **2022**, *26*, 265–274. [CrossRef]
40. Ishikawa, A.; Waseda, M.; Ishii, T.; Kaneko, M.K.; Kato, Y.; Kaneko, S. Improved anti-solid tumor response by humanized anti-podoplanin chimeric antigen receptor transduced human cytotoxic T cells in an animal model. *Genes Cells* **2022**, *27*, 549–558. [CrossRef]
41. Tamura-Sakaguchi, R.; Aruga, R.; Hirose, M.; Ekimoto, T.; Miyake, T.; Hizukuri, Y.; Oi, R.; Kaneko, M.K.; Kato, Y.; Akiyama, Y.; et al. Moving toward generalizable NZ-1 labeling for 3D structure determination with optimized epitope-tag insertion. *Acta Crystallogr. D Struct. Biol.* **2021**, *77*, 645–662. [CrossRef]
42. Kaneko, M.K.; Ohishi, T.; Nakamura, T.; Inoue, H.; Takei, J.; Sano, M.; Asano, T.; Sayama, Y.; Hosono, H.; Suzuki, H.; et al. Development of core-fucose-deficient humanized and chimeric anti-human podoplanin antibodies. *Monoclon. Antib. Immunodiagn. Immunother.* **2020**, *39*, 167–174. [CrossRef]
43. Fujii, Y.; Matsunaga, Y.; Arimori, T.; Kitago, Y.; Ogasawara, S.; Kaneko, M.K.; Kato, Y.; Takagi, J. Tailored placement of a turn-forming PA tag into the structured domain of a protein to probe its conformational state. *J. Cell Sci.* **2016**, *129*, 1512–1522. [CrossRef]
44. Abe, S.; Kaneko, M.K.; Tsuchihashi, Y.; Izumi, T.; Ogasawara, S.; Okada, N.; Sato, C.; Tobiume, M.; Otsuka, K.; Miyamoto, L.; et al. Antitumor effect of novel anti-podoplanin antibody NZ-12 against malignant pleural mesothelioma in an orthotopic xenograft model. *Cancer Sci.* **2016**, *107*, 1198–1205. [CrossRef]
45. Kaneko, M.K.; Abe, S.; Ogasawara, S.; Fujii, Y.; Yamada, S.; Murata, T.; Uchida, H.; Tahara, H.; Nishioka, Y.; Kato, Y. Chimeric anti-human podoplanin antibody NZ-12 of lambda light chain exerts higher antibody-dependent cellular cytotoxicity and complement-dependent cytotoxicity compared with NZ-8 of kappa light chain. *Monoclon. Antib. Immunodiagn. Immunother.* **2017**, *36*, 25–29. [CrossRef]
46. Ito, A.; Ohta, M.; Kato, Y.; Inada, S.; Kato, T.; Nakata, S.; Yatabe, Y.; Goto, M.; Kaneda, N.; Kurita, K.; et al. A real-time near-infrared fluorescence imaging method for the detection of oral cancers in mice using an indocyanine green-labeled podoplanin antibody. *Technol. Cancer Res. Treat.* **2018**, *17*, 1533033818767936. [CrossRef]
47. Tamura, R.; Oi, R.; Akashi, S.; Kaneko, M.K.; Kato, Y.; Nogi, T. Application of the NZ-1 fab as a crystallization chaperone for PA tag-inserted target proteins. *Protein Sci.* **2019**, *28*, 823–836. [CrossRef]
48. Shiina, S.; Ohno, M.; Ohka, F.; Kuramitsu, S.; Yamamichi, A.; Kato, A.; Motomura, K.; Tanahashi, K.; Yamamoto, T.; Watanabe, R.; et al. CAR T cells targeting podoplanin reduce orthotopic glioblastomas in mouse brains. *Cancer Immunol. Res.* **2016**, *4*, 259–268. [CrossRef]
49. Kuwata, T.; Yoneda, K.; Mori, M.; Kanayama, M.; Kuroda, K.; Kaneko, M.K.; Kato, Y.; Tanaka, F. Detection of circulating tumor cells (CTCs) in malignant pleural mesothelioma (MPM) with the "universal" CTC-chip and an anti-podoplanin antibody NZ-1.2. *Cells* **2020**, *9*, 888. [CrossRef]
50. Nishinaga, Y.; Sato, K.; Yasui, H.; Taki, S.; Takahashi, K.; Shimizu, M.; Endo, R.; Koike, C.; Kuramoto, N.; Nakamura, S.; et al. Targeted phototherapy for malignant pleural mesothelioma: Near-infrared photoimmunotherapy targeting podoplanin. *Cells* **2020**, *9*, 1019. [CrossRef]
51. Fujii, Y.; Kaneko, M.; Neyazaki, M.; Nogi, T.; Kato, Y.; Takagi, J. PA tag: A versatile protein tagging system using a super high affinity antibody against a dodecapeptide derived from human podoplanin. *Protein Expr. Purif.* **2014**, *95*, 240–247. [CrossRef]
52. Kato, Y.; Kaneko, M.K.; Kunita, A.; Ito, H.; Kameyama, A.; Ogasawara, S.; Matsuura, N.; Hasegawa, Y.; Suzuki-Inoue, K.; Inoue, O.; et al. Molecular analysis of the pathophysiological binding of the platelet aggregation-inducing factor podoplanin to the C-type lectin-like receptor CLEC-2. *Cancer Sci.* **2008**, *99*, 54–61. [CrossRef] [PubMed]

53. Kato, Y.; Vaidyanathan, G.; Kaneko, M.K.; Mishima, K.; Srivastava, N.; Chandramohan, V.; Pegram, C.; Keir, S.T.; Kuan, C.T.; Bigner, D.D.; et al. Evaluation of anti-podoplanin rat monoclonal antibody NZ-1 for targeting malignant gliomas. *Nucl. Med. Biol.* **2010**, *37*, 785–794. [CrossRef] [PubMed]

54. Miura, K.; Yoshida, H.; Nosaki, S.; Kaneko, M.K.; Kato, Y. RAP tag and PMab-2 antibody: A tagging system for detecting and purifying proteins in plant cells. *Front. Plant Sci.* **2020**, *11*, 510444. [CrossRef] [PubMed]

55. Fujii, Y.; Kaneko, M.K.; Ogasawara, S.; Yamada, S.; Yanaka, M.; Nakamura, T.; Saidoh, N.; Yoshida, K.; Honma, R.; Kato, Y. Development of RAP tag, a novel tagging system for protein detection and purification. *Monoclon. Antib. Immunodiagn. Immunother.* **2017**, *36*, 68–71. [CrossRef]

56. Fujii, Y.; Kaneko, M.K.; Kato, Y. MAP tag: A novel tagging system for protein purification and detection. *Monoclon. Antib. Immunodiagn. Immunother.* **2016**, *35*, 293–299. [CrossRef]

57. Wakasa, A.; Kaneko, M.K.; Kato, Y.; Takagi, J.; Arimori, T. Site-specific epitope insertion into recombinant proteins using the MAP tag system. *J. Biochem.* **2020**, *168*, 375–384. [CrossRef]

58. Mody, M.D.; Rocco, J.W.; Yom, S.S.; Haddad, R.I.; Saba, N.F. Head and neck cancer. *Lancet* **2021**, *398*, 2289–2299. [CrossRef]

59. Ludwig, N.; Szczepanski, M.J.; Gluszko, A.; Szafarowski, T.; Azambuja, J.H.; Dolg, L.; Gellrich, N.C.; Kampmann, A.; Whiteside, T.L.; Zimmerer, R.M. CD44(+) tumor cells promote early angiogenesis in head and neck squamous cell carcinoma. *Cancer Lett.* **2019**, *467*, 85–95. [CrossRef]

60. Boxberg, M.; Götz, C.; Haidari, S.; Dorfner, C.; Jesinghaus, M.; Drecoll, E.; Boskov, M.; Wolff, K.D.; Weichert, W.; Haller, B.; et al. Immunohistochemical expression of CD44 in oral squamous cell carcinoma in relation to histomorphological parameters and clinicopathological factors. *Histopathology* **2018**, *73*, 559–572. [CrossRef]

61. Chen, J.; Zhou, J.; Lu, J.; Xiong, H.; Shi, X.; Gong, L. Significance of CD44 expression in head and neck cancer: A systemic review and meta-analysis. *BMC Cancer* **2014**, *14*, 15. [CrossRef]

62. de Jong, M.C.; Pramana, J.; van der Wal, J.E.; Lacko, M.; Peutz-Kootstra, C.J.; de Jong, J.M.; Takes, R.P.; Kaanders, J.H.; van der Laan, B.F.; Wachters, J.; et al. CD44 expression predicts local recurrence after radiotherapy in larynx cancer. *Clin. Cancer Res.* **2010**, *16*, 5329–5338. [CrossRef]

63. Zöller, M. CD44: Can a cancer-initiating cell profit from an abundantly expressed molecule? *Nat. Rev. Cancer* **2011**, *11*, 254–267. [CrossRef]

64. Woerner, S.M.; Givehchian, M.; Dürst, M.; Schneider, A.; Costa, S.; Melsheimer, P.; Lacroix, J.; Zöller, M.; Doeberitz, M.K. Expression of CD44 splice variants in normal, dysplastic, and neoplastic cervical epithelium. *Clin. Cancer Res.* **1995**, *1*, 1125–1132.

65. Dall, P.; Heider, K.H.; Hekele, A.; von Minckwitz, G.; Kaufmann, M.; Ponta, H.; Herrlich, P. Surface protein expression and messenger RNA-splicing analysis of CD44 in uterine cervical cancer and normal cervical epithelium. *Cancer Res.* **1994**, *54*, 3337–3341.

66. Dall, P.; Hekele, A.; Ikenberg, H.; Göppinger, A.; Bauknecht, T.; Pfleiderer, A.; Moll, J.; Hofmann, M.; Ponta, H.; Herrlich, P. Increasing incidence of CD44v7/8 epitope expression during uterine cervical carcinogenesis. *Int. J. Cancer* **1996**, *69*, 79–85. [CrossRef]

67. Mereiter, S.; Martins, Á.M.; Gomes, C.; Balmaña, M.; Macedo, J.A.; Polom, K.; Roviello, F.; Magalhães, A.; Reis, C.A. O-glycan truncation enhances cancer-related functions of CD44 in gastric cancer. *FEBS Lett.* **2019**, *593*, 1675–1689. [CrossRef]

68. Matter, N.; Herrlich, P.; König, H. Signal-dependent regulation of splicing via phosphorylation of Sam68. *Nature* **2002**, *420*, 691–695. [CrossRef]

69. Weg-Remers, S.; Ponta, H.; Herrlich, P.; König, H. Regulation of alternative pre-mRNA splicing by the ERK MAP-kinase pathway. *Embo J.* **2001**, *20*, 4194–4203. [CrossRef]

70. Rothenberg, S.M.; Ellisen, L.W. The molecular pathogenesis of head and neck squamous cell carcinoma. *J. Clin. Investig.* **2012**, *122*, 1951–1957. [CrossRef]

71. Compagnone, M.; Gatti, V.; Presutti, D.; Ruberti, G.; Fierro, C.; Markert, E.K.; Vousden, K.H.; Zhou, H.; Mauriello, A.; Anemone, L.; et al. ΔNp63-mediated regulation of hyaluronic acid metabolism and signaling supports HNSCC tumorigenesis. *Proc. Natl. Acad. Sci. USA* **2017**, *114*, 13254–13259. [CrossRef] [PubMed]

72. Stransky, N.; Egloff, A.M.; Tward, A.D.; Kostic, A.D.; Cibulskis, K.; Sivachenko, A.; Kryukov, G.V.; Lawrence, M.S.; Sougnez, C.; McKenna, A.; et al. The mutational landscape of head and neck squamous cell carcinoma. *Science* **2011**, *333*, 1157–1160. [CrossRef]

73. Menke-van der Houven van Oordt, C.W.; Gomez-Roca, C.; van Herpen, C.; Coveler, A.L.; Mahalingam, D.; Verheul, H.M.; van der Graaf, W.T.; Christen, R.; Rüttinger, D.; Weigand, S.; et al. First-in-human phase I clinical trial of RG7356, an anti-CD44 humanized antibody, in patients with advanced, CD44-expressing solid tumors. *Oncotarget* **2016**, *7*, 80046–80058. [CrossRef] [PubMed]

74. Riechelmann, H.; Sauter, A.; Golze, W.; Hanft, G.; Schroen, C.; Hoermann, K.; Erhardt, T.; Gronau, S. Phase I trial with the CD44v6-targeting immunoconjugate bivatuzumab mertansine in head and neck squamous cell carcinoma. *Oral. Oncol.* **2008**, *44*, 823–829. [CrossRef] [PubMed]

75. Tijink, B.M.; Buter, J.; de Bree, R.; Giaccone, G.; Lang, M.S.; Staab, A.; Leemans, C.R.; van Dongen, G.A. A phase I dose escalation study with anti-CD44v6 bivatuzumab mertansine in patients with incurable squamous cell carcinoma of the head and neck or esophagus. *Clin. Cancer Res.* **2006**, *12*, 6064–6072. [CrossRef]

76. Mitsunaga, M.; Ogawa, M.; Kosaka, N.; Rosenblum, L.T.; Choyke, P.L.; Kobayashi, H. Cancer cell-selective in vivo near infrared photoimmunotherapy targeting specific membrane molecules. *Nat. Med.* **2011**, *17*, 1685–1691. [CrossRef]

77. Maruoka, Y.; Wakiyama, H.; Choyke, P.L.; Kobayashi, H. Near infrared photoimmunotherapy for cancers: A translational perspective. *EBioMedicine* **2021**, *70*, 103501. [CrossRef]

78. Kato, T.; Wakiyama, H.; Furusawa, A.; Choyke, P.L.; Kobayashi, H. Near infrared photoimmunotherapy; A review of targets for cancer therapy. *Cancers* **2021**, *13*, 2535. [CrossRef]

79. Ogawa, M.; Tomita, Y.; Nakamura, Y.; Lee, M.J.; Lee, S.; Tomita, S.; Nagaya, T.; Sato, K.; Yamauchi, T.; Iwai, H.; et al. Immunogenic cancer cell death selectively induced by near infrared photoimmunotherapy initiates host tumor immunity. *Oncotarget* **2017**, *8*, 10425–10436. [CrossRef]

80. Cognetti, D.M.; Johnson, J.M.; Curry, J.M.; Kochuparambil, S.T.; McDonald, D.; Mott, F.; Fidler, M.J.; Stenson, K.; Vasan, N.R.; Razaq, M.A.; et al. Phase 1/2a, open-label, multicenter study of RM-1929 photoimmunotherapy in patients with locoregional, recurrent head and neck squamous cell carcinoma. *Head Neck* **2021**, *43*, 3875–3887. [CrossRef]

81. Nagaya, T.; Nakamura, Y.; Okuyama, S.; Ogata, F.; Maruoka, Y.; Choyke, P.L.; Allen, C.; Kobayashi, H. Syngeneic mouse models of oral cancer are effectively targeted by anti-CD44-based NIR-PIT. *Mol. Cancer Res.* **2017**, *15*, 1667–1677. [CrossRef]

82. Li, G.; Suzuki, H.; Ohishi, T.; Asano, T.; Tanaka, T.; Yanaka, M.; Nakamura, T.; Yoshikawa, T.; Kawada, M.; Kaneko, M.K.; et al. Antitumor activities of a defucosylated anti-EpCAM monoclonal antibody in colorectal carcinoma xenograft models. *Int. J. Mol. Med.* **2023**, *51*, 1–14. [CrossRef]

83. Nanamiya, R.; Takei, J.; Ohishi, T.; Asano, T.; Tanaka, T.; Sano, M.; Nakamura, T.; Yanaka, M.; Handa, S.; Tateyama, N.; et al. Defucosylated anti-epidermal growth factor receptor monoclonal antibody (134-mG(2a)-f) exerts antitumor activities in mouse xenograft models of canine osteosarcoma. *Monoclon. Antib. Immunodiagn. Immunother.* **2022**, *41*, 1–7. [CrossRef]

84. Kawabata, H.; Suzuki, H.; Ohishi, T.; Kawada, M.; Kaneko, M.K.; Kato, Y. A defucosylated mouse anti-CD10 monoclonal antibody (31-mG(2a)-f) exerts antitumor activity in a mouse xenograft model of CD10-overexpressed tumors. *Monoclon. Antib. Immunodiagn. Immunother.* **2022**, *41*, 59–66. [CrossRef]

85. Kawabata, H.; Ohishi, T.; Suzuki, H.; Asano, T.; Kawada, M.; Suzuki, H.; Kaneko, M.K.; Kato, Y. A defucosylated mouse anti CD10 monoclonal antibody (31-mG(2a)-f) exerts antitumor activity in a mouse xenograft model of renal cell cancers. *Monoclon. Antib. Immunodiagn. Immunother.* **2022**, *41*, 320–327. [CrossRef]

86. Asano, T.; Tanaka, T.; Suzuki, H.; Li, G.; Ohishi, T.; Kawada, M.; Yoshikawa, T.; Kaneko, M.K.; Kato, Y. A defucosylated anti. EpCAM monoclonal antibody (EpMab-37-mG(2a)-f) exerts antitumor activity in xenograft model. *Antibodies* **2022**, *11*, 74. [CrossRef]

87. Tateyama, N.; Nanamiya, R.; Ohishi, T.; Takei, J.; Nakamura, T.; Yanaka, M.; Hosono, H.; Saito, M.; Asano, T.; Tanaka, T.; et al. Defucosylated anti-epidermal growth factor receptor monoclonal antibody 134-mG(2a)-f exerts antitumor activities in mouse xenograft models of dog epidermal growth factor receptor-overexpressed cells. *Monoclon. Antib. Immunodiagn. Immunother.* **2021**, *40*, 177–183. [CrossRef]

88. Takei, J.; Ohishi, T.; Kaneko, M.K.; Harada, H.; Kawada, M.; Kato, Y. A defucosylated anti-PD-L1 monoclonal antibody 13-mG(2a)-f exerts antitumor effects in mouse xenograft models of oral squamous cell carcinoma. *Biochem. Biophys. Rep.* **2020**, *24*, 100801. [CrossRef]

Article

IFIT2 Depletion Promotes Cancer Stem Cell-like Phenotypes in Oral Cancer

Kuo-Chu Lai [1,2,3,*,†], Prabha Regmi [4,†], Chung-Ji Liu [5], Jeng-Fan Lo [6] and Te-Chang Lee [4,*]

1 Department of Physiology and Pharmacology, College of Medicine, Chang Gung University, Taoyuan City 33302, Taiwan
2 Graduate Institute of Biomedical Sciences, College of Medicine, Chang Gung University, Taoyuan City 33302, Taiwan
3 Division of Hematology and Oncology, Department of Internal Medicine, New Taipei Municipal TuCheng Hospital, New Taipei City 23652, Taiwan
4 Institute of Biomedical Sciences, Academia Sinica, Taipei 11529, Taiwan
5 Department of Oral and Maxillofacial Surgery, Mackay Memorial Hospital, Taipei 10421, Taiwan
6 Institute of Oral Biology, National Yang Ming Chiao Tung University, Taipei 11221, Taiwan
* Correspondence: kuochu@mail.cgu.edu.tw (K.-C.L.); bmtcl@ibms.sinica.edu.tw (T.-C.L.)
† These authors contributed equally to this work.

Abstract: (1) Background: Cancer stem cells (CSCs) are a small cell population associated with chemoresistance, metastasis and increased mortality rate in oral cancer. Interferon-induced proteins with tetratricopeptide repeats 2 (IFIT2) depletion results in epithelial to mesenchymal transition, invasion, metastasis, and chemoresistance in oral cancer. To date, no study has demonstrated the effect of IFIT2 depletion on the CSC-like phenotype in oral cancer cells. (2) Methods: Q-PCR, sphere formation, Hoechst 33,342 dye exclusion, immunofluorescence staining, and flow cytometry assays were performed to evaluate the expression of the CSC markers in IFIT2-depleted cells. A tumorigenicity assay was adopted to assess the tumor formation ability. Immunohistochemical staining was used to examine the protein levels of IFIT2 and CD24 in oral cancer patients. (3) Results: The cultured IFIT2 knockdown cells exhibited an overexpression of ABCG2 and CD44 and a downregulation of CD24 and gave rise to CSC-like phenotypes. Clinically, there was a positive correlation between IFIT2 and CD24 in the patients. IFIT2^{high}/CD24^{high}/CD44^{low} expression profiles predicted a better prognosis in HNC, including oral cancer. The TNF-α blockade abolished the IFIT2 depletion-induced sphere formation, indicating that TNF-α may be involved in the CSC-like phenotypes in oral cancer. (4) Conclusions: The present study demonstrates that IFIT2 depletion promotes CSC-like phenotypes in oral cancer.

Keywords: interferon-induced proteins with tetratricopeptide repeats 2; oral cancer; cancer stem cells

Citation: Lai, K.-C.; Regmi, P.; Liu, C.-J.; Lo, J.-F.; Lee, T.-C. IFIT2 Depletion Promotes Cancer Stem Cell-like Phenotypes in Oral Cancer. *Biomedicines* **2023**, *11*, 896. https://doi.org/10.3390/biomedicines11030896

Academic Editor: Vui King Vincent-Chong

Received: 4 February 2023
Revised: 5 March 2023
Accepted: 9 March 2023
Published: 14 March 2023

1. Introduction

Oral cancer is the sixth most prevalent malignancy worldwide, and its incidence varies depending on the geographic region [1,2]. Oral squamous cell carcinoma (OSCC) is highly malignant and is associated with cervical nodal and distant metastasis in 25–65% of patients [3]. Despite the advancements in diagnosis and therapeutics, the overall five year survival rate of oral cancer patients has remained at 50–60% for decades [4,5]. Treatment failure and eventually mortality in oral cancer patients is in general due to the recurrence, metastasis, and chemo/radioresistance that are associated with a small number of cancer stem cells (CSCs) or tumor-initiating cells [6].

CSCs are capable of sustaining tumor growth in situ and possess self-renewal and differentiation abilities and seed tumors in xenograft models [7]. CSCs have been reported in both primary oral cancer and cell lines and are resistant to various chemotherapeutic agents, such as cisplatin, docetaxel, etoposide, gemcitabine, paclitaxel, carboplatin, and

5-fluorouracil (5-FU) [6,8]. CSCs are capable of cellular plasticity, where they can switch between the epithelial to mesenchymal transition (EMT) differentiation program and the reverse program [9]. Consequently, various studies have reported the role of phenotypic plasticity in cancer initiation, progression, metastasis, and resistance to therapy in various primary and metastatic neoplasms, such as lung, prostate, pancreatic, and head and neck cancers (HNCs) [10–16]. Numerous studies also suggest that cells undergoing EMT have the potential to acquire the CSC characteristics of self-renewal and to reprogram the gene expression associated with stemness [9,17,18]. However, studies on the pathogenesis and molecular mechanism of CSCs in oral cancer are limited and require further investigation [19].

There are four members in the interferon-induced proteins with the tetratricopeptide repeat (IFIT) family (IFIT1, IFIT2, IFIT3 and IFIT5) that display a broad spectrum of antiviral functions and are involved in various biological processes, such as proliferation, migration, translational initiation, and RNA signaling [20,21]. Emerging studies have shown the importance of IFITs in the signaling pathways involved in cancer progression, metastasis, and drug resistance [22]. Intriguingly, although IFITs share conserved structural motifs, their isoforms display differential expression and nonredundant functions [23,24]. Elevated levels of IFIT1, IFIT3 and IFIT5 have been shown to play significant roles in cancer progression, while the decreased expression of IFIT2 has been reported to enhance invasion, tumor progression, and drug resistance in various cancer types [25–32]. Furthermore, increased IFIT1 and IFIT3 expression but decreased IFIT2 expression is correlated with poor survival in OSCC patients [25,33]. On the contrary, increased IFIT2 expression inhibits cell proliferation and triggers apoptosis in various cancer types [34–37]. The IFIT family may play a critical role in tumor progression in various types of cancer.

We have previously shown that IFIT2 knockdown leads to the activation of atypical protein kinase C (PKC) signaling, accompanied by EMT and an increase in migration, invasion, distant metastasis, chemoresistance, and TNF-α expression, and eventually results in angiogenesis and poor survival in oral cancer [31–33,38]. The accumulating evidence shows that cells undergoing EMT have CSC characteristics and that EMT is an important bridge between metastasis, drug resistance, and CSCs [9,39,40]. Whether IFIT2-depleted oral cancer cells exhibit a CSC-like phenotype is still unknown. In the present study, we aimed to assess the CSC properties in IFIT2 knockdown and control cells. Examining the relationship between IFIT2 depletion and CSCs may be helpful for proposing an approach to block cancer progression, recurrence, and metastasis, to overcome drug resistance, and to stratify patients based on optimal treatment regimens.

2. Materials and Methods

2.1. Cell Culture and Reagents

Sh-control and stable IFIT2-depleted (sh-IFIT2-1 and sh-IFIT2-2) cells were generated in the human oral cancer cell line CAL27 as previously described [31]. The cells were screened for mycoplasma contamination following the manufacturer's instructions (abm Mycoplasma PCR detection kit; Cat no. G238 Applied Biological Materials Inc. Richmond BC, Canada) and cultured as previously described [32]. We also tried to knockdown IFIT2 in other cell lines. However, based on the IFIT2 expression levels, CAL27 cells were apparently the most convenient cells for this study.

2.2. Western Blot Analysis

Western blot analysis was performed to determine the expression level of proteins as previously described [32]. The primary antibodies, anti-IFIT2 (sc-390724), anti-OCT3/4 (sc-5279), anti-NANOG (sc-293121), and anti-ABCG2 (sc-58222), were obtained from Santa Cruz Biotechnology (Santa Cruz, CA, USA); anti-TNFα (#6945) from Cell Signaling Technology (Danvers, MA, USA); anti-GAPDH (#60004-1-Ig) and anti-β-actin (66009-1-Ig) from Proteintech (Chicago, IL, USA). Anti-rabbit-HRP (#ab97051) and anti-mouse-HRP (#205719) were purchased from Abcam (Cambridge, UK).

2.3. Anchorage-Independent Growth Assay

The soft agar assay is a common method to examine anchorage-independent cell growth to validate transformed cells [41]. To evaluate the clonogenicity of IFIT2 knockdown cells on soft agar, the bottom layer of the 60 mm dish was coated with 0.7% agar prepared in 2× DMEM supplemented with 4% peptone and FBS. After gel solidification, 3 mL of 0.3% top agar containing 1000 cells per well was added and incubated at 37 °C for 6 weeks. During the course of incubation, 500 µL of fresh medium was added every 3–4 days. The cells were stained with 750 µL of 1 mg/mL iodonitrotetrazolium (INT) for 2–3 days. Images were taken under the microscope at 10× magnification, and the colonies were counted.

2.4. Sphere Formation Assay

To evaluate the sphere growth and formation, the cells were plated at a density of 1000 cells per well in ultralow attachment 6-well plates (Corning Inc., Corning, NY, USA) and cultured in serum-free DMEM-F12 medium supplemented with 1% B27 supplement, 20 ng/mL epidermal growth factor (EGF) and 20 ng/mL basic fibroblast growth factor (bFGF) for 2 weeks. EGF and bFGF were purchased from PeproTech, Rocky Hill, NJ, USA. The sphere formation was assessed by counting the spheres (>50 µm) under a microscope.

2.5. Side Population (SP) Assay

The SP was quantified by a Hoechst dye exclusion assay [42]. The OSCC cells (10^6/mL) were incubated with Hoechst 33,342 dye (5 µg/mL) in the presence or absence of verapamil (50 µM) and incubated at 37 °C in a water bath for 90 min with intermittent shaking. Then, the cells were washed with ice-cold phosphate-buffered saline (PBS) with 2% FBS and resuspended in 2 µg/mL propidium iodide prepared in ice-cold PBS containing 2% FBS. The SP population was quantified using fluorescence-activated cell sorting (FACS) in the IFIT2 knockdown cells and compared to vector control cells, as well as verapamil-treated and nontreated cells. The Hoechst 33,342 dye was excited, and its fluorescence at dual wavelengths was analyzed (blue, 402–446 nm; red, 650–670 nm). Hoechst dye was purchased from BD Bioscience (San Jose, CA, USA).

2.6. Quantitative Real-Time PCR (QPCR)

Q-PCR was performed as previously described [32]. The primer sequences are listed in Supplementary Table S1.

2.7. Flow Cytometry Analysis

The growing cells cultured in 100 mm dishes were washed with 1× PBS, trypsinized, centrifuged, and resuspended in 2% FBS in Hank's balanced salt solution (HBSS). The viable cell number was counted using an automatic cell counter (trypan blue exclusion), and the cell density was adjusted to 1×10^6 cells/100 µL. For CD24 and CD44 staining, 20 µL of antibodies (anti-CD24, anti-CD44 and PE-mouse IgG isotype control and APC-mouse IgG isotype control) were added to 100 µL of cell suspension and incubated at 4 °C for 30 min per the manufacturer's protocol. However, for ABCG2 staining, 5 µL of APC Mouse Anti-Human CD338 and APC-mouse-IgG isotype control were added to 100 µL cell suspension and incubated at 4 °C for 30 min, according to the manufacturer's instructions. After incubation, the cells were washed three times with HBSS and centrifuged at 2000 rpm for 3 min. Then, the cells were resuspended in 300 µL 1× PBS containing 1 µg/mL DAPI and were passed through tube filters to avoid cell aggregation. Finally, the cells were analyzed using flow cytometry. PE mouse anti-human CD24, APC mouse anti-human CD44, PE mouse IgG2α, κ isotype control (555574), APC mouse IgG2α, and κ isotype control antibodies were purchased from BD Bioscience (San Jose, CA, USA), and anti-mouse IgG$_1$ isotype control (MAB002) was purchased from R & D Systems, Inc. (Minneapolis, MN, USA).

2.8. Immunofluorescence and Confocal Microscopy

The intracellular location of ABCG2, CD24 and CD44 was examined by immunofluorescence staining. The cultured sh-control and stable IFIT2 knockdown cells were fixed with 100% ice-cold methanol and permeabilized with 0.2% Triton X-100 and incubated with the ABCG2, CD24, and CD44 antibodies. Anti-mouse IgG secondary antibody Alexa Fluor 555 (Molecular Probes, Eugene, OR, USA) was then added and incubated for 1 h. The cell nuclei were stained with 4',6-diamidino-2-phenylindole (DAPI). After mounting the slides with 50% glycerol in PBS, images were acquired under a laser scanning confocal microscope (Carl Zeiss MicroImaging Inc., Thornwood, NY, USA) and Axio Vision software.

2.9. Tumorigenicity Assay

In vivo tumorigenesis was performed according to the guidelines of experimental animals and approved by the Institutional Animal Care and Utilization Committee (IACUC) of Academia Sinica. The mice were housed in a specific pathogen-free environment under 12 h light-dark cycles in an animal core facility at the Institute of Biomedical Sciences, Academia Sinica, Taipei, Taiwan. To determine the tumorigenicity, the sh-control and stable IFIT2 -depleted cells were subcutaneously implanted into the dorsal flank region of 5-week-old male BALB/c nude mice. Different numbers of cells (10^5, 5×10^4, 10^4, 5×10^3, 10^3, 500, 100, 50, 25 and 10 cells) were mixed with Matrigel (1:1 volume) to generate a 100 µL cell-Matrigel mixture. Matrigel was purchased from Corning (Two Oak Park, Bedford, MA, USA). The mice were housed under observation until the tumor was seen. Tumor volume was measured using the formula: Volume = 1/2 (length $\times$ width^2). Tumor formation was confirmed using hematoxylin and eosin (H & E) stained sections.

2.10. Immunohistochemical Staining

Oral cancer patient tissue samples were obtained and analyzed, abiding the rules and approval of Mackay Memorial Hospital's institutional review board and the IRB of the Institute of Biomedical Research, Academia Sinica. In brief, the slides with tissue sections were deparaffinized in xylene for 7 min, twice, and then rehydrated in graded ethanol serially from 100% to 70%, followed by rinsing with distilled water. The deparaffinization process was conducted by the Pathology Core of IBMS, Academia Sinica. Subsequently, the slides were immersed in citrate buffer (0.01 M, pH 6.0) and boiled for 50 min in a cooker for antigen retrieval, followed by cooling for 30 min at ambient temperature. Then, the slides were rinsed in distilled water again before immunohistochemistry was performed. The staining procedure was conducted using the Novolink Polymer Detection System (Leica Biosystems, Wetzlar, Germany) following the manufacturer's instructions. The primary antibody dilution for both the IFIT2 and CD24 was used at a ratio of 1:100. Anti-CD24 antibody was purchased from Santa Cruz Biotechnology (Santa Cruz, CA, USA). Then, the immunostained tissue section slides were scanned using a Panoramic 250 Flash II whole slide scanner at high magnification. The expression levels of IFIT2 and CD24 were analyzed using the 3DHISTECH Panoramic viewer software (3DHISTECH Ltd., Budapest, Hungary).

2.11. Enzyme-Linked Immunoassay (ELISA)

A total of 3×10^5 and 5×10^5 cells were seeded in 6-well plates and cultured in serum-free medium for 48–72 h. The collected conditioned medium was used to perform an ELISA as per the manufacturer's instructions. The absorbance of the standard was compared to that of the test sample to quantify the TNF-α concentration. The Human TNF-α Quantikine ELISA kit (DTA00C) was purchased from R & D Systems, Inc. (Minneapolis, MN, USA).

2.12. Statistical Analysis

The data analysis was performed using the GraphPad software (version 6.0). To identify significant differences between the two different groups, Student's *t*-test was used. The body weight and tumor volume between the animal groups were compared using a one-way ANOVA. *p* values < 0.05 were considered to indicate statistical significance. The correlation and significance of differences for immunohistochemistry were determined by the Pearson correlation coefficient test.

3. Results

3.1. IFIT2 Knockdown Cells Exhibit CSC-like Properties

Self-renewal and differentiation are the main features of CSCs and are commonly assessed by analyzing the anchorage-independent growth and spheroid abilities. We first confirmed the low expression of IFIT2 in the IFIT2-depleted cells using Western blotting (Figure 1A). Similarly, a clonogenicity assay showed that the IFIT2 knockdown increased the anchorage-independent growth compared to the sh-control cells (Figure 1A). Moreover, we analyzed the spheroid formation ability of the IFIT2 knockdown cells using a sphere formation assay. The Sh-IFIT2-1 and sh-IFIT2-2 cells showed a significantly higher number of spheroids than the sh-control cells, indicating an enhanced self-renewal capacity in the IFIT2 knockdown cells (Figure 1B). CSCs have a higher drug efflux capacity, mainly due to their high expression levels and the activity of drug transporters such as ABCB1 and ABCG2 [43,44]. Using the Hoechst 33,342 dye exclusion assay, we observed an enriched SP in the stable IFIT2-depleted cells compared to the sh-control cells. Approximately 2% of the stained population was designated as the SP group in the IFIT2 knockdown cells (Figure 1C).

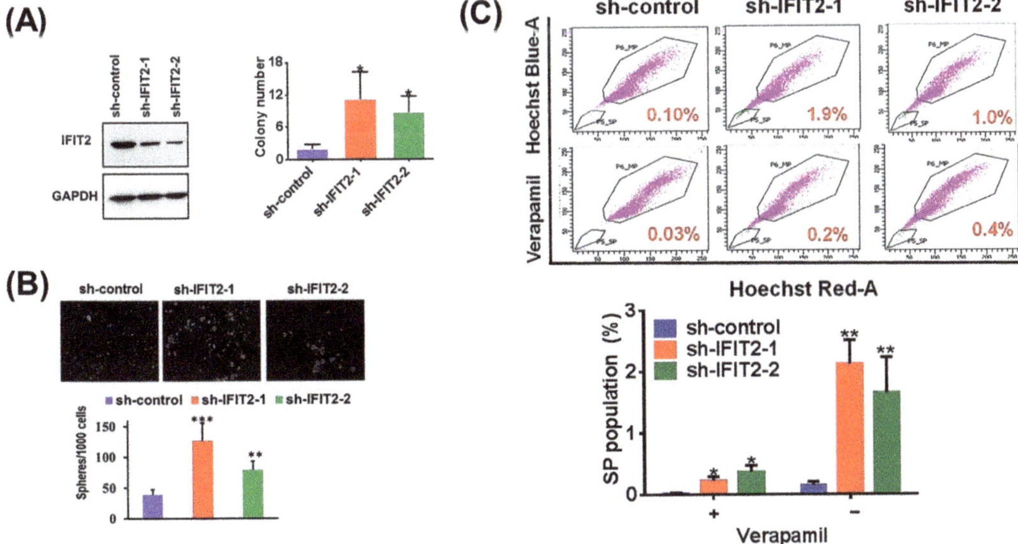

Figure 1. In vitro soft agar colony and spheroid formation in stable IFIT2-depleted cells. (**A**) The IFIT2 protein levels in sh-control and stable IFIT2 knockdown cells were determined by Western blotting. GAPDH expression was used as a loading control (left panel). The results of the anchorage-independent colony formation assay were shown in right panel; (**B**) Representative image of spheroid formation and quantitated data of spheroid numbers counted in three independent experiments in sh-control and stable IFIT2 knockdown cells; (**C**) Representative flow cytometry of SP assay with or without verapamil treatment (upper). The quantitative analysis of the SP population was average of three independent experiments (bottom). The symbols *, ** and *** indicate statistically significant differences compared to sh-control at $p < 0.05$, $p < 0.01$ and $p < 0.001$, respectively.

3.2. Characterization of CSC Markers in IFIT2 Knockdown Cells

Our unpublished cDNA microarray data showed that several probes targeting CD44 or ABCG2 were significantly enhanced, whereas CD24 was decreased, in the IFIT2 knockdown cells compared with the control cells. Other CSC markers, such as NANOG, SOX2, and KLF4, were not different among them (Supplementary Figure S1). Accordingly, Q-PCR and Western blot analysis were conducted to characterize the CSC features in the IFIT2 knockdown cell. We confirmed that the mRNA levels of ABCG2 and CD44 were significantly increased in the IFIT2 knockdown cells compared with the sh-control cells, whereas the mRNA of CD24 were significantly lower in the IFIT2-depleted cells. The protein levels of CD44 and ABCG2 were consistent with the relative mRNA expression (Figure 2A). The mRNA and protein levels of Nanog and Oct-4 were not significantly different between these cells (Figure 2A and Supplementary Figure S2). Furthermore, we evaluated the population of ABCG2, CD44, and CD24 expression in these cells using flow cytometry. The ABCG2-positive populations accounted for 49.5% and 32% of the sh-IFIT2-1 and sh-IFIT2-2 cells, respectively, which was an increase compared to the proportion in the sh-control cells (12.5%). Similarly, double staining for CD44 and CD24 was performed to analyze the $CD44^+/CD24^-$, $CD44^+/CD24^+$, $CD44^-/CD24^-$, and $CD44^-/CD24^+$ populations. The proportions of $CD44^{high}/CD24^{low}$ cells was 5.48% in the sh-control cells; however, $CD44^{high}/CD24^{low}$ cells were 33.2% and 24.3%, respectively, observed in the sh-IFIT2-1 and sh-IFIT2-2 cells (Figure 2B). These results depict the enrichment of CD44 high and CD24 low populations in the IFIT2-depleted cells, showing the potential role of IFIT2 knockdown in harboring CSC-like properties. In addition, the intracellular localization of CD44, CD24 and ABCG2 was visualized using immunofluorescence staining. As expected, ABCG2 and CD44 were highly expressed in the stable IFIT2-depleted cells compared to the sh-control cells, whereas CD24 expression showed minimal staining in the IFIT2 knockdown cells (Figure 2C). These results validate that IFIT2 depletion may enhance the CSC population in oral cancer cells.

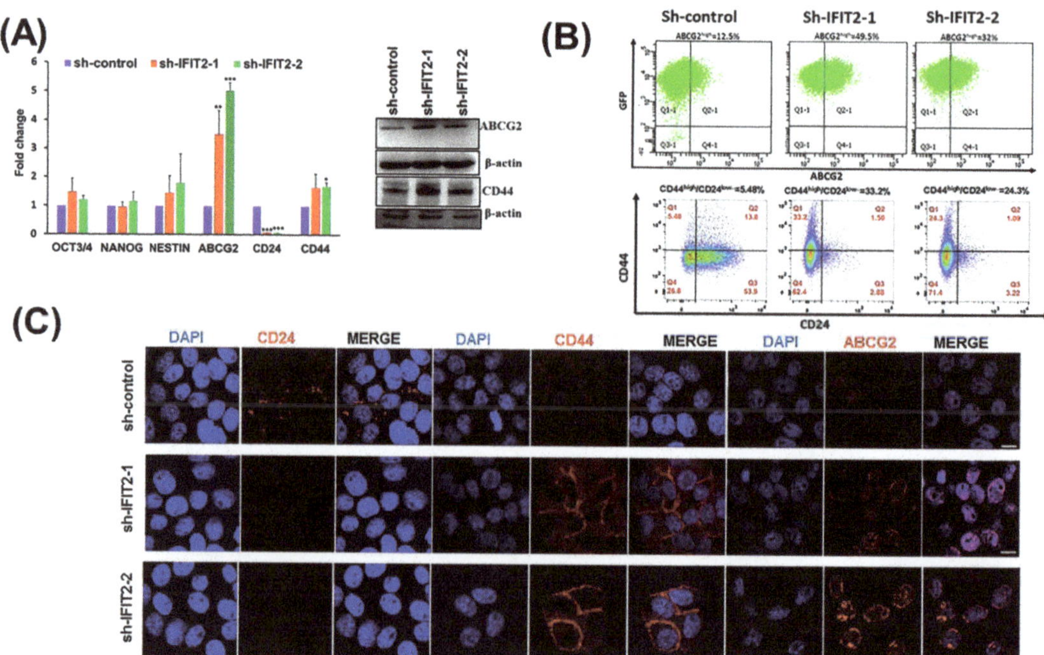

Figure 2. Expression of CSC markers in IFIT2 knockdown cells. (**A**) The mRNA expression levels of CSC markers in sh-control and stable IFIT2 knockdown cells were determined by Q-PCR, while the

protein levels were analyzed by Western blotting. GAPDH was used as an internal control in Q-PCR and β-actin was used as an internal control in Western blotting; (**B**) Representative flow cytometry analysis of ABCG2, CD44, and CD24 in sh-control and IFIT2-depleted cells. Experiments were performed four times; (**C**) Representative image of immunofluorescence of ABCG2, CD44 and CD24 staining in sh-control and stable IFIT2 knockdown cells. Rhodamine-conjugated secondary antibody for ABCG2, CD24 and CD44 (red) were applied, and nuclei were counterstained with DAPI (blue). Scale is 10 μm. The symbols *, ** and *** indicate statistically significant differences compared to sh-control at $p < 0.05$, $p < 0.01$ and $p < 0.001$, respectively.

3.3. Tumorigenicity in IFIT2 Knockdown Cells

The in vivo tumor-initiating properties of cells are also hallmarks of CSCs; hence, we performed an in vivo tumorigenicity assay by subcutaneously transplanting the stable IFIT2 knockdown and sh-control cells into the flanks of the nude mice. As shown in Table 1 and Figure 3A, these three sublines had a 100% success rate when more than 1000 cells were transplanted subcutaneously into the mice. The body weights and tumor volumes were not significantly different between these sublines (Figure 3B). The success rate decreased when the number of cells was reduced, but there was no apparent difference between the three sublines until the seeding numbers reached at least 25 and 10 cells. In particular, when the cell inoculation number was at least ten cells, there was no tumor growth in the sh-control group, but tumor formation was still observed in the IFIT2 knockdown cells (Figure 3C). The results show enriched tumorigenesis in the IFIT2 knockdown cells.

Table 1. In vivo tumorigenicity of IFIT2-depleted oral cancer cells.

Cell Number	Subline	Latency (Days)	Tumor Incidence (%)
	sh-control	21	100 (10/10)
100,000	sh-IFIT2-1	21	100 (10/10)
	sh-IFIT2-2	21	100 (10/10)
	sh-control	21	100 (10/10)
50,000	sh-IFIT2-1	21	100 (10/10)
	sh-IFIT2-2	21	100 (10/10)
	sh-control	30	100 (10/10)
10,000	sh-IFIT2-1	28	100 (10/10)
	sh-IFIT2-2	28	100 (10/10)
	sh-control	36	100 (10/10)
1000	sh-IFIT2-1	28	100 (10/10)
	sh-IFIT2-2	32	100 (10/10)
	sh-control	55	60 (6/10)
100	sh-IFIT2-1	48	70 (7/10)
	sh-IFIT2-2	48	80 (8/10)
	sh-control	68	50 (5/10)
50	sh-IFIT2-1	48	70 (7/10)
	sh-IFIT2-2	48	70 (7/10)
	sh-control	75	16.7 (1/6)
25	sh-IFIT2-1	68	66.7 (4/6)
	sh-IFIT2-2	68	50 (3/6)
	sh-control	114	0 (0/8)
10	sh-IFIT2-1	114	25 (2/8)
	sh-IFIT2-2	114	25 (2/8)

3.4. Clinical Signature of IFIT2 and CSCs in HNC

IFIT2 knockdown strongly promoted the properties of CSCs; however, the clinical correlation of IFIT2 and CSCs is still unknown. With the aid of IHC, the IFIT2 and CD24 protein levels in the 47 primary OSCC tissues were evaluated and showed a positive correlation between IFIT2 and CD24 ($p < 0.001$) (Figure 4A). In addition, we estimated the prognostic effect of IFIT2, CD24, and CD44 in patients with HNC from the TCGA database. Cox regression analysis was performed using the RNA-seq expression of IFIT2, CD24, and

CD24 in HNC patients. The correlations between IFIT2, CD24, and CD24 and survival were computed using Cox proportional hazards regression and by plotting Kaplan–Meier survival plots [45]. In the 499 HNC patients, the expressions of IFIT2 or CD24 did not show an association with overall survival; however, CD44 was associated with the overall survival ($p = 0.017$; Figure 4B). The IFIT2^{high}/CD24^{high} expression profile was associated with better overall survival in HNC patients (the medium follow-up overall survival was 61.27 months compared to 46.47 months; $p = 0.072$); the IFIT2^{high}/CD44^{low} group had a significantly better overall survival rate (the medium follow-up overall survival was 58.73 compared to 36.43 months; $p < 0.01$); and the IFIT2^{high}/CD24^{high}/CD44^{low} expression also predicted a better survival rate (the medium follow-up overall survival was 58.73 compared to 46.6 months; $p = 0.028$). These results demonstrate that IFIT2 and CSC markers may be prognostic factors in HNC.

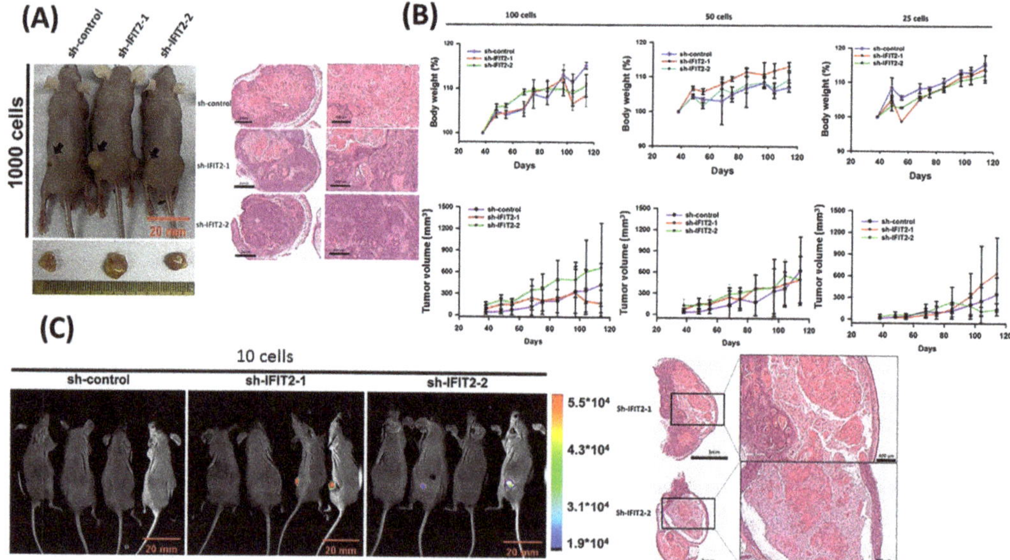

Figure 3. In vivo tumorigenicity of stable IFIT2-depleted cells. (**A**) Representative image of mice and H & E staining of primary xenograft tumors generated by injecting 1000 cells subcutaneously into nude mice; (**B**) Body weight and tumor volumes were monitored throughout the indicated period. There were no statistically significant differences in body weight and tumor volume between groups. (**C**) Images of mice bearing tumors 114 days after implantation of 10 cells. The xenograft tumors were visualized by the IVIS system, and pathology was assessed by H & E staining.

3.5. Effect of a TNF-α Inhibitor on Spheroid Formation in IFIT2 Knockdown Cells

Our previous study showed increased TNF-α expression in IFIT2-depleted metastatic and xenograft-derived sublines, and its inhibition resulted in decreased tumor growth, abolished angiogenic activity, and inhibited metastasis [38]. Inflammatory cytokines (interferons, TNF-α, IL-6, and IL-17) and inflammatory cells modulate the gene expression that regulates survival, proliferation, self-renewal, metastasis, and cancer stemness. As expected, increased TNF-α was detected in the IFIT2-depleted cells compared to the sh-control cells (Figure 5A). Furthermore, pomalidomide, a potent TNF-α inhibitor [46], suppressed the sphere formation in the IFIT2 knockdown cells at a non-toxic concentration (Supplementary Figure S3 and Figure 5B). Hence, these results suggest that blocking TNF-α may alleviate the IFIT2 depletion-induced CSC phenotype.

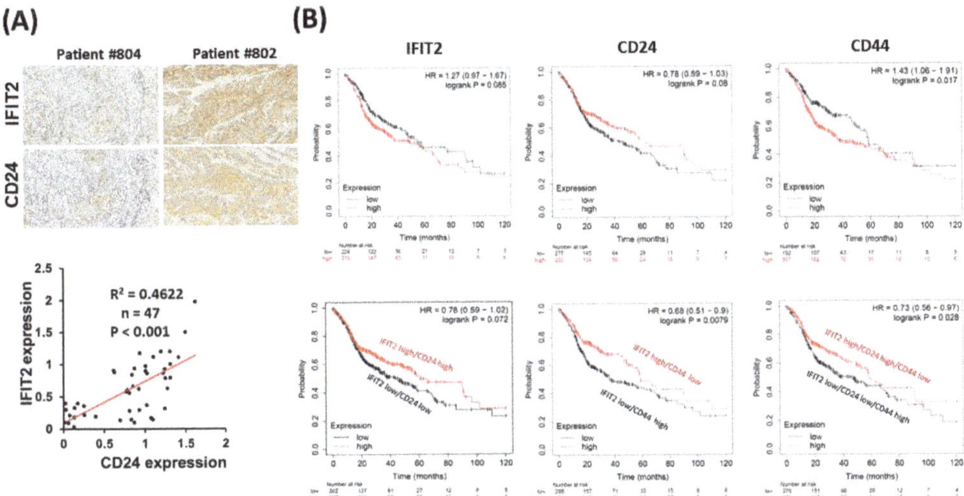

Figure 4. Clinical value of IFIT2 and CSC markers. (**A**) Representative images of low and high immunostaining for IFIT2 and CD24 proteins in OSCC patients (patient IDs: #804 and #802). Correlation between IFIT2 and CD24 expression in OSCC. The *p* value was determined by the Pearson correlation coefficient test; (**B**) Kaplan–Meier plotter analysis of overall survival in 499 HNC patients grouped according to IFIT2, CD44, and CD24 expression. The data were obtained from TCGA.

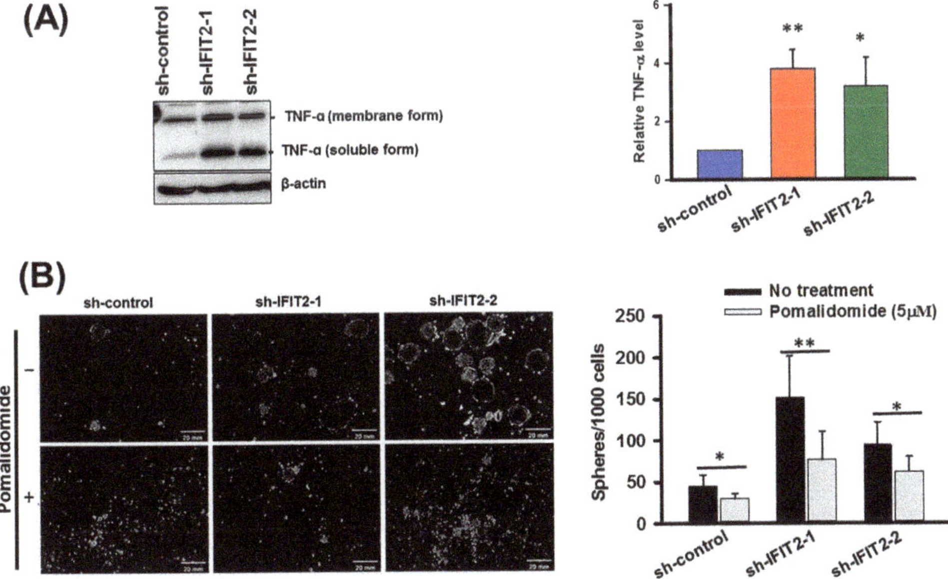

Figure 5. Effect of pomalidomide on spheroid formation in stable IFIT2 knockdown cells. (**A**) The TNF-α protein levels in sh-control and stable IFIT2 knockdown cells were determined by Western blotting (left panel) and ELISA (right panel). β-actin expression was used as an internal control in Western blotting; (**B**) Representative image of spheres with or without pomalidomide (5 μM). The quantitative data were average of seven independent experiments. The symbols * and ** indicate statistically significant differences at $p < 0.05$ and $p < 0.01$, respectively.

4. Discussion

IFIT2 depletion is associated with enhanced EMT, metastasis, and chemoresistance in oral cancer cell lines and poor survival in OSCC patients [31–33]. EMT, metastasis, drug resistance, and CSCs are intertwined. Evidently, our present results confirmed that the IFIT2-depleted cells, compared to the sh-control cells, were associated with a higher tumor sphere forming capability, anchorage-independent growth, SP cells, and self-renewal properties, which are congruent with the major hallmarks of CSCs reported in the literature [8,19,47]. The combined assessment of IFIT2 with CSC markers such as CD44 and CD24 may be applied to predict the prognosis of HNC patients. TNF-α may be involved in the IFIT2 depletion-induced CSC-like phenotype.

Tumor sphere-forming cells harbor enhanced metastatic activity, tumorigenicity, drug resistance, and expression of stemness molecules, which reflect their role in the pathogenesis and progression of cancers [48,49]. Therefore, examining the tumor sphere-forming capability is considered a robust technique to identify and isolate CSCs from heterogeneous cancer populations [50]. CSCs have a higher drug efflux capacity, primarily due to the high expression levels and activity of drug transporters such as ABCB1 and ABCG2 [43,44]. SP cells isolated from oral cancer are enriched in stem cell features and chemo/radioresistance compared to non-SP cells [51,52]. In this study, we also observed the presence of SP and an increase in the expression levels of ABCG2, which further exemplifies the stemness property of IFIT2 knockdown cells. Moreover, CSCs enriched in the SP are known to stay in a quiescent state, have intrinsic chemoresistance, evade apoptosis, and acquire chemoresistance upon drug treatment, and our previous study on IFIT2 knockdown cells showed a resistance to multiple drugs, as well as decreased apoptosis upon 5-FU treatment [32].

Emerging studies have started to stratify the CSC population by isolating CD44^{high} and CD24^{low} populations in various cancer types [53,54]. The CD44^{high}/CD24^{low} population isolated from OSCC has been reported to overexpress stem cell-related markers, exhibit EMT characteristics, and confer drug resistance via the increased expression of drug transporters, which is consistent with our study, where the IFIT2 knockdown cells present both CD44^{high} and CD24^{low} populations and increased Oct4, Nanog, Nestin, and ABCG2 expression [53,54]. CSC phenotypes are associated with the expression of pluripotency genes and the acquisition of mesenchymal traits, i.e., the loss of E-cadherin and gain of vimentin, which have already been established in IFIT2 knockdown cells [31]. Similarly, increased Oct4 expression in OSCC can modulate tumor-initiating properties via EMT [55]. In breast cancer stem cells, CD24 expression is downregulated by Twist, an EMT molecule, and it is therefore speculated that CD24 downregulation could be an EMT regulator [56]. Moreover, CD24 expression in breast cancer has shown a positive correlation with tumor grading. Our immunohistochemical analysis of CD24 and IFIT2 in OSCC patients showed a strong correlation in their expression pattern, suggesting an important role of CD24 in the CSC-like properties in IFIT2 knockdown cells. Furthermore, emerging evidence also supports the role of IFIT2 depletion in CSCs. The upregulation of IFIT2 can suppress the CSC-like characteristics of radio/chemoresistant breast cancer cells [28].

The IFIT2 knockdown cells showed higher tumorigenicity in an in vivo assay than control cells. These effects were likely due to the lower CD24 expression as the proportion of CD44 expression in the sh-control was comparatively high. CD24 is reported to regulate the gene expression and distribution of tight junction proteins, such as Zonula occludens (ZO)-1, ZO-2, occludin, claudin-7, and par-3, which are associated with the marginal barrier function of epithelial cells and E-cadherin expression in oral epithelial cells derived from human OSCC [57]. The association of E-cadherin and the actin cytoskeleton mediated by catenin and the partitioning defective-3 (par-3)/par-6/atypical PKC polarity complex that is activated by the small GTPase Cdc42 complex mediate the formation of tight junctions initiated by ZO-1 and ZO-2 [58]. Intriguingly, IFIT2-depleted cells showed decreased E-cadherin expression and increased atypical PKC activation, suggesting a possible correlation between CD24 and IFIT2.

Emerging studies have highlighted the role of TNF-α in CSCs. TNF-α treatment could switch the non-CSC (CD44$^+$/CD24$^+$) population to a CSC-like (CD44$^+$/CD24$^-$) population by increasing the expression of EMT factors in breast cancer cells [59]. TNF-α stimulates the Snail-driven EMT transition and increases the stemness properties in cholangiocarcinoma and renal cell carcinoma cells [60,61]. Additionally, TNF-α is known to regulate CD44 and its isoform expression in different breast cancer cell lines via a different pathway to promote cell migration [62]. Additionally, TNF-α enhanced the CSC properties in the OSCC via the Notch-Hes1 activation pathway [63]. However, the effect of TNF-α on the CSC phenotype in IFIT2 knockdown requires further investigation. Therefore, the role of enhanced TNF-α expression in regulating an increase in the CSC population/phenotype and facilitating CSC to overcome the immune escape mechanism by elongating cell survival and promoting the malignancy of OSCC requires further validation.

We had tried to overexpress IFIT2 in low IFIT2 expressing cells. Unfortunately, it was difficult to obtain IFIT2-overexpressed cells because the overexpression of IFIT2 triggers apoptosis [64]. Several reports have made similar observations. Altogether, the IFIT2 knockdown cells indeed exhibited an enhanced expression of putative and pluripotent CSC markers and exhibited self-renewal and tumorigenic abilities in the OSCC cells. We are the first to assess the clinical significance of IFIT2 and CSC markers in HNC, including oral cancer patients. Our previous studies have indicated that TNF-α inhibitors can block angiogenesis and metastasis. Here, we further demonstrated that TNF-α may alter the sphere forming ability. These results suggest that TNF-α may be a therapeutic target in advanced oral cancer patients with low IFIT2 expression.

5. Conclusions

The present study demonstrates that IFIT2 depletion promotes CSC-like phenotypes in oral cancer cells. Clinically, IFIT2 and CSC markers may be prognostic factors in HNC patients.

Supplementary Materials: The following supporting information can be downloaded at: https://www.mdpi.com/article/10.3390/biomedicines11030896/s1. Table S1: Q-PCR primer; Figure S1: cDNA microarray results of CSC markers in sh-control and stable IFIT2-depleted cells; Figure S2: The protein levels of Nanog and Oct-4 in sh-control and stable IFIT2-depleted cells were determined by Western blotting; Figure S3. Cytotoxicity of pomalidomide in sh-control and stable IFIT2-depleted cells.

Author Contributions: Conceptualization, K.-C.L., P.R. and T.-C.L.; data curation, P.R.; formal analysis, K.-C.L. and P.R.; funding acquisition, K.-C.L. and T.-C.L.; investigation, K.-C.L., P.R., C.-J.L., J.-F.L. and T.-C.L.; methodology, K.-C.L., P.R., C.-J.L. and J.-F.L.; project administration, T.-C.L.; resources, C.-J.L. and J.-F.L.; supervision, T.-C.L., validation, P.R., C.-J.L. and T.-C.L.; visualization, K.-C.L. and T.-C.L., writing—original draft preparation, K.-C.L. and P.R.; writing—review and editing, K.-C.L. and T.-C.L. All authors have read and agreed to the published version of the manuscript.

Funding: This work was supported by grants from the Ministry of Science and Technology (NSC 103-2321-B-001-019; MOST 107-2320-B-001-007; MOST 110-2320-B-182-023 MY3), Taiwan.

Institutional Review Board Statement: The study was conducted in accordance with the Declaration of Helsinki, and approved by the Institutional Review Board of Mackay Memorial Hospital and the Institute of Biomedical Research, Academia Sinica.

Informed Consent Statement: Not applicable.

Data Availability Statement: The data that support the findings of this study are available from the corresponding author upon reasonable request.

Acknowledgments: The authors would like to thank the Academia Sinica Core Facility for their flow cytometry and confocal microscopy services, the Institute of Biomedical Sciences and the SPF Animal Facility Academia Sinica, Taipei, Taiwan.

Conflicts of Interest: None of the authors have a financial conflict of interest.

References

1. Shield, K.D.; Ferlay, J.; Jemal, A.; Sankaranarayanan, R.; Chaturvedi, A.K.; Bray, F.; Soerjomataram, I. The global incidence of lip, oral cavity, and pharyngeal cancers by subsite in 2012. *CA Cancer J. Clin.* **2017**, *67*, 51–64. [CrossRef]
2. Bray, F.; Ferlay, J.; Soerjomataram, I.; Siegel, R.L.; Torre, L.A.; Jemal, A. Global cancer statistics 2018: GLOBOCAN estimates of incidence and mortality worldwide for 36 cancers in 185 countries. *CA Cancer J. Clin.* **2018**, *68*, 394–424. [CrossRef] [PubMed]
3. Shingaki, S.; Takada, M.; Sasai, K.; Bibi, R.; Kobayashi, T.; Nomura, T.; Saito, C. Impact of lymph node metastasis on the pattern of failure and survival in oral carcinomas. *Am. J. Surg.* **2003**, *185*, 278–284. [CrossRef]
4. Iyer, N.G.; Tan, D.S.; Tan, V.K.; Wang, W.; Hwang, J.; Tan, N.C.; Sivanandan, R.; Tan, H.K.; Lim, W.T.; Ang, M.K.; et al. Randomized trial comparing surgery and adjuvant radiotherapy versus concurrent chemoradiotherapy in patients with advanced, nonmetastatic squamous cell carcinoma of the head and neck: 10-year update and subset analysis. *Cancer* **2015**, *121*, 1599–1607. [CrossRef] [PubMed]
5. Jerjes, W.; Upile, T.; Petrie, A.; Riskalla, A.; Hamdoon, Z.; Vourvachis, M.; Karavidas, K.; Jay, A.; Sandison, A.; Thomas, G.J.; et al. Clinicopathological parameters, recurrence, locoregional and distant metastasis in 115 T1-T2 oral squamous cell carcinoma patients. *Head Neck Oncol.* **2010**, *2*, 9. [CrossRef]
6. Baillie, R.; Tan, S.T.; Itinteang, T. Cancer Stem Cells in Oral Cavity Squamous Cell Carcinoma: A Review. *Front. Oncol.* **2017**, *7*, 112. [CrossRef] [PubMed]
7. Reya, T.; Morrison, S.J.; Clarke, M.F.; Weissman, I.L. Stem cells, cancer, and cancer stem cells. *Nature* **2001**, *414*, 105–111. [CrossRef]
8. Chiou, S.H.; Yu, C.C.; Huang, C.Y.; Lin, S.C.; Liu, C.J.; Tsai, T.H.; Chou, S.H.; Chien, C.S.; Ku, H.H.; Lo, J.F. Positive correlations of Oct-4 and Nanog in oral cancer stem-like cells and high-grade oral squamous cell carcinoma. *Clin. Cancer Res.* **2008**, *14*, 4085–4095. [CrossRef] [PubMed]
9. Mani, S.A.; Guo, W.; Liao, M.J.; Eaton, E.N.; Ayyanan, A.; Zhou, A.Y.; Brooks, M.; Reinhard, F.; Zhang, C.C.; Shipitsin, M.; et al. The epithelial-mesenchymal transition generates cells with properties of stem cells. *Cell* **2008**, *133*, 704–715. [CrossRef]
10. Gupta, P.B.; Pastushenko, I.; Skibinski, A.; Blanpain, C.; Kuperwasser, C. Phenotypic Plasticity: Driver of Cancer Initiation, Progression, and Therapy Resistance. *Cell Stem Cell* **2019**, *24*, 65–78. [CrossRef] [PubMed]
11. Visvader, J.E.; Lindeman, G.J. Cancer stem cells in solid tumours: Accumulating evidence and unresolved questions. *Nat. Rev. Cancer* **2008**, *8*, 755–768. [CrossRef] [PubMed]
12. Eramo, A.; Lotti, F.; Sette, G.; Pilozzi, E.; Biffoni, M.; Di Virgilio, A.; Conticello, C.; Ruco, L.; Peschle, C.; De Maria, R. Identification and expansion of the tumorigenic lung cancer stem cell population. *Cell Death Differ.* **2008**, *15*, 504–514. [CrossRef]
13. Maitland, N.J.; Collins, A.T. Prostate cancer stem cells: A new target for therapy. *J. Clin. Oncol.* **2008**, *26*, 2862–2870. [CrossRef] [PubMed]
14. Li, C.; Heidt, D.G.; Dalerba, P.; Burant, C.F.; Zhang, L.; Adsay, V.; Wicha, M.; Clarke, M.F.; Simeone, D.M. Identification of pancreatic cancer stem cells. *Cancer Res.* **2007**, *67*, 1030–1037. [CrossRef] [PubMed]
15. Han, J.; Fujisawa, T.; Husain, S.R.; Puri, R.K. Identification and characterization of cancer stem cells in human head and neck squamous cell carcinoma. *BMC Cancer* **2014**, *14*, 173. [CrossRef]
16. Pohl, A.; Lurje, G.; Kahn, M.; Lenz, H.J. Stem cells in colon cancer. *Clin. Color. Cancer* **2008**, *7*, 92–98. [CrossRef]
17. Batlle, E.; Sancho, E.; Franci, C.; Dominguez, D.; Monfar, M.; Baulida, J.; Garcia De Herreros, A. The transcription factor snail is a repressor of E-cadherin gene expression in epithelial tumour cells. *Nat. Cell Biol.* **2000**, *2*, 84–89. [CrossRef] [PubMed]
18. Cano, A.; Perez-Moreno, M.A.; Rodrigo, I.; Locascio, A.; Blanco, M.J.; del Barrio, M.G.; Portillo, F.; Nieto, M.A. The transcription factor snail controls epithelial-mesenchymal transitions by repressing E-cadherin expression. *Nat. Cell Biol.* **2000**, *2*, 76–83. [CrossRef]
19. Baniebrahimi, G.; Mir, F.; Khanmohammadi, R. Cancer stem cells and oral cancer: Insights into molecular mechanisms and therapeutic approaches. *Cancer Cell Int.* **2020**, *20*, 113. [CrossRef]
20. Fensterl, V.; Sen, G.C. The ISG56/IFIT1 gene family. *J. Interferon Cytokine Res.* **2011**, *31*, 71–78. [CrossRef]
21. Diamond, M.S.; Farzan, M. The broad-spectrum antiviral functions of IFIT and IFITM proteins. *Nat. Rev. Immunol.* **2013**, *13*, 46–57. [CrossRef]
22. Pidugu, V.K.; Pidugu, H.B.; Wu, M.M.; Liu, C.J.; Lee, T.C. Emerging Functions of Human IFIT Proteins in Cancer. *Front. Mol. Biosci.* **2019**, *6*, 148. [CrossRef]
23. Zhou, X.; Michal, J.J.; Zhang, L.; Ding, B.; Lunney, J.K.; Liu, B.; Jiang, Z. Interferon induced IFIT family genes in host antiviral defense. *Int. J. Biol. Sci.* **2013**, *9*, 200–208. [CrossRef] [PubMed]
24. Terenzi, F.; White, C.; Pal, S.; Williams, B.R.; Sen, G.C. Tissue-specific and inducer-specific differential induction of ISG56 and ISG54 in mice. *J. Virol.* **2007**, *81*, 8656–8665. [CrossRef] [PubMed]
25. Pidugu, V.K.; Wu, M.M.; Yen, A.H.; Pidugu, H.B.; Chang, K.W.; Liu, C.J.; Lee, T.C. IFIT1 and IFIT3 promote oral squamous cell carcinoma metastasis and contribute to the anti-tumor effect of gefitinib via enhancing p-EGFR recycling. *Oncogene* **2019**, *38*, 3232–3247. [CrossRef] [PubMed]
26. Lo, U.G.; Bao, J.; Cen, J.; Yeh, H.C.; Luo, J.; Tan, W.; Hsieh, J.T. Interferon-induced IFIT5 promotes epithelial-to-mesenchymal transition leading to renal cancer invasion. *Am. J. Clin. Exp. Urol.* **2019**, *7*, 31–45. [PubMed]
27. Zhao, Y.; Altendorf-Hofmann, A.; Pozios, I.; Camaj, P.; Daberitz, T.; Wang, X.; Niess, H.; Seeliger, H.; Popp, F.; Betzler, C.; et al. Elevated interferon-induced protein with tetratricopeptide repeats 3 (IFIT3) is a poor prognostic marker in pancreatic ductal adenocarcinoma. *J. Cancer Res. Clin. Oncol.* **2017**, *143*, 1061–1068. [CrossRef] [PubMed]

28. Koh, S.Y.; Moon, J.Y.; Unno, T.; Cho, S.K. Baicalein Suppresses Stem Cell-Like Characteristics in Radio- and Chemoresistant MDA-MB-231 Human Breast Cancer Cells through Up-Regulation of IFIT2. *Nutrients* **2019**, *11*, 624. [CrossRef]
29. Chen, L.; Zhai, W.; Zheng, X.; Xie, Q.; Zhou, Q.; Tao, M.; Zhu, Y.; Wu, C.; Jiang, J. Decreased IFIT2 Expression Promotes Gastric Cancer Progression and Predicts Poor Prognosis of the Patients. *Cell. Physiol. Biochem.* **2018**, *45*, 15–25. [CrossRef]
30. Su, W.; Xiao, W.; Chen, L.; Zhou, Q.; Zheng, X.; Ju, J.; Jiang, J.; Wang, Z. Decreased IFIT2 Expression In Human Non-Small-Cell Lung Cancer Tissues Is Associated With Cancer Progression And Poor Survival Of The Patients. *OncoTargets Ther.* **2019**, *12*, 8139–8149. [CrossRef] [PubMed]
31. Lai, K.C.; Liu, C.J.; Chang, K.W.; Lee, T.C. Depleting IFIT2 mediates atypical PKC signaling to enhance the migration and metastatic activity of oral squamous cell carcinoma cells. *Oncogene* **2013**, *32*, 3686–3697. [CrossRef] [PubMed]
32. Regmi, P.; Lai, K.C.; Liu, C.J.; Lee, T.C. SAHA Overcomes 5-FU Resistance in IFIT2-Depleted Oral Squamous Cell Carcinoma Cells. *Cancers* **2020**, *12*, 3527. [CrossRef] [PubMed]
33. Lai, K.C.; Chang, K.W.; Liu, C.J.; Kao, S.Y.; Lee, T.C. IFN-induced protein with tetratricopeptide repeats 2 inhibits migration activity and increases survival of oral squamous cell carcinoma. *Mol. Cancer Res.* **2008**, *6*, 1431–1439. [CrossRef] [PubMed]
34. Zhang, Z.; Li, N.; Liu, S.; Jiang, M.; Wan, J.; Zhang, Y.; Wan, L.; Xie, C.; Le, A. Overexpression of IFIT2 inhibits the proliferation of chronic myeloid leukemia cells by regulating the BCRABL/AKT/mTOR pathway. *Int. J. Mol. Med.* **2020**, *45*, 1187–1194. [PubMed]
35. Chen, L.; Liu, S.; Xu, F.; Kong, Y.; Wan, L.; Zhang, Y.; Zhang, Z. Inhibition of Proteasome Activity Induces Aggregation of IFIT2 in the Centrosome and Enhances IFIT2-Induced Cell Apoptosis. *Int. J. Biol. Sci.* **2017**, *13*, 383–390. [CrossRef] [PubMed]
36. Feng, X.; Wang, Y.; Ma, Z.; Yang, R.; Liang, S.; Zhang, M.; Song, S.; Li, S.; Liu, G.; Fan, D.; et al. MicroRNA-645, up-regulated in human adencarcinoma of gastric esophageal junction, inhibits apoptosis by targeting tumor suppressor IFIT2. *BMC Cancer* **2014**, *14*, 633. [CrossRef] [PubMed]
37. Ohsugi, T.; Yamaguchi, K.; Zhu, C.; Ikenoue, T.; Furukawa, Y. Decreased expression of interferon-induced protein 2 (IFIT2) by Wnt/beta-catenin signaling confers anti-apoptotic properties to colorectal cancer cells. *Oncotarget* **2017**, *8*, 100176–100186. [CrossRef]
38. Lai, K.C.; Liu, C.J.; Lin, T.J.; Mar, A.C.; Wang, H.H.; Chen, C.W.; Hong, Z.X.; Lee, T.C. Blocking TNF-alpha inhibits angiogenesis and growth of IFIT2-depleted metastatic oral squamous cell carcinoma cells. *Cancer Lett.* **2016**, *370*, 207–215. [CrossRef]
39. Du, F.; Liu, H.; Lu, Y.; Zhao, X.; Fan, D. Epithelial-to-Mesenchymal Transition: Liaison between Cancer Metastasis and Drug Resistance. *Crit. Rev. Oncog.* **2017**, *22*, 275–282. [CrossRef]
40. Niess, H.; Camaj, P.; Renner, A.; Ischenko, I.; Zhao, Y.; Krebs, S.; Mysliwietz, J.; Jackel, C.; Nelson, P.J.; Blum, H.; et al. Side population cells of pancreatic cancer show characteristics of cancer stem cells responsible for resistance and metastasis. *Target Oncol.* **2015**, *10*, 215–227. [CrossRef]
41. Hamburger, A.W.; Salmon, S.E. Primary bioassay of human tumor stem cells. *Science* **1977**, *197*, 461–463. [CrossRef]
42. Song, J.; Chang, I.; Chen, Z.; Kang, M.; Wang, C.Y. Characterization of side populations in HNSCC: Highly invasive, chemoresistant and abnormal Wnt signaling. *PLoS ONE* **2010**, *5*, e11456. [CrossRef]
43. Greve, B.; Kelsch, R.; Spaniol, K.; Eich, H.T.; Gotte, M. Flow cytometry in cancer stem cell analysis and separation. *Cytometry A* **2012**, *81*, 284–293. [CrossRef] [PubMed]
44. Hirschmann-Jax, C.; Foster, A.E.; Wulf, G.G.; Nuchtern, J.G.; Jax, T.W.; Gobel, U.; Goodell, M.A.; Brenner, M.K. A distinct "side population" of cells with high drug efflux capacity in human tumor cells. *Proc. Natl. Acad. Sci. USA* **2004**, *101*, 14228–14233. [CrossRef] [PubMed]
45. Nagy, Á.; Munkácsy, G.; Győrffy, B. Pancancer survival analysis of cancer hallmark genes. *Sci. Rep.* **2021**, *11*, 6047. [CrossRef] [PubMed]
46. Wang, F.; Liu, W.; Jiang, Q.; Gong, M.; Chen, R.; Wu, H.; Han, R.; Chen, Y.; Han, D. Lipopolysaccharide-induced testicular dysfunction and epididymitis in mice: A critical role of tumor necrosis factor alphadagger. *Biol. Reprod.* **2019**, *100*, 849–861. [CrossRef] [PubMed]
47. Shrivastava, S.; Steele, R.; Sowadski, M.; Crawford, S.E.; Varvares, M.; Ray, R.B. Identification of molecular signature of head and neck cancer stem-like cells. *Sci. Rep.* **2015**, *5*, 7819. [CrossRef] [PubMed]
48. Reynolds, D.S.; Tevis, K.M.; Blessing, W.A.; Colson, Y.L.; Zaman, M.H.; Grinstaff, M.W. Breast Cancer Spheroids Reveal a Differential Cancer Stem Cell Response to Chemotherapeutic Treatment. *Sci. Rep.* **2017**, *7*, 10382. [CrossRef]
49. Morata-Tarifa, C.; Jimenez, G.; Garcia, M.A.; Entrena, J.M.; Grinan-Lison, C.; Aguilera, M.; Picon-Ruiz, M.; Marchal, J.A. Low adherent cancer cell subpopulations are enriched in tumorigenic and metastatic epithelial-to-mesenchymal transition-induced cancer stem-like cells. *Sci. Rep.* **2016**, *6*, 18772. [CrossRef] [PubMed]
50. Cao, L.; Zhou, Y.; Zhai, B.; Liao, J.; Xu, W.; Zhang, R.; Li, J.; Zhang, Y.; Chen, L.; Qian, H.; et al. Sphere-forming cell subpopulations with cancer stem cell properties in human hepatoma cell lines. *BMC Gastroenterol.* **2011**, *11*, 71. [CrossRef]
51. Zhang, P.; Zhang, Y.; Mao, L.; Zhang, Z.; Chen, W. Side population in oral squamous cell carcinoma possesses tumor stem cell phenotypes. *Cancer Lett.* **2009**, *277*, 227–234. [CrossRef] [PubMed]
52. Yanamoto, S.; Kawasaki, G.; Yamada, S.; Yoshitomi, I.; Kawano, T.; Yonezawa, H.; Rokutanda, S.; Naruse, T.; Umeda, M. Isolation and characterization of cancer stem-like side population cells in human oral cancer cells. *Oral Oncol.* **2011**, *47*, 855–860. [CrossRef] [PubMed]
53. Todoroki, K.; Ogasawara, S.; Akiba, J.; Nakayama, M.; Naito, Y.; Seki, N.; Kusukawa, J.; Yano, H. CD44v3+/CD24− cells possess cancer stem cell-like properties in human oral squamous cell carcinoma. *Int. J. Oncol.* **2016**, *48*, 99–109. [CrossRef] [PubMed]

54. Ghuwalewala, S.; Ghatak, D.; Das, P.; Dey, S.; Sarkar, S.; Alam, N.; Panda, C.K.; Roychoudhury, S. CD44(high)CD24(low) molecular signature determines the Cancer Stem Cell and EMT phenotype in Oral Squamous Cell Carcinoma. *Stem Cell Res* **2016**, *16*, 405–417. [CrossRef]
55. Tsai, L.L.; Hu, F.W.; Lee, S.S.; Yu, C.H.; Yu, C.C.; Chang, Y.C. Oct4 mediates tumor initiating properties in oral squamous cell carcinomas through the regulation of epithelial-mesenchymal transition. *PLoS ONE* **2014**, *9*, e87207. [CrossRef]
56. Vesuna, F.; Lisok, A.; Kimble, B.; Raman, V. Twist modulates breast cancer stem cells by transcriptional regulation of CD24 expression. *Neoplasia* **2009**, *11*, 1318–1328. [CrossRef]
57. Ye, P.; Nadkarni, M.A.; Hunter, N. Regulation of E-cadherin and TGF-beta3 expression by CD24 in cultured oral epithelial cells. *Biochem. Biophys. Res. Commun.* **2006**, *349*, 229–235. [CrossRef]
58. Schneeberger, E.E.; Lynch, R.D. The tight junction: A multifunctional complex. *Am. J. Physiol. Cell Physiol.* **2004**, *286*, C1213–C1228. [CrossRef]
59. Bhat-Nakshatri, P.; Appaiah, H.; Ballas, C.; Pick-Franke, P.; Goulet, R., Jr.; Badve, S.; Srour, E.F.; Nakshatri, H. SLUG/SNAI2 and tumor necrosis factor generate breast cells with $CD44^{+}/CD24^{-}$ phenotype. *BMC Cancer* **2010**, *10*, 411. [CrossRef]
60. Techasen, A.; Namwat, N.; Loilome, W.; Bungkanjana, P.; Khuntikeo, N.; Puapairoj, A.; Jearanaikoon, P.; Saya, H.; Yongvanit, P. Tumor necrosis factor-alpha (TNF-alpha) stimulates the epithelial-mesenchymal transition regulator Snail in cholangiocarcinoma. *Med. Oncol.* **2012**, *29*, 3083–3091. [CrossRef]
61. Zhang, L.; Jiao, M.; Wu, K.; Li, L.; Zhu, G.; Wang, X.; He, D.; Wu, D. TNF-alpha induced epithelial mesenchymal transition increases stemness properties in renal cell carcinoma cells. *Int. J. Clin. Exp. Med.* **2014**, *7*, 4951–4958. [PubMed]
62. Li, J.; Zha, X.M.; Wang, R.; Li, X.D.; Xu, B.; Xu, Y.J.; Yin, Y.M. Regulation of CD44 expression by tumor necrosis factor-alpha and its potential role in breast cancer cell migration. *Biomed. Pharmacother.* **2012**, *66*, 144–150. [CrossRef] [PubMed]
63. Lee, S.H.; Hong, H.S.; Liu, Z.X.; Kim, R.H.; Kang, M.K.; Park, N.H.; Shin, K.H. TNFα enhances cancer stem cell-like phenotype via Notch-Hes1 activation in oral squamous cell carcinoma cells. *Biochem. Biophys. Res. Commun.* **2012**, *424*, 58–64. [CrossRef] [PubMed]
64. Reich, N.C. A death-promoting role for ISG54/IFIT2. *J. Interferon Cytokine Res.* **2013**, *33*, 199–205. [CrossRef]

Article

Deep-Learning-Based Automated Identification and Visualization of Oral Cancer in Optical Coherence Tomography Images

Zihan Yang [1], Hongming Pan [1], Jianwei Shang [2], Jun Zhang [3] and Yanmei Liang [1,*]

[1] Institute of Modern Optics, Tianjin Key Laboratory of Micro-Scale Optical Information Science and Technology, Nankai University, Tianjin 300350, China

[2] Department of Oral Pathology, Tianjin Stomatological Hospital, Hospital of Stomatology, Nankai University, Tianjin 300041, China

[3] Department of Oral-Maxillofacial Surgery, Tianjin Stomatological Hospital, Hospital of Stomatology, Nankai University, Tianjin 300041, China

* Correspondence: ymliang@nankai.edu.cn

Abstract: Early detection and diagnosis of oral cancer are critical for a better prognosis, but accurate and automatic identification is difficult using the available technologies. Optical coherence tomography (OCT) can be used as diagnostic aid due to the advantages of high resolution and non-invasion. We aim to evaluate deep-learning-based algorithms for OCT images to assist clinicians in oral cancer screening and diagnosis. An OCT data set was first established, including normal mucosa, precancerous lesion, and oral squamous cell carcinoma. Then, three kinds of convolutional neural networks (CNNs) were trained and evaluated by using four metrics (accuracy, precision, sensitivity, and specificity). Moreover, the CNN-based methods were compared against machine learning approaches through the same dataset. The results show the performance of CNNs, with a classification accuracy of up to 96.76%, is better than the machine-learning-based method with an accuracy of 92.52%. Moreover, visualization of lesions in OCT images was performed and the rationality and interpretability of the model for distinguishing different oral tissues were evaluated. It is proved that the automatic identification algorithm of OCT images based on deep learning has the potential to provide decision support for the effective screening and diagnosis of oral cancer.

Keywords: optical coherence tomography; oral cancer; identification; deep learning; machine learning

Citation: Yang, Z.; Pan, H.; Shang, J.; Zhang, J.; Liang, Y. Deep-Learning-Based Automated Identification and Visualization of Oral Cancer in Optical Coherence Tomography Images. *Biomedicines* **2023**, *11*, 802. https://doi.org/10.3390/biomedicines11030802

Academic Editors: Vui King Vincent-Chong and Wolfgang J. Weninger

Received: 18 January 2023
Revised: 15 February 2023
Accepted: 4 March 2023
Published: 6 March 2023

1. Introduction

Oral cancer is one of the most common cancers in the head and neck [1]. In terms of the pathogenesis of oral cancer, the predominant type of oral cancer is oral squamous cell carcinoma (OSCC) with a long preclinical stage [2]. In addition, precancerous lesions (oral potentially malignant disorder), such as homogeneous leukoplakia and nonhomogeneous leukoplakia, are at risk of malignant transformation [3]. Despite the advancement in targeted cancer therapy, survival rates for oral cancer have remained flat over the last 50 years [4]. Fortunately, the patient's survival can be improved if the OSCC can be detected and diagnosed early for appropriate treatments [5]. The study indicated that the 5-year survival rate can increase from less 30% to 83% with early detection [6]. Therefore, it is critical that oral cancer can be diagnosed and treated in the pre- or early cancerous stages.

The conventional visual examination is the most commonly screening procedure for oral lesions, but its sensitivity and specificity vary greatly [7]. Auxiliary methods, such as, toluidine blue, auto-fluorescence, or non-linear microscopy have been studied [8–12]. However, there are some limitations, such as the safety assessment of chemiluminescence methods, the lack of three-dimensional (3D) information of fluorescence, or the limited field of view and depth of microscopic methods. While histopathology is still the gold standard, this processing is invasive and time-consuming.

The study has shown that the thickness of oral mucosa (epithelium and lamina propria) is less than 1 mm [13]. For oral cavity imaging with microscopic techniques, their penetration depth is limited, which may not be deep enough to investigate the existence of basement membrane. In contrast, optical coherence tomography (OCT) has the advantages of high-resolution (1–20 μm), real-time and large-depth (1–2 mm) imaging which is suitable for imaging oral mucosa. OCT has been applied in biomedical fields since it was first introduced in 1991 [14], such as ophthalmology [15], cardiology [16], gastroenterology [17], and dermatology [18]. In the oral cavity, studies based on OCT have been attempted to differentiate benign and OSCC by different structural or optical indicators, including the thickness of the epithelium, the intactness of basement membrane, or optical scattering properties [19–21]. It has been proved that OCT can enable imaging of oral mucosa and identification of the morphological structures.

Automatic image recognition and classification play an important role in biomedicine. To identify oral lesions automatically, texture feature-based methods were proposed. Krishnan et al. made use of high-order spectra, local binary pattern and laws texture energy from histopathological images to identify oral sub-mucous fibrosis [22]. Thomas et al. used the grey level co-occurrence matrix and grey level run-length for classification of oral cancer in digital camera images [23]. Recently, our laboratory has studied the use of texture features to distinguish salivary gland tumors [24], as well as OSCC [25] in OCT images.

In addition, deep learning has been surprisingly successful in recent years [26–28]. In the field of biomedicine, deep learning has been developed for disease classification, object segmentation and image enhancement. Aubreville et al. presented and evaluated an automatic approach for OSCC diagnosis using deep learning on confocal laser endomicroscopy images [29]. Welikala et al. assessed two deep-learning-based computer vision approaches for the automated detection and classification of oral lesions in photographs [30]. However, there is no research on deep-learning-based automatic recognition of oral cancer in OCT images.

The goal of this study is to explore the potential of automatic recognition of oral cancer based on deep learning in OCT images and evaluate the effectiveness by identifying precancerous and cancerous tissues. In addition, feature visualization is also studied to evaluate the rationality and interpretability of the network. It has the great potential to assist clinicians in screening and diagnosis of oral cancer and precancerous lesions.

2. Materials and Methods

2.1. Sample Preparation and Data Acquisition

Fresh tissue samples investigated in this study were obtained from the Tianjin Stomatology Hospital, China. All procedures performed in this study were in accordance with the ethical standards of the Ethics Committee of Tianjin Stomatological Hospital. These samples came from 19 patients who were diagnosed with oral diseases, including leukoplakia with hyperplasia (LEH) and OSCC. The normal and diseased oral tissues were sequentially scanned, and then were fixed and stained with H&E. The slices were evaluated by an experienced pathologist. The details about the OCT system and imaging protocol were described in the previous work [31,32].

2.2. Establishment of the Data Set

Different morphological features of oral tissues were marked in Figure 1. Figure 1a,d show the OCT image and the corresponding histopathologic image of normal mucosa. It can be found that the epithelium (EP) and the lamina propria (LP) are clearly distinguishable due to the different optical scattering intensity, which corresponds well to the histopathological image (Figure 1d). The boundary of EP and LP is called the basement membrane (BM), as shown by the white dashed curve. The typical OCT image and the corresponding histopathologic image of LEH are shown in Figure 1b,e. We can see a boundary (BM) similar to that of normal mucosa from Figure 1b,e. It is worth noting that the thickness of EP is increased and the stratum corneum (SC) can also be observed. In contrast, the epithelial cells of OSCC proliferate maliciously, resulting in the destruction of BM. Moreover, due to the aggregation of cancer cells, the distribution of optical scattering signal appears as cord-like in the OCT image, as indicated by the red arrows in Figure 1c,f.

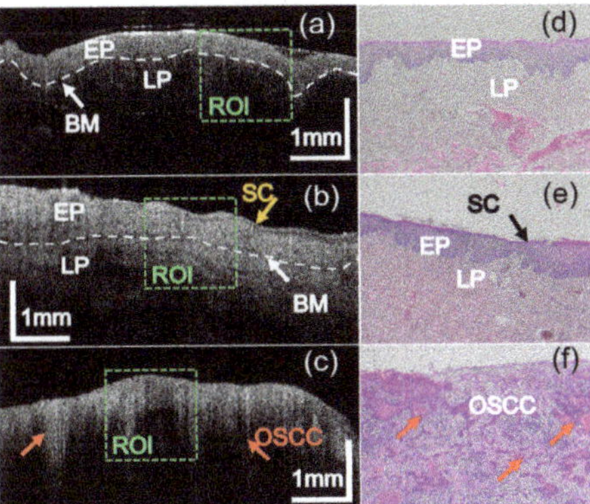

Figure 1. Morphological characteristics and statistical analysis of oral tissues. The representative OCT images of normal mucosa (**a**), LEH (**b**), and OSCC (**c**) and corresponding histopathological images (**d–f**). The ROI indicates 256 × 256 pixels.

According to the above analysis of morphological characteristics of different oral tissues in OCT images, OCT images matched with the histopathological images were manually segmented to the appropriate size (256 × 256 pixels) as the regions of interest (ROIs), which contain information unique to different tissues, as shown in the green square boxes in Figure 1. After the segmentation, a total of 13,799 OCT images of ROIs were used to establish the data set. In order to avoid data bias, we randomly selected OCT images from some patients for training and others for test, as described in detail in Table 1.

Table 1. Information of the patients and the partitioning of the data set.

Dataset	Normal *	LEH	OSCC	Total
Patients' number	-	5	14	19
Age (median [range])	-	62 (37–73)	60 (29–69)	
Gender (male/female)	-	3/2	7/7	10/9
Training set				
Patients' number	-	3	10	13
OCT images	2151	3639	3947	9737
Test set				
Patients' number	-	2	4	6
OCT images	1043	1601	1418	4062

* OCT images of normal mucosa were captured from the normal part of the abnormally excised tissues. The normal area was determined to be at least 1 cm away from abnormal area under the guidance of an experienced surgeon.

2.3. CNN Architecture

Three CNNs, including LeNet-5, VGG16, and ResNet18, were used for the classification and identification of these oral tissues (Figure S1). As one of the most basic and earliest proposed deep learning networks, LeNet-5 has a simple network structure and a small number of parameters [33]. There are two convolution layers and three fully connected layers in LeNet-5. The rectified linear unit (ReLU) activation function and max pooling operation are used after each convolutional layer. VGG16 is composed of 13 convolution layers and 3 full connection layers, in which ReLU is used as activation functions after every two convolutional layer and full connected layer [34]. VGG16 changed the convolution mode, set multiple convolution kernels, increased the channel, and reduced the matrix width and height through pooling. With the deepening of layers of VGG16, the amount of computation increases. ResNet18 consists of 18 layers with weights, including the convolutional layers and the fully connected layers. ResNet18 avoided the vanishing gradient and reduced computation amount by skip connections [35].

2.4. Training and Classification

Figure 2 is the flowchart of our experiment. CNNs were firstly trained by using random initialization parameters. To reduce the risk of overfitting, 10-fold cross-validation was performed. Then, the independent test set (never seen by the network before) was used to test the classification performance of different CNNs. In this study, the CNNs were implemented under the PyTorch framework. The batch size is set to 32, the cross-entropy is used as the loss function, and Adam is used as the optimizer with a learning rate of 0.0001, momentum of 0.9.

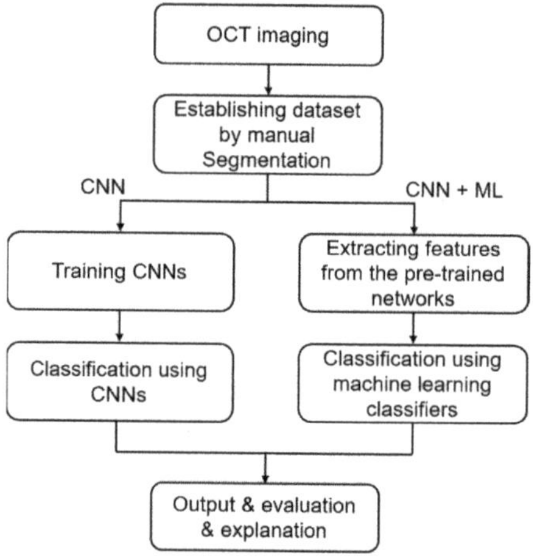

Figure 2. Flowchart of the oral tissue classification experiment.

Further, considering that CNN can perform feature extraction, we used CNN as the feature extractor and machine learning (ML) as the classifier to evaluate the effectiveness in oral tissues classification (CNN + ML). Here, the features were extracted from the last layer before the classification layer (the last fully connected layer) of the pre-trained networks. After that, the feature dimensionality reduction was carried out by the principal component analysis (PCA) algorithm [25]. Finally, three kinds of ML classifiers were used, including decision tree (DT), random forest (RF), and support vector machine (SVM) [25].

In this method, transfer learning was applied and the networks were trained on the ImageNet dataset in advance. Transfer learning can transfer the acquired powerful skills to relevant problems, thus saving time and computing costs [36]. For the classifiers we used, DT model is a kind of tree structure, which is composed of a series of nodes, and each node represents a feature. RF is an algorithm that integrates multiple decision trees through ensemble learning. The random vector is used to generate the ensemble of trees and control the growth of each tree in the ensemble, which can significantly improve the classification accuracy. The number and the depth of tree nodes were used to optimize the best results. Multi-class SVM classifiers with Gaussian radial basis function as the kernel function were employed and the non-linear decision boundary was obtained. The penalty factors C and gamma were optimized for SVM.

In order to apply this method to a common scenario, the algorithms were executed on a desktop computer with an eight-core Intel Xeon 3.5 GHz (E5-1620) processor and a 24 GB random-access memory using the Python programming software (Version 3.7.3).

2.5. Evaluation Indicators

To evaluate the performance of different CNNs and approaches in distinguishing oral tissues, four metrics including sensitivity (*Sen*), specificity (*Spe*), precision (*Pre*), and accuracy (*Acc*) were calculated.

$$Sen = \frac{TP}{TP + FN} \tag{1}$$

$$Spe = \frac{TN}{TN + FP} \tag{2}$$

$$Pre = \frac{TP}{TP + FP} \tag{3}$$

$$Acc = \frac{TP + TN}{TP + FP + TN + FN} \tag{4}$$

TP: true positives, *FP*: false positives, *FN*: false negatives, *TN*: true negatives

In addition, receiver operating characteristic curves (ROCs) were plotted and areas under ROC (AUCs) were also calculated. ROCs and AUCs can be used to describe the classification performance of models objectively. The degree of convergence of the networks was determined by the loss value obtained by the loss function (cross entropy loss).

2.6. Visualization

To more directly display classification performance, the predictions of the CNNs were calculated. We extracted 384 overlapping patches from each image. Each patch was input the trained network one by one, and the prediction results were visualized using pseudo-color map.

To enhance interpretability of networks, gradient weighted class activation mapping (Grad-CAM) technique was used to highlight the important regions in the OCT images of oral tissues, which creates the visual explanation for CNNs and helps determine more information about the models when performing detection or prediction work [37].

3. Results

3.1. Identification Using CNN Alone

All the CNNs were trained and tested using PyTorch, which is a deep learning framework enabling fast implementation. After 40 epochs of training, loss values of three kinds of CNNs tend to converge (Figure S2). The results of three CNNs using 10-fold cross-validation were shown in Table S1 and the accuracy of the CNN models was verified.

The performances of three kinds of CNNs were presented in Figure 3a–c, respectively. It is observed that three kinds of CNNs are capable of distinguishing each type of tissue, especially for LEH (AUC = 0.99 for all CNNs). The classification accuracies were further calculated, as shown in Figure 3d. For LeNet-5, the classification accuracies of LEH, normal mucosa, and OSCC are 99.56%, 97.51%, and 93.37%, respectively. For VGG16, the accuracy of each type of tissue is 97.87%, 99.77%, and 82.79%, respectively. For ResNet18, the accuracy of each class is 99.87%, 99.32%, and 77.01%, respectively. The overall accuracies of using LeNet-5, VGG16, and ResNet18 are 96.76%, 91.94%, and 90.43%, respectively.

3.2. Identification Using CNN + ML

Given that ML-based methods often require manual feature extraction and feature selection, it brings about contingency and inconvenience for accurate recognition. To address these issues, we used CNNs as feature extractor, and then used ML as classifier to identify different tissues in OCT images. Figure 4 shows the performance of three classifiers after feature extraction using different networks. For LeNet-5 as feature extractor, the overall accuracies of SVM, DT, and RF are 92.52%, 88.23%, and 91.53%, respectively. For VGG16 as feature extractor, the overall accuracies of SVM, DT, and RF are 91.33%, 89.42%, and 90.52%, respectively. For ResNet18 as feature extractor, the overall accuracies of SVM,

DT, and RF are 89.51%, 90.12%, and 91.01%, respectively. The corresponding ROC curves show that CNNs combined with SVM can obtain the best results (Figure S3).

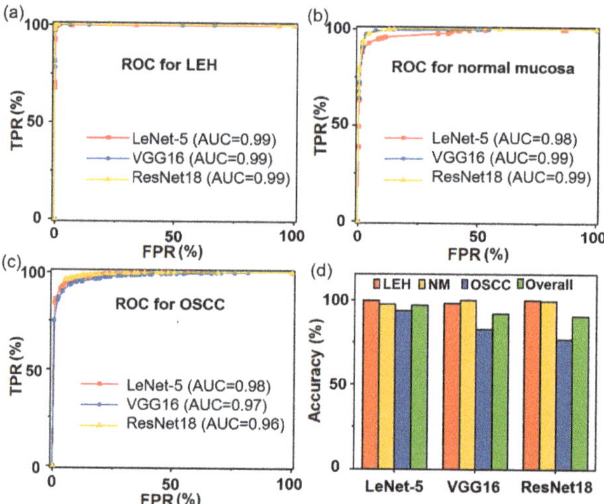

Figure 3. The classification evaluation of three kinds of CNNs. (**a**–**c**) are the ROCs and AUCs of distinguishing three types of tissues using these three CNNs, respectively. (**d**) is the classification accuracies of three types of tissues with three types of CNNs. TPR: true positive rate, FPR: false positive rate.

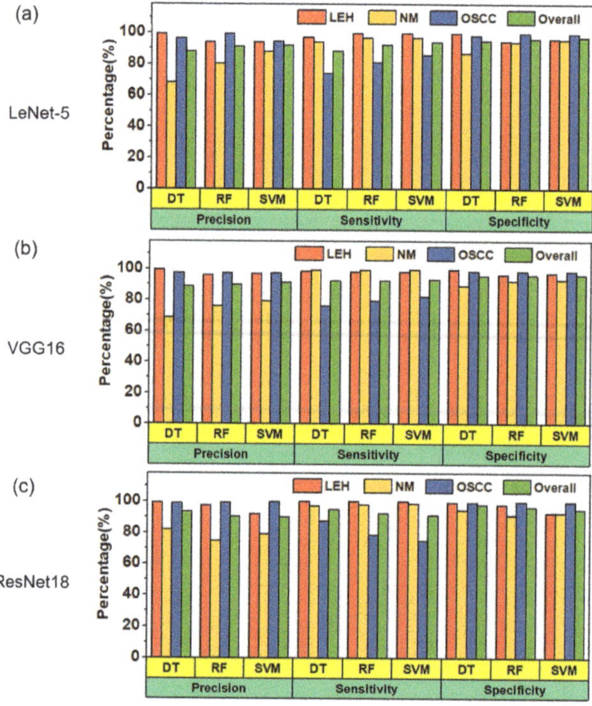

Figure 4. The performance comparison of the three classifiers (DT, RF, and SVM) when using LeNet-5 (**a**), VGG16 (**b**), and ResNet18 (**c**) as a feature extractor, respectively.

To further evaluate the best strategy, the comparison of using CNNs as feature extraction and SVM as classifier is shown in Figure 5. As a whole, it can be found that SVM combined with LeNet-5 achieved best results, whose overall classification accuracy is 92.52%. Accordingly, the precision, sensitivity, and specificity of identifying normal mucosa, LEH, and OSCC are shown in Table 2.

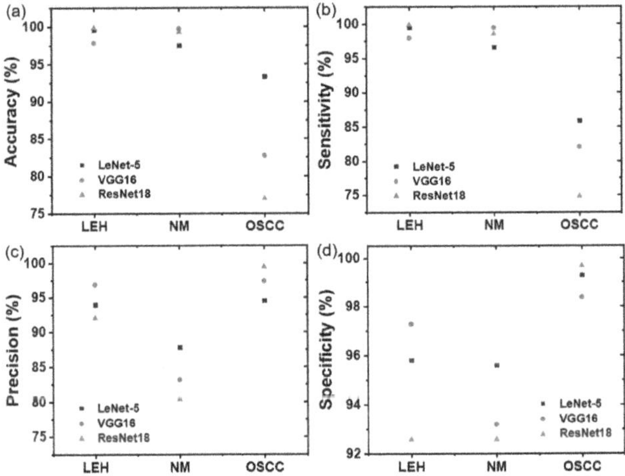

Figure 5. The performance comparison of using SVM as classifier and CNNs as feature extractor. Accuracy (**a**), sensitivity (**b**), precision (**c**), and specificity (**d**) of classification models using LeNet-5, VGG16, and ResNet18 as feature extractor, respectively.

Table 2. The precision, sensitivity, and specificity of identifying NM, LEH, and OSCC using SVM combined with LeNet-5.

Parameter	NM	LEH	OSCC
Precision (%)	87.8	94.0	94.5
Sensitivity (%)	90.7	99.5	86.0
Specificity (%)	95.6	95.8	97.3

3.3. Performance Evaluation of Two Strategies

Two classification strategies including the use of CNN alone and the use of CNN combined with ML were evaluated from accuracy (Table 3). It can be found that if only CNN is used, LeNet-5 obtained the highest accuracy of 96.76%; if CNN + ML was used, LeNet-5 combined with SVM achieved the highest accuracy of 92.52%. Therefore, the evaluation between two best strategies was implemented. The confusion matrices were shown in Figure S4. In addition, the statistics analysis of two best strategies was performed (Figure S5). Based on the two-sample student's t test, there was a statistical significance between the accuracies of LeNet-5 (CNN only) and LeNet-5 combined with SVM (CNN + ML) at $p < 0.05$.

Table 3. Overall accuracy (%) of the two classification strategies.

Model	Classifier	LeNet-5	VGG16	ResNet18
CNN alone	-	96.76	91.94	90.43
CNN + ML	DT	87.23	89.42	90.51
	RF	91.53	90.52	90.01
	SVM	92.52	91.33	89.51

In addition, the training time spent on both strategies was also assessed (Table 4). For using CNN alone, due to the difference in the number of network structure and parameters, the average time of each epoch of training LeNet-5 is much less than that of training VGG16, and ResNet18. Similarly, LeNet-5 need the least time for the network to converge. For CNN + ML, it took less time to train ML classifiers, although it required extracting features from the CNNs.

Table 4. Time of training CNNs and machine learning classifiers.

Model	LeNet-5	VGG16	ResNet18
CNN alone			
Each epoch/s	228	2891	1618
Convergence/s	9120	115,640	64,720
CNN + ML			
Feature extraction/s	86	710	481
DT/s	0.57	7.12	0.88
RF/s	0.27	1.25	0.29
SVM/s	15	22	1.56

3.4. Predictive Visualization

Figure 6 shows the predictive visualization results at the junction between normal mucosa and OSCC using a trained CNN model. Figure 6a shows the imaging area in the photograph of the excised sample. According to the histopathological image (Figure 6b), the normal area and the cancerous area are located on the left and the right sides of the image, respectively. From the corresponding OCT image, as shown in Figure 6c, there is a slight distinction between normal and cancerous regions. We can see that there is a slight BM structure on the left, but not on the right. After the patches are input into the network, the differences between the left and the right of the predicted results can be clearly seen in Figure 6d and are consistent with the histopathological image.

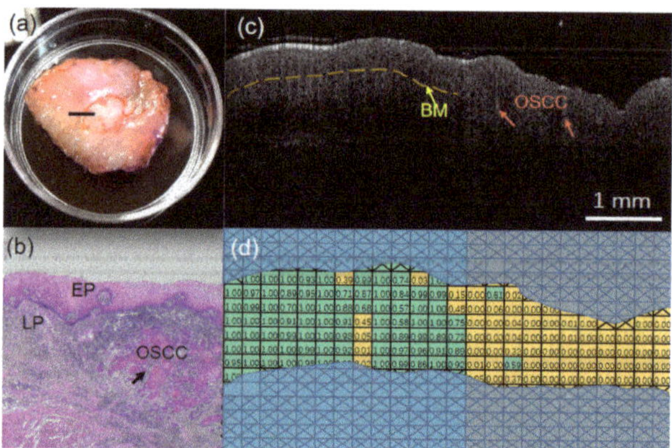

Figure 6. OCT imaging and prediction results at the junction between normal mucosa and OSCC. (a) is a photograph of the excised tissue. (b) is the corresponding histopathological image with the normal region on the left and the cancerous region on the right. (c) is the OCT image at the black line of (a). (d) is the corresponding prediction visualization.

3.5. Grad-CAM Visualization

The interpretability of neural networks using Grad-CAM was also evaluated for making efficient and confident decisions. As shown in Figure 7, different oral tissues showed different characteristics in a trained CNN model. Figure 7a is an OCT image of normal mucosa and the corresponding activation map is shown in Figure 7d. It can be found that the network primarily extracts the EP. The OCT image and corresponding activation map of LEH are shown in Figure 7b,e. It can be found that the thickened EP and LP are highlighted. The OCT image of OSCC is shown in Figure 7c. The cord-like morphological structures can be seen in the OCT image due to the accumulation of cancer nests. From Figure 7f, the highlighted area is mainly the aggregation area of cancer cells, and the neural network pays more attention to this area (the area below indicated by the yellow dotted line). These results are consistent with histological findings as described above. It demonstrated that the network learned different characteristics of oral lesions to distinguish each type of oral tissues.

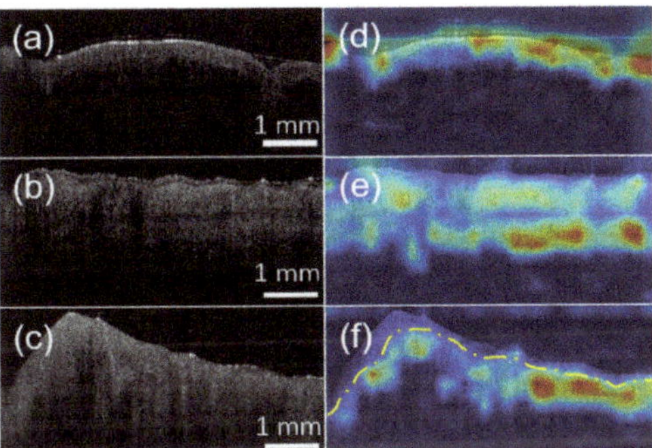

Figure 7. Visualization on OCT images of oral tissues using Grad-CAM. (**a**–**c**) are representative OCT images of normal mucosa, LEH, and OSCC, respectively. (**d**–**f**) are the corresponding activation maps with unique feature aggregations, respectively.

4. Discussion

We studied deep-learning-based identification of oral precancerous and cancerous lesions in this paper. Firstly, three basic kinds of CNNs were trained and evaluated based on oral OCT image data sets that includes normal mucosa, LEH, and OSCC. Next, to avoid the contingency and inconvenience of traditional machine learning methods when extracting features manually, CNNs were used as the feature extractors. DT, RF, and SVM were trained by using the activations of the last layer before the classification layer. Both of two strategies obtained excellent classification results. In addition, the performance of the networks was further verified by feature visualization in OCT images.

Compared to traditional ML methods, deep learning reduces the dependence on feature extraction. The methods used in deep learning are substantially effective to describe the characteristics of images than texture features. Through comparison on the same dataset, it can be found that using CNN alone especially using LeNet-5 obtained better classification results than that of using CNN + ML, whereas the training time of the former about several hundreds of times longer than that of the latter.

To speed up network training, and expand the data set, we segmented the entire image into a certain size. The images were split according to the following segmentation criteria:

(1) According to the requirements of the network on the input size, the size of ROI is determined as 256 × 256 pixels, which can speed up network training compared to the whole image being input.

(2) Each ROI must contain the unique characteristics of oral tissue, for example, the epithelium and lamina propria must be included for normal tissue. It can be found that the size of 256 × 256 pixels can not only include the features of oral tissue, but also effectively reduce the interference of background area.

(3) In order to effectively use the information in the image, we selected ROI areas in an overlapping approach.

(4) Areas with poor image quality were discarded, such as areas without focus due to large fluctuation of tissue surface.

Similar to the data augmentation used in conventional deep learning, the ROIs we obtained were used to make the network learn invariable features and prevent the network from learning irrelevant features, thus improving network performance.

According to the high-performance model, we evaluated the oral tissue OCT images to automatically predict and visualize the lesions, which will be more in line with the actual needs and suitable for intuitive judgement between normal and cancerous areas.

In addition, neural networks are often seen as black boxes in disease screening because they provide only the final diagnosis of the subject without any details of the basis of the diagnosis, which brings a major challenge to the application of artificial intelligence in clinical devices. In our study, the Grad-CAM was used to visualize the important regions in the oral tissue OCT images. It can be found that three types of oral tissues showed different characteristics in the deep learning network. Moreover, the aggregation of characteristics can reflect the unique feature of each oral tissue. Through the feature visualization, there is reasonable basis for understanding the model classification and identification.

Although microscopic or histopathological examination of tissue is the gold standard, an accurate result of biopsy may depend on the clinician's experience and confidence, and the selection of biopsy site. A more accurate diagnosis was achieved via multiple site biopsies and larger volume samples [38], which seems to be more important but makes patients more painful for oral precancerous lesions suspected to be malignant transformation.

In addition, intraoperative frozen section biopsy for surgical margin is a routine procedure after oral cancers are resected en bloc. Surgical margins are usually selected according to surgeons' estimate for suspicious sites of inadequate resection, which may result in omission of positive margins.

As auxiliary tools, imaging techniques have become indispensable in clinic, where image identification algorithms play an important role [39]. This study extends our prior work in oral cancer, which demonstrated the feasibility of OCT image-based identification of OSCC and normal mucosa by using optical parameters as markers to establish the optical attenuation model and using texture-based ML models [21,25].

It is noted that the robustness of deep-learning-based identification methods is worth further exploration with different OCT systems. Deep-learning-based identification is performed by extracting differential features from OCT images of oral tissues. That is, the OCT images of different lesions contain differentiated morphological features, which are recorded by the OCT system. In this case, if different OCT systems are employed, such as different wavelengths or different bandwidths, then the acquired OCT images contain different features. Therefore, collaboration between different devices to obtain more data and conduct robust research based on deep learning is the next direction.

Fortunately, using the powerful learning capabilities of deep learning and the advantage of high-speed and high-resolution imaging of swept-source OCT system, it laid a foundation for guiding clinicians to screen and resect tumors in real time accurately.

5. Conclusions

In conclusion, the feasibility and validity of automatic recognition strategies for OSCC based on OCT and deep learning have been demonstrated. The interpretability of disease assessment was further investigated by visualizing network feature maps. It is proved that automatic identification methods combining the powerful learning capabilities of deep learning with the advantages of OCT imaging are feasible, which is expected to provide decision support for effective screening and diagnosis of oral cancer and precancerous tissues.

Supplementary Materials: The following supporting information can be downloaded at: https://www.mdpi.com/article/10.3390/biomedicines11030802/s1, Table S1: The accuracy (%) of identifying oral tissues using 10-fold cross-validation of different CNN models; Figure S1: Schematic architectures of three typical CNNs; Figure S2: The training loss curves of three CNN models; Figure S3: The ROC curves for SVM, DT, and RF as classifiers and LeNet-5, VGG16, and ResNet18 as feature extractors; Figure S4: Confusion matrices of two strategies; Figure S5: Statistical analysis of two strategies based on student's *t* test.

Author Contributions: Conceptualization, Z.Y. and Y.L.; methodology, Y.L.; software, Z.Y. and H.P.; validation, J.S., J.Z. and Y.L.; formal analysis, Z.Y., J.S. and J.Z.; investigation, Y.L. and J.Z.; resources, J.S. and J.Z.; data curation, Z.Y. and J.S.; writing—original draft preparation, Z.Y.; writing—review and editing, Y.L., J.S. and J.Z.; visualization, Z.Y., H.P. and J.S.; supervision, Y.L.; project administration, Y.L.; funding acquisition, Y.L. All authors have read and agreed to the published version of the manuscript.

Funding: This research was funded by the National Natural Science Foundation of China, grant number 61875092, the Science and Technology Support Program of Tianjin, the grant number 17YFZCSY00740, and the Beijing-Tianjin-Hebei Basic Research Cooperation Special Program, grant number 19JCZDJC65300.

Institutional Review Board Statement: The study was conducted in accordance with the Declaration of Helsinki, and approved by the Ethics Committee of Tianjin Stomatological Hospital (ethical code: PH2017-B-003 and approval Date: 30 August 2017).

Informed Consent Statement: Informed consent was obtained from all subjects involved in the study.

Data Availability Statement: Data will be available from the corresponding author upon reasonable request.

Acknowledgments: We thank all participants for their participation.

Conflicts of Interest: The authors declare no conflict of interest.

References

1. Siegel, R.L.; Miller, K.D.; Wagle, N.S.; Jemal, A. Cancer statistics, 2023. *CA A Cancer J. Clin.* **2023**, *73*, 17–48. [CrossRef]
2. Chen, P.H.; Wu, C.H.; Chen, Y.F.; Yeh, Y.C.; Lin, B.H.; Chang, K.W.; Lai, P.Y.; Hou, M.C.; Lu, C.L.; Kuo, W.C. Combination of structural and vascular optical coherence tomography for differentiating oral lesions of mice in different carcinogenesis stages. *Biomed. Opt. Express* **2018**, *9*, 1461–1476. [CrossRef] [PubMed]
3. Amagasa, T.; Yamashiro, M.; Uzawa, N. Oral premalignant lesions: From a clinical perspective. *Int. J. Clin. Oncol.* **2011**, *16*, 5–14. [CrossRef] [PubMed]
4. Warnakulasuriya, S. Global epidemiology of oral and oropharyngeal cancer. *Oral Oncol.* **2009**, *45*, 309–316. [CrossRef]
5. Tiziani, S.; Lopes, V.; Günther, U.L. Early stage diagnosis of oral cancer using 1H NMR—Based metabolomics. *Neoplasia* **2009**, *11*, 269–276, IN7–IN10. [CrossRef]
6. Messadi, D.V. Diagnostic aids for detection of oral precancerous conditions. *Int. J. Oral Sci.* **2013**, *5*, 59–65. [CrossRef]
7. Downer, M.C.; Moles, D.R.; Palmer, S.; Speight, P.M. A systematic review of test performance in screening for oral cancer and precancer. *Oral Oncol.* **2004**, *40*, 264–273. [CrossRef] [PubMed]
8. Lingen, M.W.; Kalmar, J.R.; Karrison, T.; Speight, P.M. Critical evaluation of diagnostic aids for the detection of oral cancer. *Oral Oncol.* **2008**, *44*, 10–22. [CrossRef]
9. Tsai, M.R.; Shieh, D.B.; Lou, P.J.; Lin, C.F.; Sun, C.K. Characterization of oral squamous cell carcinoma based on higher-harmonic generation microscopy. *J. Biophotonics* **2012**, *5*, 415–424. [CrossRef]
10. Kumar, P.; Kanaujia, S.K.; Singh, A.; Pradhan, A. In vivo detection of oral precancer using a fluorescence-based, in-house-fabricated device: A Mahalanobis distance-based classification. *Lasers Med. Sci.* **2019**, *34*, 1243–1251. [CrossRef]
11. Brouwer de Koning, S.G.; Weijtmans, P.; Karakullukcu, M.B.; Shan, C.; Baltussen, E.J.M.; Smit, L.A.; van Veen, R.L.P.; Hendriks, B.H.W.; Sterenborg, H.J.C.M.; Ruers, T.J.M. Toward assessment of resection margins using hyperspectral diffuse reflection imaging (400–1700 nm) during tongue cancer surgery. *Lasers Surg. Med.* **2020**, *52*, 496–502. [CrossRef] [PubMed]
12. Scanlon, C.S.; Van Tubergen, E.A.; Chen, L.-C.; Elahi, S.F.; Kuo, S.; Feinberg, S.; Mycek, M.-A.; D'Silva, N.J. Characterization of squamous cell carcinoma in an organotypic culture via subsurface non-linear optical molecular imaging. *Exp. Biol. Med.* **2013**, *238*, 1233–1241. [CrossRef] [PubMed]
13. Kurtzman, J.; Sukumar, S.; Pan, S.; Mendonca, S.; Lai, Y.; Pagan, C.; Brandes, S. The impact of preoperative oral health on buccal mucosa graft histology. *J. Urol.* **2021**, *206*, 655–661. [CrossRef] [PubMed]
14. Huang, D.; Swanson, E.A.; Lin, C.P.; Schuman, J.S.; Stinson, W.G.; Chang, W.; Hee, M.R.; Flotte, T.; Gregory, K.; Puliafito, C.A. Optical coherence tomography. *Science* **1991**, *254*, 1178–1181. [CrossRef] [PubMed]
15. Takusagawa, H.L.; Hoguet, A.; Junk, A.K.; Nouri-Mandavi, K.; Radhakrishnan, S.; Chen, T.C. Swept-source OCT for evaluating the lamina cribrosa: A report by the american academy of ophthalmology. *Ophthalmology* **2019**, *126*, 1315–1323. [CrossRef] [PubMed]
16. Yonetsu, T.; Bouma, B.E.; Kato, K.; Fujimoto, J.G.; Jang, I.K. Optical coherence tomography-15 years in cardiology. *Circ. J.* **2013**, *77*, 1933–1940. [CrossRef]
17. Tsai, T.H.; Leggett, C.L.; Trindade, A.J.; Sethi, A.; Swager, A.F.; Joshi, V.; Bergman, J.J.; Mashimo, H.; Nishioka, N.S.; Namati, E. Optical coherence tomography in gastroenterology: A review and future outlook. *J. Biomed. Opt.* **2017**, *22*, 121716. [CrossRef]

18. Olsen, J.; Holmes, J.; Jemec, G.B. Advances in optical coherence tomography in dermatology-a review. *J. Biomed. Opt.* **2018**, *23*, 040901. [CrossRef]

19. Tsai, M.-T.; Lee, H.-C.; Lee, C.-K.; Yu, C.-H.; Chen, H.-M.; Chiang, C.-P.; Chang, C.-C.; Wang, Y.-M.; Yang, C.C. Effective indicators for diagnosis of oral cancer using optical coherence tomography. *Opt. Express* **2008**, *16*, 15847–15862. [CrossRef]

20. Adegun, O.K.; Tomlins, P.H.; Hagi-Pavli, E.; McKenzie, G.; Piper, K.; Bader, D.L.; Fortune, F. Quantitative analysis of optical coherence tomography and histopathology images of normal and dysplastic oral mucosal tissues. *Lasers Med. Sci.* **2012**, *27*, 795–804. [CrossRef]

21. Yang, Z.; Shang, J.; Liu, C.; Zhang, J.; Liang, Y. Identification of oral cancer in OCT images based on an optical attenuation model. *Lasers Med. Sci.* **2020**, *35*, 1999–2007. [CrossRef] [PubMed]

22. Krishnan, M.M.; Venkatraghavan, V.; Acharya, U.R.; Pal, M.; Paul, R.R.; Min, L.C.; Ray, A.K.; Chatterjee, J.; Chakraborty, C. Automated oral cancer identification using histopathological images: A hybrid feature extraction paradigm. *Micron* **2012**, *43*, 352–364. [CrossRef]

23. Thomas, B.; Kumar, V.; Saini, S. Texture analysis based segmentation and classification of oral cancer lesions in color images using ANN. In Proceedings of the 2013 IEEE International Conference on Signal Processing, Computing and Control (ISPCC), Solan, India, 26–28 September 2013; pp. 1–5.

24. Yang, Z.; Shang, J.; Liu, C.; Zhang, J.; Liang, Y. Classification of oral salivary gland tumors based on texture features in optical coherence tomography images. *Lasers Med. Sci.* **2021**, *37*, 1139–1146. [CrossRef] [PubMed]

25. Yang, Z.; Shang, J.; Liu, C.; Zhang, J.; Liang, Y. Identification of oral squamous cell carcinoma in optical coherence tomography images based on texture features. *J. Innov. Opt. Health Sci.* **2020**, *14*, 2140001. [CrossRef]

26. Azam, S.; Rafid, A.K.M.R.H.; Montaha, S.; Karim, A.; Jonkman, M.; De Boer, F. Automated detection of broncho-arterial pairs using CT scans employing different approaches to classify lung diseases. *Biomedicines* **2023**, *11*, 133. [CrossRef] [PubMed]

27. Ali, Z.; Alturise, F.; Alkhalifah, T.; Khan, Y.D. IGPred-HDnet: Prediction of immunoglobulin proteins using graphical features and the hierarchal deep learning-based approach. *Comput. Intell. Neurosci.* **2023**, *2023*, 2465414. [CrossRef]

28. Hassan, A.; Alkhalifah, T.; Alturise, F.; Khan, Y.D. RCCC_Pred: A novel method for sequence-based identification of renal clear cell carcinoma genes through DNA mutations and a blend of features. *Diagnostics* **2022**, *12*, 3036. [CrossRef]

29. Aubreville, M.; Knipfer, C.; Oetter, N.; Jaremenko, C.; Rodner, E.; Denzler, J.; Bohr, C.; Neumann, H.; Stelzle, F.; Maier, A. Automatic classification of cancerous tissue in laserendomicroscopy images of the oral cavity using deep learning. *Sci. Rep.* **2017**, *7*, 11979. [CrossRef]

30. Welikala, R.A.; Remagnino, P.; Lim, J.H.; Chan, C.S.; Rajendran, S.; Kallarakkal, T.G.; Zain, R.B.; Jayasinghe, R.D.; Rimal, J.; Kerr, A.R.; et al. Automated detection and classification of oral lesions using deep learning for early detection of oral cancer. *IEEE Access* **2020**, *8*, 132677–132693. [CrossRef]

31. Li, K.; Yang, Z.; Liang, W.; Shang, J.; Liang, Y.; Wan, S. Low-cost, ultracompact handheld optical coherence tomography probe for in vivo oral maxillofacial tissue imaging. *J. Biomed. Opt.* **2020**, *25*, 046003. [CrossRef]

32. Yang, Z.; Shang, J.; Liu, C.; Zhang, J.; Hou, F.; Liang, Y. Intraoperative imaging of oral-maxillofacial lesions using optical coherence tomography. *J. Innov. Opt. Health Sci.* **2020**, *13*, 2050010. [CrossRef]

33. Li, T.; Jin, D.; Du, C.; Cao, X.; Chen, H.; Yan, J.; Chen, N.; Chen, Z.; Feng, Z.; Liu, S. The image-based analysis and classification of urine sediments using a LeNet-5 neural network. *Comput. Methods Biomech. Biomed. Eng. Imaging Vis.* **2020**, *8*, 109–114. [CrossRef]

34. Krishnaswamy Rangarajan, A.; Purushothaman, R. Disease classification in eggplant using pre-trained VGG16 and MSVM. *Sci. Rep.* **2020**, *10*, 2322. [CrossRef] [PubMed]

35. Odusami, M.; Maskeliūnas, R.; Damaševičius, R.; Krilavičius, T. Analysis of features of alzheimer's disease: Detection of early stage from functional brain changes in magnetic resonance images using a finetuned ResNet18 network. *Diagnostics* **2021**, *11*, 1071. [CrossRef]

36. Weiss, K.; Khoshgoftaar, T.M.; Wang, D. A survey of transfer learning. *J. Big Data* **2016**, *3*, 9. [CrossRef]

37. Panwar, H.; Gupta, P.K.; Siddiqui, M.K.; Morales-Menéndez, R.; Bhardwaj, P.; Singh, V. A deep learning and grad-CAM based color visualization approach for fast detection of COVID-19 cases using chest X-ray and CT-Scan images. *Chaos Solitons Fractals* **2020**, *140*, 110190. [CrossRef]

38. Chen, S.; Forman, M.; Sadow, P.M.; August, M. The diagnostic accuracy of incisional biopsy in the oral cavity. *J. Oral Maxillofac. Surg.* **2016**, *74*, 959–964. [CrossRef]

39. Bisht, S.R.; Mishra, P.; Yadav, D.; Rawal, R.; Mercado-Shekhar, K.P. Current and emerging techniques for oral cancer screening and diagnosis: A review. *Prog. Biomed. Eng.* **2021**, *3*, 042003. [CrossRef]

biomedicines

Article

Psychosocial Adjustment Changes and Related Factors in Postoperative Oral Cancer Patients: A Longitudinal Study

Yi-Wei Chen [1,2,†], Ting-Ru Lin [3,†], Pei-Ling Kuo [4], Shu-Chiung Lee [5], Kuo-Feng Wu [6], Tuyen Van Duong [7,8,*] and Tsae-Jyy Wang [1,*]

1   School of Nursing, National Taipei University of Nursing and Health Sciences, Taipei City 112, Taiwan
2   Department of Cardiothoracic Surgery, Hualien Tzu Chi Hospital, Buddhist Tzu Chi Medical Foundation, Hualien City 970, Taiwan
3   Department of Nursing, Cardinal Tien College of Healthcare and Management, New Taipei City 231, Taiwan
4   School of Nursing, National Yang-Ming Chiao Tung University, Taipei City 112, Taiwan
5   Department of Nursing, Taipei Veterans General Hospital, Taipei City 112, Taiwan
6   Department of Nurse-Midwifery and Women Health, National Taipei University of Nursing and Health Science, Taipei City 112, Taiwan
7   School of Nutrition & Health Sciences, Taipei Medical University, Taipei City 110, Taiwan
8   International Master/Ph.D. Program in Medicine, College of Medicine, Taipei Medical University, Taipei City 110, Taiwan
*   Correspondence: tvduong@tmu.edu.tw (T.V.D.); tsaejyy@ntunhs.edu.tw (T.-J.W.)
†   These authors contributed equally to this work.

Abstract: Disease and treatment-related symptoms and dysfunctions can interfere with the psychosocial adjustment of patients with oral cancer. Identifying factors influencing psychosocial maladjustment is important because at-risk individuals can be targeted for early intervention. This prospective longitudinal study investigated psychosocial adjustment changes and associated factors in postoperative oral cancer patients. Data on psychosocial adjustment, facial disfigurement, symptoms, and social support were collected before surgery (T1) at one month (T2), three months (T3), and five months after discharge (T4). Fifty subjects completed the study, and their data were included in the analysis. Psychosocial maladjustment was reported in 50%, 59.2%, 66%, and 62% of subjects at T1, T2, T3, and T4, respectively. The subjects' psychosocial adjustment deteriorated after surgery. Results from generalized estimating equations indicated that financial status, cancer stage, pain, speech problems, social eating problems, and less sexuality were significant predictors of changes in psychosocial adjustment. Patients with insufficient income, stage III/IV cancer, severe pain, speech problems, social eating problems, and less sexuality were at higher risk for postoperative psychosocial maladjustment. Continued psychosocial assessment and appropriate supportive measures are needed to strengthen the psychosocial adjustment of these high-risk groups.

Keywords: psychosocial adjustment; oral cancer; symptoms; facial disfigurement

Citation: Chen, Y.-W.; Lin, T.-R.; Kuo, P.-L.; Lee, S.-C.; Wu, K.-F.; Duong, T.V.; Wang, T.-J. Psychosocial Adjustment Changes and Related Factors in Postoperative Oral Cancer Patients: A Longitudinal Study. *Biomedicines* **2022**, *10*, 3231. https://doi.org/10.3390/biomedicines10123231

Academic Editor: Vui King Vincent-Chong

Received: 6 August 2022
Accepted: 5 December 2022
Published: 12 December 2022

1. Introduction

Oral cancer is Taiwan's fifth leading cause of cancer death [1]. Compared with the worldwide data, Taiwan has a much higher incidence and prevalence of oral cancer [2]. The age-standardized incidence rates per 100,000 people in 2020 were 30.6 and 3.06 for males and females, respectively. The age-standardized mortality rates per 100,000 people in 2020 were 11.48 and 0.95 for males and females, respectively [3]. The five-year relative survival rate was 56.4% and 65.1% for males and females, respectively [4].

Psychosocial adjustment to illness refers to managing intrapsychic and social demands in response to physical disease [5]. Such adjustments may include managing the impact of physical illness on the individual's concerns about health, the work environment, family life, sexual function or relationships, relationships with extended family, social and leisure activities, and disturbing thoughts and feelings about the physical illness [5]. A reasonable psychosocial adjustment will help patients face the psychosocial challenges of oral cancer and its treatment. Oral cancer patients can become long-term survivors [6,7]. However, these survivors often face disease and treatment-related adverse effects, including facial disfigurement, dysphagia, xerostomia, and truisms [7]. These physical sequelae compromise patients' body image, verbal communication, and social interactions and negatively impact patients' psychosocial well-being [8]. Compared with other types of cancer, patients with oral cancers reported more psychosocial issues, including anxiety, depression, social isolation, and work and relationship problems [9]. Understating the factors that influence psychosocial adjustment in patients with oral cancer will help to identify risk groups for psychosocial maladjustment and develop interventions to strengthen patients' psychosocial adjustment.

Facial disfigurement has been described as a state in which a person's facial appearance has been medically severe and persistently damaged [10]. Despite advances in reconstructive surgery, oral cancer surgery can still lead to severe facial disfigurement depending on the location and stage of cancer. Facial disfigurement can negatively impact patients' psychosocial well-being [11,12]. Appearance affects an individual's body image and self-concept. Facial disfigurement makes contact with others difficult and embarrassing. The more concerned patients were about their facial deformities, the worse their body image, and the more likely they were to avoid social activities [12]. The social stigma of disfigurement can also bring psychological distress to the disfigured [13]. Hence, facial disfigurement may negatively affect the psychosocial adjustment of patients with oral cancer.

Patients with oral cancer experienced physical symptoms and side effects during and after treatment, with gradual remission over 3 to 12 months. Common symptoms in patients following oral cancer surgery include pain, dry mouth, sticky saliva, dental problems, and difficulty speaking, chewing, eating, and swallowing [14,15]. These symptoms can negatively impact a patient's psychosocial adjustment. For example, difficulty with swallowing and chewing interferes with a patient's eating, and the patient may avoid social eating to prevent embarrassment during eating. Difficulty speaking causes communication difficulties and interferes with social activities [16], affecting the patient's workability.

Social support is the support individuals can obtain from their social network when needed [17]. Sources of social support include family, friends, neighbors, colleagues, caregivers, etc. It can take the form of emotional support (e.g., trust and caring), informational support (e.g., giving advice), instrumental support (e.g., direct material assistance), or appraisal support (e.g., affirmation) [17]. Social support is an essential resource for coping with stress and psychosocial adjustment. Past research has found that social support can help cancer patients manage psychological stress, reduce anxiety and depression, and improve their quality of life [18].

In summary, previous studies showed that oral cancer patients had poor psychosocial adjustment. Facial disfigurement, symptoms, and poor social support negatively impact oral cancer patients' psychosocial adjustment. However, most of these findings were from cross-sectional studies. Few previous studies have investigated the changes in psychosocial adjustment over time in oral cancer patients. Oral cancer patients face different psychological and social challenges after reconstructive surgeries. Information on postoperative psychosocial adjustment changes and their influencing factors can help identify high-risk groups of maladjustment and develop appropriate measures to enhance psychosocial adjustment in patients with oral cancer.

Therefore, the study-specific aims were: (1) to describe the changes in psychosocial adjustment after oral cancer surgery; (2) to explore the effects of demographics, disease characteristics, facial disfigurement, social support, and symptoms on psychosocial adjustment.

2. Materials and Methods

2.1. Study Design and Subject Recruitment

This prospective longitudinal study was conducted from 2010 to 2013. Oral cancer patients who met the following eligibility criteria were recruited from the oral and maxillofacial surgical wards or the otolaryngology wards of two hospitals in Taiwan. One is a 3000-bed general hospital in Taipei, and the other is a 1000-bed general hospital in Hualien (eastern Taiwan). The inclusion criteria were (a) 20 years of age or older, (b) scheduled for reconstructive surgery for oral cancer, and (c) able to read Chinese. The exclusion criteria were: (a) with recurrence of oral cancer, (b) previously received reconstructive surgery for oral cancer, or (c) diagnosed with psychiatric illness. All subjects gave written informed consent before participating in the study.

2.2. Sample Size Determination

The required sample size was estimated using G-Power version 3.1 (Heinrich-Heine-Universität Düsseldorf, Düsseldorf, Germany). There have been no previous reports on the impacts of the study variables on postoperative psychosocial adjustment. The required sample size was estimated using the medium effect size suggested by Cohen [19]. The input parameters are an F-test, four repeated measures, a within-factor design, a correlation of 0.5 among repeated measures, a medium effect size (f = 0.25), a power of 80%, and a significance level of 0.05. A sample of 24 was required to analyze psychosocial adjustment changes over time. To explore potential predictors of psychosocial adjustment, an estimated 45 subjects were required. Given the longitudinal nature of the study, a dropout rate of 25% was estimated. Therefore, a sample of 62 subjects was recruited to overcome potential dropout issues. The final sample in the analysis included 50 patients.

2.3. Data Collection

Data collection occurred in each patient's room (for baseline data) and a quiet room at outpatient clinics (for follow-up data). One of the investigators (Y.-W.C.) and a research assistant collected data from each subject using self-reported questionnaires. The research assistant with a bachelor's degree in public health was trained in research protocols and data collection procedures. Subjects self-administered the study questionnaire. For subjects who had difficulty reading or comprehending the questionnaire, the data collector read each question to the subjects. Data on psychosocial adjustment, symptoms, and social support were collected before surgery (T1) and one month (T2), three months (T3), and five months after hospital discharge (T4). Data on demographics, disease variables, and facial disfigurement were collected at T1 only.

2.4. Instruments

In this study, demographics, disease characteristics, facial disfigurement, social support, and symptoms were the independent variables, and psychosocial adjustment was the dependent variable. Demographic data were collected from each subject, including age, gender, education level, marital status, employment status, and financial status. Disease characteristics, including cancer location, cancer stage, and adjuvant therapy, were collected from each subject's medical records. Facial disfigurement was measured using a patient-rated facial disfigurement analogue scale [20]. Subjects rated the degree to which their facial appearance had changed due to the surgery on a visual analog scale of 0 (not at all) to 100 (worst possible). The scale has shown good psychometric properties in previous studies [11,20].

The following instruments were used to collect data for the study variables at all four data collection time points. Social support was measured using a social support scale developed in Chinese [21]. The scale has 16 items that measure four dimensions of social support: appraisal support, informational support, emotional support, and instrumental support. Subjects rated each item on a Likert scale (0 (never) to 3 (always)) to indicate how often they had received support from family members or essential others in the past month.

The sum of all items is the social support score, with a possible range of 0–48. The higher the score, the greater the perceived support. The scale has shown good psychometric properties in previous studies on heart transplants [22], dialysis [23], and cancer patients [24]. In this study, Cronbach's alpha for this scale was 0.94.

Symptoms were measured using the Chinese version of the European Organization for Research and Treatment of Cancer Quality of Life Questionnaire Head and Neck Cancer Module (EORTC QLQ-H&N35) [22,25]. The QLQ-H&N35 consists of 35 items, including 7 multi-item scales and 11 single-item scales. Seven multi-item scales were scored on a 4-point Likert scale (1 (not at all problem) to 4 (very much)) to assess pain, swallowing, taste and smell problems, speech problems, trouble with social eating, trouble with social contact, and less sexuality. Of the 11 single-item scales, 6 items used a 4-point Likert scale to assess teeth, opening the mouth, dry mouth, thick saliva, cough, and feeling unwell; 5 items were scored as yes or no to assess pain medication, nutritional supplements, feeding tube, weight loss, and weight gain. Only the 7 multi-item scales and the 6 single-item scales with a 4-point scale for scoring were used in this study. For all items and scales, high scores indicated more severe symptoms. Scores on all composite scales were calculated as the average of all items in these scales. These scores were then converted to normalized scores ranging from 0 to 100 using a linear transformation according to the scoring procedure. A high score indicates a higher level of symptomatology/problems. In previous studies, this scale has shown favorable psychometric properties in the HNC population [25]. In the current study, the internal consistency of the scale was acceptable, with Cronbach's alpha coefficients of 0.70 (pain), 0.74 (swallowing), 0.61 (taste and smell problems), 0.81 (speech problems), 0.84 (social eating), 0.78 (social contact), and 0.85 (sexuality).

Psychosocial adjustment was measured using the Chinese version of the Psychosocial Adjustment to Illness Scale-Self Report (PAIS-SR) [26]. The 46-item scale measures seven major adjustment domains: healthcare orientation (8 items), vocational environment (6 items), domestic environment (8 items), sexual relationships (6 items), extended family relationships (5 items), social environment (6 items), and psychological distress (7 items). Each item was scored on a 4-point scale (0 (no problem) to 3 (a lot of difficulty)), with higher scores indicating worse adjustment. Scores for all domain scales were calculated as the total score of all items in those scales. These scores were then converted to standardized T-scores, with a possible range of 0–100 for each domain [27]. The score for PAIS-SR was calculated by summing the seven domain T-scores to provide global adjustment information, with a possible range of 0–700. A higher score indicates more difficulty experienced. A cutoff score of 393 was considered the clinical level of maladjustment [27]. In past studies involving cancer populations, the scale has shown acceptable reliability and validity [11,27]. In the current study, Cronbach's alpha coefficients ranged from 0.52–0.87 on the seven domain scales and 0.91 on the full scale. The PAIS-SR score of all 46 items was used to represent the psychosocial adjustment for all analyses in this study.

2.5. Data Analysis

All statistical analyses were conducted using the Statistical Package for Social Sciences 20.0. Chi-square and Mann–Whitney tests were used to compare differences in demographics and disease characteristics between subjects who completed the study and those who were lost to follow-up. Descriptive statistics were used to describe study variables. A univariate generalized estimating equation (GEE) was used to analyze changes in psychosocial adjustment over time (from T1 to T4). Paired *t*-tests further explored differences in psychosocial adjustment scores between time points. A multivariate GEE with an exchangeable correlation structure was used to analyze the effects of demographics, disease variables, facial disfigurement, social support, and symptoms on psychosocial adaptation. The psychosocial adjustment was entered as the dependent variable. Time, demographics, disease variables, facial disfigurement, social support, and symptoms were entered as the independent variables.

3. Results

3.1. Subjects' Characteristics

Seventy-nine potential subjects were approached. Five individuals did not meet the eligibility criteria, and 12 refused to participate. Sixty-two eligible individuals signed informed consent and participated in the study. At T2, eight subjects were lost to follow-up due to disease status ($n = 2$), failure to return or loss of contact ($n = 2$), and disinterest or inability to cooperate ($n = 4$). At T3, four subjects were lost to follow-up due to disease status ($n = 1$), failure to return or loss of contact ($n = 2$), and disinterest or inability to cooperate ($n = 2$). Fifty subjects completed the study, and their data were included in the analysis. Chi-square and Mann–Whitney tests showed no significant difference in demographics and disease characteristics between subjects who completed the study and those who were lost to follow-up.

We followed postoperative oral cancer patients for up to five months with an acceptable dropout rate of 19.4%. Most subjects were middle-aged men with high school education and poor financial status. Furthermore, 46% of subjects had buccal mucosa cancer, 36% had stage IV cancer, and 60% received adjuvant therapy (Table 1). This demographic and disease profile is similar to the epidemiological data for oral cancer in Taiwan [3].

Table 1. Demographics, disease characteristics, and perceived facial disfigurement ($N = 50$).

Variables	Frequency	%	Mean (SD)	Range
Hospital				
A	31	62		
B	19	38		
Age			50.04 (10.53)	32–78
Gender				
Male	50	100		
Education level				
Primary school and below	9	18		
Middle school	29	58		
College and above	12	24		
Marital status				
Single	16	32		
Married	34	68		
Employment status				
No	40	80		
Yes	9	18		
Financial status				
Not enough	22	44		
Enough	22	44		
More than enough	6	12		
Tumor Location				
Buccal mucosa	23	46		
Tongue, mouth floor	17	34		
Gingiva, lips	9	18		
Cancer stage				
I	9	18		
II	8	16		
III	14	28		
IV	18	36		
Adjuvant therapy				
No	20	40		
Yes	30	60		
Facial disfigurement			45.14 (32.25)	0–100

SD, standard deviation.

3.2. Social Support

The mean (standard deviation, SD) support for T1, T2, T3, and T4 was 39.66 (9.43), 36. 9 (11.17), 36.35 (11.24), and 34.76 (10.67), respectively. Univariate GEE analysis results showed that social support had significant time effects (Table 2), indicating that both sources of support change significantly over time. Social support was highest at T1 and gradually decreased over time.

Table 2. Psychosocial adjustment, social support, and symptoms at pre-operation, and one month, three months, and five months after discharge (*n* = 50).

Variable	Time	Mean	SD	Range	B	X^2	*p*-Value
Psychosocial adjustment	T4	409.3	52.4	275–487	10.9	4.5	0.033 *
	T3	415.9	52.2	275–490	15.6	9.3	0.002 **
	T2	410.9	50.8	301–515	12.8	6.5	0.011 *
	T1	398.0	43.1	309–486	0		
Social support	T4	34.8	10.7	0–48	−4.8	10.0	0.002 **
	T3	36.4	11.2	10–48	−3.0	3.9	0.048 *
	T2	36.9	11.2	2–48	−3.0	4.2	0.042 *
	T1	39.7	9.4	9–48	0		
		Median	**IQR**				
Pain	T4	25.0	8.3–33.3	0–75	4.2	2.0	0.161
	T3	33.3	12.5–37.5	0–75	6.1	4.1	0.042 *
	T2	25.0	8.3–33.3	0–91.67	5.9	4.1	0.043 *
	T1	16.7	8.3–27.1	0–66.67	0		
Swallowing	T4	41.7	20.8–58.3	0–100	26.2	49.6	<0.001 ***
	T3	41.7	25.0–70.8	0–100	28.3	57.7	<0.001 ***
	T2	33.3	16.7–66.7	0–100	26.6	53.4	<0.001 ***
	T1	8.3	0–25	0–50	0		
Senses problems	T4	16.7	0–33.3	0–100	12.6	8.5	0.004 **
	T3	16.7	0–50	0–66.67	16.1	13.8	<0.001 ***
	T2	0	0–33.3	0–100	9.2	5.1	0.024 *
	T1	0	0–0	0–100	0		
Speech problems	T4	33.3	22.2–61.1	0–77.78	22.0	37.2	<0.001***
	T3	33.3	16.7–55.6	0–88.89	20.1	31.0	<0.001 ***
	T2	33.3	11.1–55.6	0–100	19.2	29.3	<0.001 ***
	T1	11.1	0–22.2	0–100	0		
Trouble with social eating	T4	50.0	33.3–83.3	0–100	28.3	53.1	<0.001 ***
	T3	66.7	33.3–75.0	0–100	29.7	58.5	<0.001 ***
	T2	54.2	33.3–72.9	0–100	27.7	52.8	<0.001 ***
	T1	16.7	0–41.7	0–100	0		
Trouble with social contact	T4	33.3	16.7–40.0	0–100	19.5	33.8	<0.001 ***
	T3	33.3	16.7–40.0	0–100	16.9	25.5	<0.001 ***
	T2	26.7	6.7–46.7	0–100	16.9	26.4	<0.001 ***
	T1	6.7	0–26.7	0–73.33	0		
Less sexuality	T4	33.3	16.7–50.0	0–100	17.4	13.6	<0.001 ***
	T3	33.3	0–66.7	0–100	19.8	17.8	<0.001 ***
	T2	33.3	16.7–33.3	0–100	13.5	8.6	0.003 **
	T1	16.7	0–33.3	0–100	0		
Teeth problems	T4	33.3	33.3–66.7	0–100	−2.1	0.13	0.720
	T3	66.7	33.3–100	0–100	11.3	3.8	0.052
	T2	33.3	33.3–100	0–100	−0.7	0.0	0.904
	T1	50	0–75	0–100	0		
Opening the mouth	T4	66.7	33.3–100	0–100	17.0	11.9	0.001 **
	T3	66.7	33.3–100	0–100	20.4	17.2	<0.001 ***
	T2	33.3	33.3–100	0–100	13.2	7.5	0.006 **
	T1	33.3	0–66.7	0–100	0		

Table 2. *Cont.*

Variable	Time	Mean	SD	Range	B	X^2	*p*-Value
Dry mouth	T4	33.3	33.3–100	0–100	22.7	20.5	<0.001 ***
	T3	33.3	33.3–100	0–100	31.4	39.1	<0.001 ***
	T2	33.3	33.3–66.7	0–100	17.3	12.3	<0.001 ***
	T1	33.3	0–33.3	0–100	0		
Sticky saliva	T4	33.3	33.3–66.7	0–100	11.9	5.7	0.017 *
	T3	66.7	33.3–100	0–100	22.6	20.5	<0.001 ***
	T2	33.3	33.3–66.7	0–100	16.6	11.5	0.001 **
	T1	33.3	0–33.3	0–100	0		
Cough	T4	33.3	0–33.3	0–100	12.1	11.2	0.001 **
	T3	33.3	33.3–33.3	0–100	15.9	19.4	<0.001 ***
	T2	33.3	0–33.3	0–100	10.2	8.2	0.004 **
	T1	33.3	0–33.3	0–100	0		
Felt ill	T4	33.3	33.3–33.3	0–100	5.7	1.6	0.202
	T3	33.3	33.3–33.3	0–100	6.0	1.8	0.180
	T2	33.3	33.3–66.7	0–100	11.7	7.0	0.008 **
	T1	33.3	0–33.3	0–100	0		

Note. Generalized estimating equations for repeated measurements and an exchangeable correlation structure were used. T1, preoperative; T2, one month after discharge; T3, three months after discharge; T4, five months after discharge; SD, standard division; IQR, interquartile range; B, unstandardized coefficient; X^2, value of Wald chi-square; * $p < 0.05$; ** $p < 0.01$; *** $p < 0.001$.

3.3. Symptoms

At T1, the most severe symptom experienced was teeth problems (median = 50, SD = 37.64). At T2, the most severe symptom was trouble with social eating (median = 54.2, interquartile range (IQR): 33.3–72.9). At T3, the most severe symptoms were trouble with social eating (median = 66.7, IQR: 33.3–75.0), teeth problems (median = 66.7, IQR: 33.3–100), opening the mouth (median = 66.7, IQR: 33.3–100), and sticky saliva (median = 66.7, IQR: 33.3–100). At T4, the three most severe symptoms were opening the mouth (median = 66.7, IQR: 33.3–100) and trouble with social eating (median = 50, IQR: 33.3–83.3, Table 2).

Univariate GEE analysis showed that all symptoms, except the teeth problems, had a significant time effect, indicating that symptoms change significantly over time (Table 2). Most of these symptoms were aggravated after surgery, the most severe at T3, maintained or slightly improved from T3 to T4 (Figure 1).

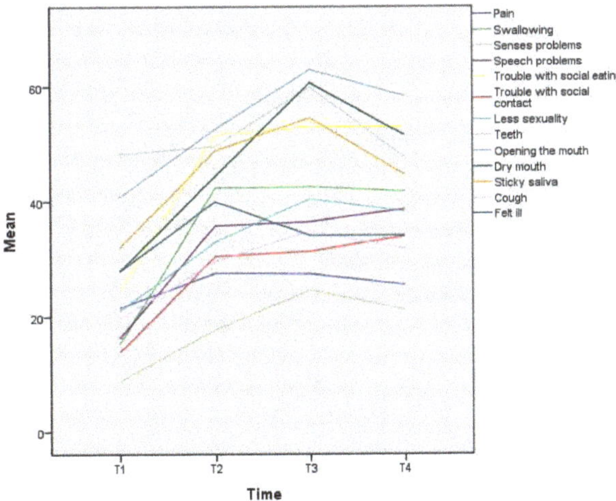

Figure 1. Oral cancer patients' symptoms change over time.

3.4. Psychosocial Adjustment

The mean (SD) PAIS-SR scores for T, T2, T3, and T4 were 398.0 (43.1), 410.9 (50.8), 415.9 (52.2), and 409.3 (52.4) (Table 2). Univariate GEE analysis results showed a significant time effect. Subjects' PAIS-SR scores were significantly higher at T2 (unstandardized coefficient (B) = 12.8), T3 (B = 15.6), and T4 (B = 10.9) than at T1, indicating that subjects experienced more psychosocial adjustment challenges after surgery. Paired *t*-tests further explored differences in PAIS-SR scores between time points. The results showed a statistically significant difference between T1 and T2 (mean difference (MD) = 12.4, 95% confidence interval (CI): 1.7~23.0); between T3 and T4 (MD = −6.4, 95%CI: −12.3~−0.5). The difference between T2 and T3 was not statistically significant. These results suggest that subjects' psychosocial adjustment challenges increased from T1 to T2, maintained from T2 to T3, and decreased from T3 to T4. Taking 393 as the cut-off point, 50%, 59.2%, 66%, and 62% of patients reported psychosocial maladjustment at T1, T2, T3, and T4, respectively. Of the seven domains of psychosocial adjustment, subjects experienced the most significant challenges in extended family relationships, vocational environment, and healthcare orientation (Figure 2).

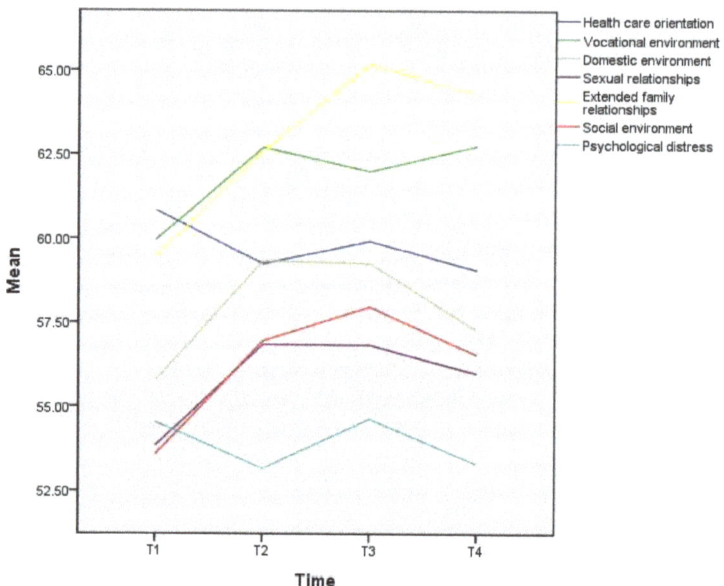

Figure 2. Oral cancer patients' seven domains of psychosocial adjustment.

3.5. Factors Associated with Psychosocial Adjustment

Results from GEE indicated that financial status, cancer stage, pain, speech problems, social eating problems, and decreased sexuality were significantly associated with psychosocial adjustment. Patients with enough income reported better psychosocial adjustment than patients without (B = −28.31, *p* = 0.041). Patients with stage I cancer had better psychosocial adjustment than patients with stage IV (B = 25.33, *p* = 0.047) and stage III cancers (B = 31.13, *p* = 0.026). Patients with severer pain (B = 0.38, *p* = 0.025), more speech problems (B = 0.39, *p* = 0.040), more trouble with social eating (B = 0.35, *p* = 0.013), and less sexuality (B = 0.25, *p* = 0.028) had worse psychosocial adjustment (Table 3). The following variables were not significantly associated with psychosocial adjustment: age, education level, marital status, employment status, tumor location, adjuvant therapy, facial disfigurement, social support, senses problems, trouble with social contact, teeth problems, opening the mouth, dry mouth, sticky saliva, cough, and felt ill.

Table 3. Factors associated with psychosocial adjustment ($n = 50$).

Variables	B	SE	95% CI		Wald X^2	p-Value
			Lower	Upper		
Intercept	395.75	29.72	337.50	454.00	177.32	<0.001 ***
Time						
T4 vs. T1	−10.52	7.36	−24.94	3.91	2.04	0.153
T3 vs. T1	−8.83	7.28	−23.1	5.44	1.47	0.225
T2 vs. T1	−6.24	6.60	−19.18	6.69	0.90	0.344
Age	−0.20	0.44	−1.07	0.67	0.21	0.648
Hospital (B vs. A)	−14.79	10.07	−34.52	4.95	2.16	0.142
Education level						
College vs. Primary school	−0.30	15.16	−30.02	29.4	0.00	0.984
Middle vs. Primary school	2.62	11.84	−20.58	25.82	0.05	0.825
Marital (Married vs. Single)	−5.29	9.51	−23.93	13.35	0.31	0.578
Employment (Yes vs. No)	−20.84	13.52	−47.33	5.65	2.38	0.123
Financial status						
More than enough vs. Not enough	−28.31	13.88	−55.51	−1.11	4.16	0.041 *
Enough vs. Not enough	−3.70	8.86	−21.06	13.66	0.17	0.676
Tumor Location						
Gingiva, lips vs. Buccal	19.77	10.77	−1.35	40.88	3.37	0.067
Tongue, mouth floor vs. Buccal mucosa	−5.28	9.65	−24.18	13.63	0.30	0.584
Cancer stage						
IV vs I	25.33	12.76	0.33	50.33	3.94	0.047 *
III vs. I	31.12	13.96	3.76	58.49	4.97	0.026 *
II vs. I	13.72	11.60	−9.03	36.46	1.4	0.237
Adjuvant therapy (yes vs. no)	−14.94	8.98	−32.53	2.66	1.78	0.182
Facial disfigurement	0.19	0.13	−0.76	0.45	1.94	0.164
Social support	−0.29	0.33	−0.93	0.36	0.76	0.382
Pain	0.38	0.17	0.05	0.70	5.02	0.025 *
Swallowing	0.16	0.15	−0.14	0.46	1.07	0.301
Senses problems	0.15	0.13	−0.11	0.40	1.29	0.257
Speech problems	0.39	0.19	−0.75	−0.02	4.23	0.040 *
Trouble with social eating	0.35	0.14	0.07	0.63	6.17	0.013 *
Trouble with social contact	0.34	0.19	−0.03	0.70	3.28	0.070
Less sexuality	0.25	0.11	0.03	0.47	4.85	0.028 *
Teeth problems	0.04	0.08	−0.12	0.20	0.21	0.645
Opening the mouth	−0.13	0.09	−0.31	0.06	1.76	0.185
Dry mouth	0.02	0.10	−0.18	0.22	0.02	0.879
Sticky saliva	−0.12	0.11	−0.33	0.09	1.19	0.275
Cough	0.19	0.12	−0.05	0.42	2.42	0.119
Felt ill	0.16	0.13	−0.1	0.41	1.48	0.223

Note. Generalized estimating equations for repeated measurements and an exchangeable correlation structure were used. T1, preoperative; T2, one month after discharge; T3, three months after discharge; T4, five months after discharge. B, unstandardized beta; SE, standard error; X^2, value of Wald chi-square; * $p < 0.05$, ***, $p < 0.001$.

4. Discussion

4.1. Psychosocial Adjustment

Our results show that patients with oral cancer had poor psychosocial adjustment. At each of the four data collection time points, more than half of the subjects had psychosocial maladjustment. Overall, the psychosocial adjustment of subjects worsened over time, and they reported the worst psychosocial adjustment three months after surgery. The findings suggest that oral cancer patients face significant psychosocial adjustment challenges after reconstructive surgery, especially in the first three months. Healthcare professionals should pay close attention to the psychosocial needs of this population and provide support to help patients cope with psychosocial challenges. Attention should be paid to adapting to changes in extended family relationships, vocational environment, and healthcare orientation, as subjects reported these were the most challenging areas.

4.2. Factors Associated with Psychosocial Adjustment

We found that financial status, cancer stages, pain, speech problems, social eating problems, and less sexuality were significant predictors of psychosocial adjustment after reconstructive surgery in patients with oral cancer. Like what was reported in cross-sectional studies, patients with not enough income [21], cancer stage III/IV [28], severer pain [29], more speech problems [30], more social eating problems, and less sexuality report poorer psychosocial adjustment. Patients with these characteristics or symptoms are at higher risk for postoperative psychosocial maladjustment. These findings are interpreted in terms of *p*-values, which may not be clinically meaningful findings. However, the study's finding provides preliminary data for understanding changes in psychosocial adjustment after oral cancer surgery and the potential impact of facial disfigurement, social support, and symptoms on psychosocial adjustment. Given the high prevalence and worsening of psychosocial maladjustment in this population over time, continued psychosocial assessment and appropriate supportive measures are needed to enhance psychosocial adjustment in these at-risk groups. Speech and social eating problems should be noted. Many oral cancer patients experience social avoidance due to concerns about speech or eating, which negatively impacts patients' psychosocial adjustment [29]. In addition, postoperative oral cancer patients require ongoing assessment and appropriate interventions to help patients manage symptoms, especially during the three months following surgery. Particular attention should be paid to teeth problems, trouble with social eating, opening the mouth, and sticky saliva.

Unlike previous research reports [24], we found that psychosocial adjustment was independent of age, education, marital status, employment status, cancer location, and type of cancer treatment. Different patient populations and study designs may partially explain this discrepancy. Our subjects were recruited from two hospitals in Taiwan, and their characteristics may differ from patients in other clinical settings. Instead of a cross-sectional design, we used a longitudinal study design and followed patients with oral cancer for up to 5 months after discharge.

Furthermore, unlike previous research reports, we found no significant effect of facial disfigurement on psychosocial adjustment. Our results fail to support the research hypothesis that facial disfigurement negatively affects patients' psychosocial adjustment over time. This difference may be because the perceived disfigurement of the face differs from the actual impairment of oral function, which may significantly impact psychosocial adjustment. Moreover, the effects of facial disfigurement on psychosocial adjustment may be primarily mediated by functional impairment or social isolation (e.g., speech or social eating problems) rather than direct effects [30,31]. Our findings suggest that patients with oral cancer experience many uncomfortable symptoms. Among them, pain, speech problems, social eating problems, and less sexuality significantly affected changes in psychosocial adjustment over time. The effects of other symptoms on psychosocial adjustments, such as sensory problems, social contact, teeth problems, opening the mouth, dry mouth, sticky saliva, cough, and felt ill, were not statistically significant.

4.3. Study Limitations

The study was limited by its small sample size, the use of a convenient sample, and the recruitment of male subjects only. We recruited a convenience sample of oral cancer patients from two medical centers in Taiwan. Characteristics of subjects may differ from patients in other clinical settings. Moreover, women may be more sensitive to facial changes. Although we did not exclude women, only men participated in this study. It is difficult to recruit a representative sample of female patients due to the low incidence of oral cancer in women. Therefore, our findings may not generalize beyond this sample. When considering the impact of perceived financial stress on psychosocial adjustment, we asked subjects to indicate how they viewed their income as meeting their needs, rather than their actual income. However, having enough or enough income can be subjective and vary from person to person. Using better scales could provide better information about subjects' financial status.

5. Conclusions

Postoperative oral cancer patients experience poor psychosocial adjustment that worsens over time. Patients with poor financial status, cancer stage III/IV, severe pain, more speech problems, more social eating problems, and less sexuality were at the most significant risk for psychosocial maladjustment. Identifying these risk factors is essential because at-risk individuals can be targeted for early psychosocial assessment and intervention. Enhancing the psychosocial adjustment of oral cancer patients requires ongoing support and team-based and multidisciplinary collaborative care.

Author Contributions: Conceptualization, Y.-W.C., P.-L K. and T.-J.W.; data curation, Y.-W.C. and S.-C.L.; formal analysis, T.-R.L., K.-F.W. and T.-J.W.; funding acquisition, T.-J.W.; investigation, P.-L.K. and S.-C.L.; methodology, T.-R.L., P.-L.K. and T.-J.W.; project administration, Y.-W.C., S.-C.L. and T.-J.W.; supervision, T.V.D. and T.-J.W.; validation, T.-R.L. and P.-L.K.; visualization, K.-F.W.; writing—original draft, Y.-W.C. and P.-L.K.; writing—review and editing, T.-R.L., T.V.D. and T.-J.W. All authors have read and agreed to the published version of the manuscript.

Funding: This research was funded by the National Science and Technology Council, Taiwan, ROC, grant number NSC 97-2314-B-227-002-MY2.

Institutional Review Board Statement: The study was approved by the Institutional Review Board of the Hualien Tzu Chi Hospital (IRB no. ACT-IRB098-07).

Informed Consent Statement: Informed consent was obtained from all subjects involved in the study.

Data Availability Statement: Data will be available from the corresponding author upon reasonable request.

Acknowledgments: We thank all participants for their participation.

Conflicts of Interest: The authors declare no conflict of interest.

References

1. Taiwan Ministry of Health and Welfare. Cancer Incidence in Taiwan in 2019. Available online: https://www.mohw.gov.tw/cp-5264-65547-1.html (accessed on 1 July 2022).
2. International Agency Research on Cancer/World Health Organization. Globocan 2020: Estimated Cancer Incidence, Mortality, and Prevalence Worldwide in 2020. Available online: http://globocan.iarc.fr/Pages/fact_sheets_population.aspx (accessed on 1 July 2022).
3. Taiwan Health Promotion Administration. Cancer Registration Online Inquiry System. Available online: https://cris.hpa.gov.tw/pagepub/Home.aspx?itemNo=cr.a.10 (accessed on 1 July 2022).
4. Taiwan Health Promotion Administration. Cancer Registry Annual Report. 2019. Available online: https://www.hpa.gov.tw/Pages/Detail.aspx?nodeid=269&pid=14913 (accessed on 1 July 2022).
5. Derogatis, L.R.; Fleming, M.P.; Sudler, N.C.; Della, P.L. Psychological assessment. In *Managing Chronic Illness: A Biopsychosocial Perspective*; Nicassio, P.A., Smith, T.W., Eds.; American Psychological Association: Washington, DC, USA, 1995; pp. 59–115.
6. National Comprehensive Cancer Network. NCCN Clinical Practice Guidelines in Oncology (NCCN Guidelines®) Head and Neck Cancers. Available online: https://www.nccn.org/professionals/physician_gls/pdf/head-and-neck.pdf (accessed on 1 July 2022).
7. Ringash, J.; Bernstein, L.J.; Devins, G.; Dunphy, C.; Giuliani, M.; Martino, R.; McEwen, S. Head and Neck Cancer Survivorship: Learning the Needs, Meeting the Needs. *Semin. Radiat. Oncol.* **2018**, *28*, 64–74. [CrossRef] [PubMed]
8. Doss, J.G.; Ghani, W.M.N.; Razak, I.A.; Yang, Y.H.; Rogers, S.N.; Zain, R.B. Changes in health-related quality of life of oral cancer patients treated with curative intent: Experience of a developing country. *Int. J. Oral Maxillofac. Surg.* **2017**, *46*, 687–698. [CrossRef] [PubMed]
9. Kim, J.H.; McMahon, B.T.; Hawley, C.; Brickham, D.; Gonzalez, R.; Lee, D.H. Psychosocial adaptation to chronic illness and disability: A virtue based model. *J. Occup. Rehabil.* **2016**, *26*, 45–55. [CrossRef] [PubMed]
10. Katz, M.R.; Irish, J.C.; Devins, G.M.; Rodin, G.M.; Gullane, P.J. Reliability and Validity of an Observer-Rated Disfigurement Scale for Head and Neck Cancer Patients. *Head Neck* **2000**, *22*, 132–141. [CrossRef]

11. Wang, T.-J.; Lu, M.-H.; Kuo, P.-L.; Chen, Y.-W.; Lee, S.-C.; Liang, S.-Y. Influences of Facial Disfigurement and Social Support for Psychosocial Adjustment among Patients with Oral Cancer in Taiwan: A Cross Sectional Study. *BMJ Open* **2018**, *8*, e023670. [CrossRef]
12. Rifkin, W.J.; Kantar, R.S.; Ali-Khan, S.; Plana, N.M.; Diaz-Siso, J.R.; Tsakiris, M.; Rodriguez, E.D. Facial Disfigurement and Identity: A Review of the Literature and Implications for Facial Transplantation. *AMA J. Ethics* **2018**, *20*, 309–323. [CrossRef]
13. Chaudhary, F.A.; Ahmad, B. The relationship between psychosocial distress and oral health status in patients with facial burns and mediation by oral health behaviour. *BMC Oral Health* **2021**, *21*, 172. [CrossRef]
14. Khawaja, S.N.; Jamshed, A.; Hussain, R.T. Prevalence of Pain in Oral Cancer: A retrospective study. *Oral Dis.* **2021**, *27*, 1806–1812. [CrossRef]
15. Ou, M.; Wang, G.; Yan, Y.; Chen, H.; Xu, X. Perioperative Symptom Burden and Its Influencing Factors in Patients with Oral Cancer: A Longitudinal Study. *Asia-Pac. J. Oncol. Nurs.* **2022**, *9*, 100073. [CrossRef]
16. Constantinescu, G.; Rieger, J.; Winget, M.; Paulsen, C.; Seikaly, H. Patient Perception of Speech Outcomes: The Relationship between Clinical Measures and Self-Perception of Speech Function Following Surgical Treatment for Oral Cancer. *Am. J. Speech-Lang. Pathol.* **2017**, *26*, 241–247. [CrossRef]
17. Cohen, S.; Wills, T.A. Stress, Social Support, and the Buffering Hypothesis. *Psychol. Bull.* **1985**, *98*, 310–357. [CrossRef] [PubMed]
18. Jagannathan, A.; Juvva, S. Emotions and Coping of Patients with Head and Neck Cancers after Diagnosis: A Qualitative Content Analysis. *J. Postgrad. Med.* **2016**, *62*, 143. [CrossRef]
19. Cohen, J. *Statistical Power Analysis for the Behavioral Sciences*, 2nd ed.; Lawrence Erlbaum Associates Publishers: Hillsdale, NJ, USA, 1988.
20. Lueg, E.A.; Irish, J.C.; Katz, M.R.; Brown, D.H.; Gullane, P.J. A Patient- and Observer-Rated Analysis of the Impact of Lateral Rhinotomy on Facial Aesthetics. *Arch. Facial Plast. Surg.* **2001**, *3*, 241–244. [CrossRef] [PubMed]
21. Wang, S.M.; Ku, N.P.; Lin, H.T.; Wei, J. The Relationships of Symptom Distress, Social Support and Self-Care Behaviors in Heart Transplant Recipients. *J. Nurs. Res.* **1998**, *6*, 4–18.
22. Chie, W.C.; Hong, R.L.; Lai, C.C.; Ting, L.L.; Hsu, M.M. Quality of life in patients of nasopharyngeal carcinoma: Validation of the Taiwan Chinese Version of the EORTC QLQ-C30 and the EORTC QLQ-H&N35. *Qual. Life Res.* **2003**, *12*, 93–98. [CrossRef] [PubMed]
23. Wang, T.-J.; Lin, M.; Liang, S.-Y.; Wu, S.-F.V.; Tung, H.-H.; Tsay, S.-L. Factors Influencing Peritoneal Dialysis Patients' Psychosocial Adjustment. *J. Clin. Nurs.* **2014**, *23*, 82–90. [CrossRef] [PubMed]
24. Huang, S.M.; Tseng, L.M.; Lai, J.C.Y.; Tsai, Y.F.; Lien, P.J.; Chen, P.H. Impact of Symptom and Social Support on Fertility Intention in Reproductive-Age Women with Breast Cancer. *Clin. Nurs. Res.* **2018**, *2018*, 1–16. [CrossRef]
25. Bjordal, K.; Hammerlid, E.; Ahlner-Elmqvist, M.; Graeff, A.; de Boysen, M.; Evensen, J.F.; Biörklund, A.; Leeuw, J.R.; de Fayers, P.M.; Jannert, M.; et al. Quality of life in head and neck cancer patients: Validation of the European Organization for Research and Treatment of Cancer Quality of Life Questionnaire-H&N35. *J. Clin. Oncol.* **1999**, *17*, 1008–1019.
26. Derogatis, L.R. The Psychosocial Adjustment to Illness Scale (PAIS). *J. Psychosom. Res.* **1986**, *30*, 77–91. [CrossRef]
27. Derogatis, L.R.; Derogatis, M.A. PAIS & PAIS-SR: Administration, Scoring & Procedures Manual-II. *Clin. Psychom. Res.* **1990**, *30*, 77–91.
28. Tsui, W.Y.; Lin, K.C.; Huang, H.C. An Investigation of Body Image Changes and Coping Behaviors in Oral Cancer Patients Following Surgery. *J. Nurs. Healthc. Res.* **2013**, *9*, 127–138.
29. Chen, S.C. Life Experiences of Taiwanese Oral Cancer Patients During the Postoperative Period. *Scand. J. Caring Sci.* **2012**, *26*, 98–103. [CrossRef]
30. Fingeret, M.C.; Yuan, Y.; Urbauer, D.; Weston, J.; Nipomnick, S.; Weber, R. The Nature and Extent of Body Image Concerns Among Surgically Treated Patients with Head and Neck Cancer. *Psycho-Oncology* **2012**, *21*, 836–844. [CrossRef] [PubMed]
31. Pateman, K.A.; Ford, P.J.; Batstone, M.D.; Farah, C.S. Coping with an Altered Mouth and Perceived Supportive Care Needs Following Head and Neck Cancer Treatment. *Support. Care Cancer* **2015**, *23*, 2365–2373. [CrossRef] [PubMed]

Communication

Lymph Node Ratio in Head and Neck Cancer with Submental Flap Reconstruction

Hidenori Suzuki *, Shintaro Beppu, Daisuke Nishikawa, Hoshino Terada, Michi Sawabe and Nobuhiro Hanai

Department of Head and Neck Surgery, Aichi Cancer Center Hospital, Nagoya 464-8681, Aichi, Japan
* Correspondence: hi.suzuki@aichi-cc.jp

Abstract: This study aimed to investigate the relationship between the lymph node ratio (LNR) and survival results of patients with head and neck squamous cell carcinoma (HNSCC) reconstructed by a submental artery flap (SMAF) to limit tumor size. This study retrospectively recruited 49 patients with HNSCC who underwent both primary resection and neck dissection with SMAF reconstruction. The LNR was the ratio of the number of metastatic lymph nodes to the sum number of examined lymph nodes. A LNR of 0.04 was the best cut-off value for HNSCC-specific death on receiver operating curve analysis. Patients with LNRs > 0.04 were univariately related to cancer-specific, disease-free, distant metastasis-free, and locoregional recurrence-free survival than those with LNRs $\leq$ 0.04 by log-rank test. In a Cox's proportional hazards model with hazard ratio (HR) and 95% confidence interval (CI) adjusting for pathological stage, extranodal extension and or surgical margins, the LNR (>0.04/$\leq$0.04) predicted multivariate shorter cancer-specific (HR = 9.24, 95% CI = 1.49–176), disease-free (HR = 3.44, 95% CI = 1.23–10.3), and distant metastasis-free (HR = 9.76, 95% CI = 1.57–187) survival. In conclusion, LNR for patients of HNSCC with SMAF reconstruction for limited tumor size was a prognostic factor for survival outcomes.

Keywords: lymph node ratio; squamous cell carcinoma; head and neck; submental artery flap; survival

Citation: Suzuki, H.; Beppu, S.; Nishikawa, D.; Terada, H.; Sawabe, M.; Hanai, N. Lymph Node Ratio in Head and Neck Cancer with Submental Flap Reconstruction. *Biomedicines* **2022**, *10*, 2923. https://doi.org/10.3390/biomedicines10112923

Academic Editor: Vui King Vincent-Chong

Received: 17 October 2022
Accepted: 10 November 2022
Published: 14 November 2022

Publisher's Note: MDPI stays neutral with regard to jurisdictional claims in published maps and institutional affiliations.

1. Background

Pathological metastasis of lymph nodes was recognized as a prognostic factor of survival outcomes in various types of carcinoma [1,2]. The lymph node ratio (LNR), which was defined as the ratio of the number of lymph node metastasis to the number of resected lymph nodes, was a pathologically simple continuous variable with the reflection of surgery, sampling, and staging [3]. The LNR, regardless of several patterns for neck dissection, has been widely adopted as a survival predictor for head and neck squamous cell carcinoma (HNSCC) [4]. Moreover, the LNR in our institution also predicted survival results for 46 cases of hypopharyngeal squamous cell carcinoma (SCC) from 2000 to 2015 [3] and 35 cases of oral SCC from 2008 to 2013 [5].

The submental artery flap (SMAF) is a regional flap, which was firstly described in 1993 [6], was globally developed as a useful flap for medium-sized surgical defects for HNSCC from retrospective and prospective studies [7,8]. The SMAF with both less invasive procedures and good oncologic results was evaluated as a game-changer reconstruction without microvascular anastomosis in comparison to free-flap reconstruction [9]. The subsite of head and neck cancer is heterogenous. To date, the prognostic value of LNR should be assessed for individuals with SMAF for HNSCC.

Therefore, this research purposed to investigate the association between LNR and survival outcomes for patients of HNSCC treated by surgery with SMAF reconstruction.

2. Methods

This retrospective observational study at the Department of Head and Neck Surgery in our hospital, following the Declaration of Helsinki, was carried out and approved by our hospital review board (receipt number of 2019-1-427). Of the 53 patients with HNSCC who were newly diagnosed without distant metastasis and underwent tumor resection with SMAF reconstruction from March 2009 to March 2020, four patients who received no neck dissection were excluded. Therefore, 49 patients who had pathological diagnoses of lymph nodes for interventions and examinations with informed consent were recruited. The treatment strategy using SMAF in this cohort mainly applied to small or intermediate defects in patients with advanced age or exhibition of comorbidity.

3. Submental Artery Flap

The SMAF was made by head and neck surgeons and is similar to the supraclavicular artery flap as previously described [10]. The SMAF was designed by a pinch test at the submental area of the primary tumor side. The SMAF was designed with both the anterior belly of the digastric muscle and the partial mylohyoid muscle elevated by preserving the submental artery, submental vein, and facial marginal nerve. Primary tumor resection as well as neck dissection were performed by preserving the elevated SMAF. The defect for primary tumor resection was carefully covered without tension by the SMAF. Figure 1 shows a representative image for elevated SMAF.

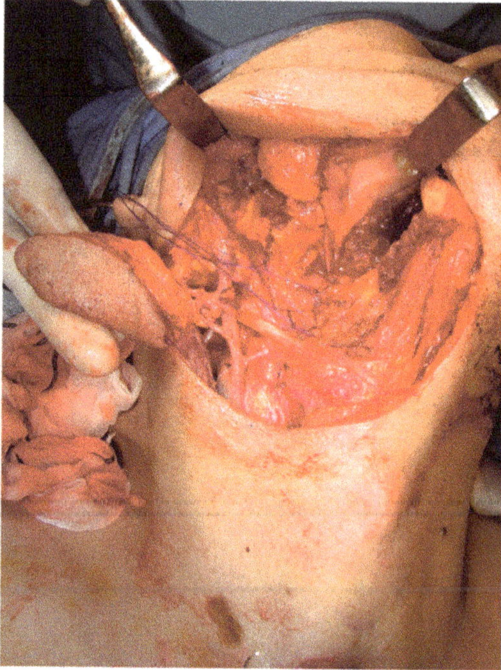

Figure 1. Submental artery flap in right side.

4. Clinicopathological Parameters

The median $\pm$ standard deviation of age was 67 ± 12.0 years old. Clinical Tumor, Node, Metastasis (TNM) staging was diagnosed by appropriate images as previously reported [3]. Bilateral neck dissection was recommended for clinical metastasis of bilateral metastases of lymph node or floor of the mouth as primary tumor subsites. The primary sites in the head and neck were oral cavity ($n = 37$), oropharynx ($n = 8$), and hypopharynx ($n = 4$). There were two patients of positive status and six patients of unknown status for human papilloma virus in oropharyngeal cancer. Each pathological restaging of SCC in the primary site was conducted following the seventh edition of the International Union Against Cancer [11]. Experienced pathologists determined the pathological TNM diagnosis with both surgical margins for resected primary tumor and extranodal extension for the metastatic lymph node. The calculation for the LNR was the number of involved lymph nodes relative to the total number of dissected lymph nodes [3]. The median $\pm$ standard deviation of primary tumor sizes was 23 ± 13.3 mm based on maximum size from pathological and surgical reports. The main regimen of preoperative chemotherapy was 5-fluorouracil and cisplatin. The main purpose for using induction chemotherapy by 5-fluorouracil and cisplatin was for maximum organ preservation as previously described [12]. Postoperative treatment was recommended by the presence of multiple metastases of lymph node, positive surgical margins, and extranodal extension from pathological reports. Locoregional recurrence for follow-up was performed by salvage treatment as possible.

5. Statistical Analysis

The Kaplan–Meier method was applied to calculate survival duration from SMAF reconstruction to a target outcome or last date of contact. The target outcome for each survival type was death from HNSCC to cancer-specific survival (CSS), recurrence or metastasis to disease-free survival (DFS), local or regional recurrence to locoregional recurrence-free survival (LRRFS), distant metastasis to distant metastasis-free survival (DMFS), and death to overall survival (OS). Versatile cut-off values for the LNR were assessed for HNSCC specific death by a receiver operating curve (ROC) analysis with the area under the curve (AUC), as performed by other groups previously [13]. All patients were distinguished into two categories (those with LNR of ≤ 0.04 vs. > 0.04). The comparisons between the two categories in clinicopathological parameters (age, sex, pathological T and N classification, pathological stage, primary tumor size, primary site, positive surgical margin, extranodal extension, type of neck dissection, postoperative treatment, preoperative chemotherapy, smoking history, and extranodal extension and or positive surgical margin) or survival results were assessed by Fisher's exact test or the log-rank test, respectively. A Cox proportional hazards models with hazard ratio (HR) and 95% confidence interval (95% CI) was used to evaluate multivariate analyses of CSS, DFS, DMFS, and LRRFS. The interaction between LNR and extranodal extension was assessed by the Mann–Whitney U test. Statistical analyses were executed using the JMP software (version 9, SAS: Cary, NC, USA), and p-values < 0.05 were considered significant.

6. Results

The median number $\pm$ standard deviation of positive lymph nodes and the sum of harvested lymph nodes was 1 ± 2.55 and 29 ± 13.8, respectively. The mean and median $\pm$ standard deviation of LNR was 0.03, and 0 ± 0.08, respectively. Table 1 presents the associations between LNR and clinicopathological parameters.

The median follow-up $\pm$ standard deviation at last contact in the study was 5.04 ± 2.45 years for whole cases, 5.13 ± 2.13 years for the 35 survivors, 2.74 ± 2.52 years for the 14 cases who died, and 1.75 ± 1.32 years for the 10 cases who died from HNSCC. Local recurrence was observed in 10 patients, regional recurrence in 14, and distant metastasis in 9. The 5-year rates of CSS, DFS, LRRFS, DMFS, and OS were 78.0%, 60.5%, 65.7%, 80.6%, and 75.8%, respectively.

Figure 2 shows the ROC, the AUC of the ROC for death from HNSCC, 1-specificity, and sensitivity. The optimal cut-off values for LNR to find HNSCC specific death was 0.04 (AUC = 0.68, $p = 0.01$). The sensitivity and specificity in this ROC model were 0.5 and 0.15, respectively. Patients were separated into two categories based on the LNR of 0.04.

Table 1. Association between clinicopathologic parameters and LNR.

Parameter		Number	LNR (Mean ± Standard Deviation)
Age	<67 year	25	0.04 ± 0.05
	≥67 year	24	0.04 ± 0.11
Sex	Male	31	0.05 ± 0.10
	Female	18	0.01 ± 0.02
Pathological T classification	T1	16	0.01 ± 0.02
	T2	16	0.06 ± 0.13
	T3	9	0.03 ± 0.02
	T4	8	0.05 ± 0.08
Pathological N classification	N0	27	0
	N1	10	0.05 ± 0.03
	N2a	1	0.03
	N2b	10	0.12 ± 0.15
	N2c	1	0.03
Pathological stage	I	13	0
	II	7	0
	III	10	0.03 ± 0.02
	IVA	19	0.07 ± 0.12
Primary tumor size	<23 mm	21	0.01 ± 0.02
	≥23 mm	28	0.05 ± 0.10
Primary site	Oral	37	0.03 ± 0.09
	Oropharynx	8	0.03 ± 0.01
	Hypopharynx	4	0.08 ± 0.09
Positive surgical margin	Presence	7	0.13 ± 0.19
	Absence	42	0.02 ± 0.03
Extranodal extension	Presence	5	0.14 ± 0.21
	Absence	44	0.02 ± 0.04
Type of neck dissection	Unilateral	45	0.04 ± 0.08
	Bilateral	4	0
Postoperative treatment	Radiation	2	0.27 ± 0.35
	Chemoradiation	4	0.06 ± 0.04
	Chemotherapy	3	0.13 ± 0.09
	Absence	40	0.01 ± 0.02
Preoperative chemotherapy	Presence	6	0.06 ± 0.08
	Absence	43	0.03 ± 0.08
Smoking history	Presence	27	0.03 ± 0.05
	Absence	22	0.04 ± 0.11
Extranodal extension and or positive surgical margin	Presence	11	0.10 ± 0.15
	Absence	38	0.02 ± 0.02

LNR = lymph node ratio.

Figure 3 presents the Kaplan–Meier curves of the two categories for LNR. The log-rank test significantly showed that the group with LNR of >0.04 ($n = 11$) was related to shorter CSS ($p = 0.03$), DFS ($p < 0.01$), DMFS ($p = 0.01$), and LRRFS ($p < 0.01$) in comparison to the group with LNR of ≤0.04 ($n = 38$). Conversely, no significant relationship was found in OS between the two groups for LNR ($p = 0.32$).

Table 2 shows the relationship between clinicopathological parameters and the two categories. Pathological N1-N2 ($p < 0.01$), pathological stage III-IVA ($p < 0.01$), and the presence of postoperative treatment ($p < 0.01$) were more frequently in LNR of >0.04 compared with LNR of ≤0.04.

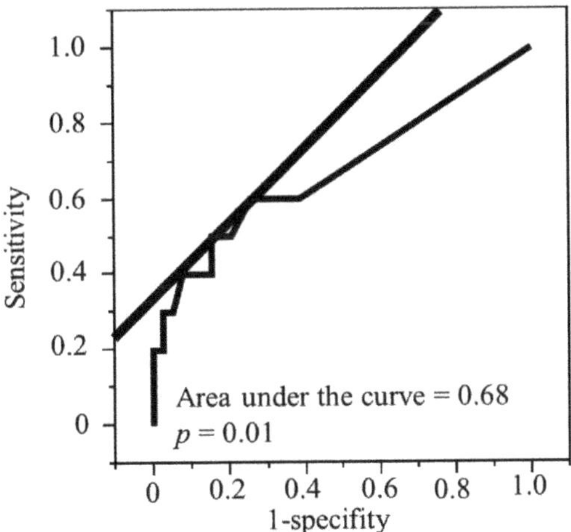

Figure 2. Receiver operating curves for 49 patients with head and neck squamous cell carcinoma.

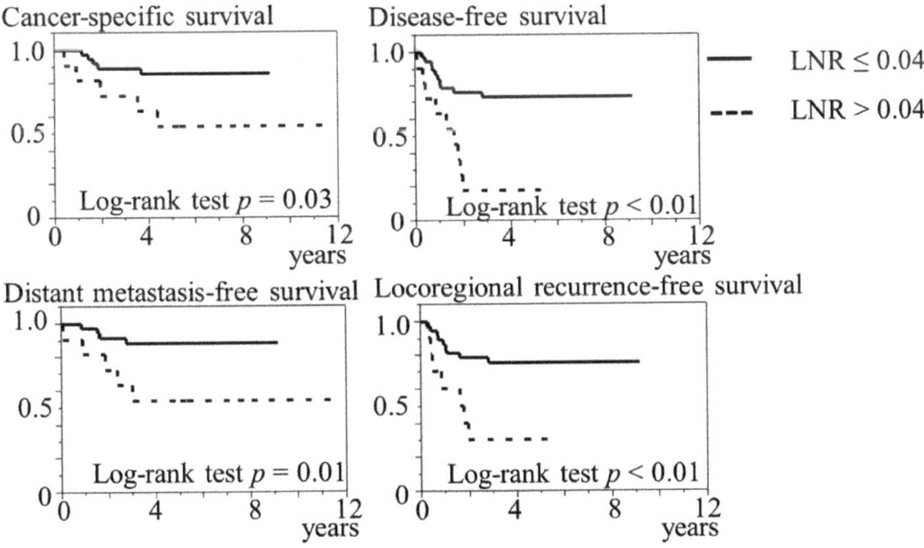

Figure 3. Kaplan–Meier curves in 49 patients were divided into two groups of lymph node ratios. LNR = lymph node ratio.

Table 3 presents the multivariate analyses. The LNR (>0.04/≤0.04) were significantly shorter CSS ($p = 0.02$, HR = 9.24, 95% CI = 1.23–176), DFS ($p = 0.02$, HR = 3.44, 95% CI: 1.23–10.3), and DMFS ($p = 0.01$, HR: 9.76, 95% CI: 1.57–187). No significant associations were found between LNR (>0.04/≤0.04) and LRRFS. Neither pathological Stage (III-IVA/I-II) nor extranodal extension and or positive surgical margin (presence/absence) were associated with survival results.

The LNR of patients with the presence of extranodal extension was a significantly higher value than those with the absence of extranodal extension ($p < 0.01$).

Table 2. Association between clinicopathologic parameters and LNR by Fisher's exact test.

Parameter		LNR ≤ 0.04 (*n* = 38)	LNR > 0.04 (*n* = 11)	*p*-Value
Age	<67 year	19	6	
	≥67 year	19	5	1
Sex	Male	22	9	
	Female	16	2	0.18
Pathological T classification	T1–2	25	7	
	T3–4	13	4	1
Pathological N classification	N0	35	2	
	N1–2	3	9	<0.01
Pathological stage	I–II	20	0	
	III–IVA	18	11	<0.01
Primary tumor size	<23 mm	18	3	
	≥23 mm	20	8	0.31
Primary site	Oral	30	7	
	Pharynx	8	4	0.43
Positive surgical margin	Presence	4	3	
	Absence	34	8	0.18
Extranodal extension	Presence	3	2	
	Absence	35	9	0.31
Type of neck dissection	Unilateral	34	11	
	Bilateral	4	0	0.56
Postoperative treatment	Presence	3	6	
	Absence	35	5	<0.01
Preoperative chemotherapy	Presence	4	2	
	Absence	34	9	0.61
Smoking history	Presence	22	5	
	Absence	16	6	0.51
Extranodal extension and or positive surgical margin	Presence	7	4	
	Absence	31	7	0.24

LNR = lymph node ratio.

Table 3. Multivariate survival analyses by Cox's hazard proportional model.

Parameter	CSS	DFS	DMFS	LRRFS
LNR (>0.04/≤0.04)				
HR	9.24	3.44	9.76	2.77
95% CI	1.49–176	1.23–10.3	1.57–187	0.91–8.65
p-value	0.02	0.02	0.01	0.07
Pathological Stage (III–IVA/I–II)				
HR	0.15	1.63	0.26	2.59
95% CI	0.01–1.22	0.43–6.71	0.01–2.29	0.65–12.7
p-value	0.08	0.47	0.24	0.18
Extranodal extension and or positive surgical margin (Presence/Absence)				
HR	3.67	1.16	2.00	0.71
95% CI	0.77–19.3	0.39–3.20	0.38–9.75	0.19–2.17
p-value	0.10	0.78	0.39	0.56

CSS = cancer-specific survival, DFS = disease-free survival, LRRFS = locoregional recurrence-free survival, DMFS = distant metastasis-free survival, LNR = lymph node ratio, HR = Hazard ratio, CI = confidence interval.

7. Discussion

The present study demonstrated using both univariate and multivariate survival analyses, adjusted with pathological stage and extranodal extension and or positive surgical margin, that a significant association existed between higher LNR and shorter CSS, DFS, and DMFS in patients with HNSCC who underwent by surgery with SMAF reconstruction.

The LNR, as a significant predictor for survival outcomes, was reported in HNSCC by meta-analyses and some individual institutions [3–5] and was evaluated for patients with focusing on the surgical procedure [3–5]. For example, the LNR in 79 patients after primary total laryngopharyngectomy was an independent predictor for OS, CSS, and DFS in univariate and multivariate analyses [13]. Furthermore, the LNR for 327 patients following minimally invasive esophagectomy also predicted OS [14]. Patients with focusing SMAF reconstruction in HNSCC revealed a significant relationship between survival outcomes and LNR, and are similar to previous results [13,14].

Several prognostic factors following SMAF reconstruction in HNSCC were investigated [9,15,16]. Among patients with both tumor resection and SMAF reconstruction, pathological metastasis of lymph node was related to shorter OS and CSS in 160 cases with T1-2 oral SCC [15], the pathological stage was related to DFS in 1169 cases [9], and N stage and pathological differentiation were related to locoregional recurrence in 229 cases [16]. The relationships between LNR and survival results in patients with SMAF reconstruction were not fully investigated because these studies did not investigate LNR [9,15,16]. Therefore, the present study is thought to contribute to additional research. Although one major problem certainly is dissection of level Ia in patients receiving SMAF reconstruction, we focused on both LNR and SMAF with interesting topics in this special cohort. The SMAF in OSCC is often used for patients with comorbid disorders to avoid long anesthesia or microvascular reconstruction due to safeness. This work combines two interesting topics for head and neck surgeons.

Extranodal extension and or the surgical margin and the pathological stage for multivariate analysis in the present study were selected due to both being possible confounding and comprehensive prognostic factors. As one of approaches derived from the pathophysiological significant relationship between the LNR and survival outcomes in both univariate and multivariate analyses of the present and previous results including meta-analysis [3–5], the LNR at operation with SMAF for HNSCC was considered for a pathological indicator for postoperative chemoradiaton or radiation.

The present study contains several limitations. Limited sample size was retrospectively observed by a single institution. Therefore, more utile results with more statistical points should be prospectively assessed by a larger cohort from multi-institutions. Because the tumor staging system used in this study was the International Union Against Cancer of the 7th edition, future study is advised to apply that from the American Joint Committee on Cancer of the 8th edition.

8. Conclusions

A high-level LNR in HNSCC was a prognostic factor for survival outcomes after operation with SMAF reconstruction.

Author Contributions: H.S. performed conceptualization, study design, data collection and statistical analysis, and writing. S.B., D.N., H.T., M.S. and N.H. performed data collection and review. All authors have read and agreed to the published version of the manuscript.

Funding: Support for this study was the Japan Society for the Promotion of Science (KAKENHI Grant Number: 21K09575).

Institutional Review Board Statement: This retrospective observational study at the Department of Head and Neck Surgery in our hospital, following the Declaration of Helsinki, was carried out and approved by our hospital review board (receipt number of 2019-1-427).

Informed Consent Statement: Patients had the chance to decline participation on institutional website of Aichi Cancer Center according to institutional review board.

Data Availability Statement: The datasets of this study are available based on reasonable request of the corresponding author.

Acknowledgments: We thank the staff of our institution for their patient's care.

Conflicts of Interest: The authors declared no conflict of interest.

Abbreviations

LNR: lymph node ratio; SCC: squamous cell carcinoma; HNSCC: head and neck squamous cell carcinoma; SMAF: submental artery flap; CSS: cancer-specific survival; DFS: disease-free survival; LRRFS: locoregional recurrence-free survival; DMFS: distant metastasis-free survival; OS: overall survival; ROC: receiver operating curve; AUC: area under the curve; HR: hazard ratio; CI: confidence interval.

References

1. de Boer, M.; van Deurzen, C.H.; van Dijck, J.A.; Borm, G.F.; van Diest, P.J.; Adang, E.M.; Nortier, J.W.; Rutgers, E.J.; Seynaeve, C.; Menke-Pluymers, M.B.; et al. Micrometastases or isolated tumor cells and the outcome of breast cancer. *N. Engl. J. Med.* **2009**, *361*, 653–663. [CrossRef] [PubMed]
2. Brierley, J.; Gospodarowicz, M.D.; Wittkeind, C. *TNM Classification of Malignant Tumors*, 8th ed.; International Union Against Cancer; Wiley-Blackwell: Oxford, UK, 2016.
3. Suzuki, H.; Matoba, T.; Hanai, N.; Nishikawa, D.; Fukuda, Y.; Koide, Y.; Hasegawa, Y. Lymph node ratio predicts survival in hypopharyngeal cancer with positive lymph node metastasis. *Eur. Arch. Otorhinolaryngol.* **2016**, *273*, 4595–4600. [CrossRef] [PubMed]
4. Gartagani, Z.; Doumas, S.; Kyriakopoulou, A.; Economopoulou, P.; Psaltopoulou, T.; Kotsantis, I.; Sergentanis, T.N.; Psyrri, A. Lymph node ratio as a prognostic factor in neck dissection in oral cancer patients: A systematic review and meta-analysis. *Cancers* **2022**, *14*, 4456. [CrossRef] [PubMed]
5. Suzuki, H.; Beppu, S.; Hanai, N.; Hirakawa, H.; Hasegawa, Y. Lymph node density predicts lung metastases in oral squamous cell carcinoma. *Br. J. Oral Maxillofac. Surg.* **2016**, *54*, 213–218. [CrossRef] [PubMed]
6. Martin, D.; Pascal, J.F.; Baudet, J.; Mondie, J.M.; Farhat, J.B.; Athoum, A.; Gaillard, P.; Peri, G. The submental island flap: A new donor site. Anatomy and clinical applications as a free or pedicled flap. *Plast. Reconstr. Surg.* **1993**, *92*, 867–873. [CrossRef] [PubMed]
7. Mishra, A.; Mishra, N.; Pati, D.; Samal, D.; Kar, I.B.; Mohapatra, D.; Sarkar, D.F. Oncologic safety of submental island flap reconstruction in clinically node-negative oral cancer patients: A prospective comparative study. *Int. J. Oral Maxillofac. Surg.* **2022**, *51*, 159–165. [CrossRef] [PubMed]
8. Gabrysz-Forget, F.; Tabet, P.; Rahal, A.; Bissada, E.; Christopoulos, A.; Ayad, T. Free versus pedicled flaps for reconstruction of head and neck cancer defects: A systematic review. *J. Otolaryngol. Head Neck Surg.* **2019**, *48*, 13. [CrossRef] [PubMed]
9. Pradhan, S.A.; Kannan, R.; Tiwari, N.; Jain, S.; Khan, S.; Rodrigues, D.; Doctor, A.; Jatale, R.G.; Agrawal, K.; Shaikh, M. Submental flap: Game changer in oral cancer reconstruction-A study of 1169 cases. *J. Surg. Oncol.* 2020, *in press*. [CrossRef] [PubMed]
10. Suzuki, H.; Iwaki, S.; Higaki, E.; Abe, T.; Sawabe, M.; Beppu, S.; Kobayashi, Y.; Nishikawa, D.; Terada, H.; Hanai, N. Supraclavicular artery flap for oral reconstruction prior to esophagectomy during the COVID-19 pandemic: A case report. *In Vivo* **2021**, *35*, 3597–3601. [CrossRef] [PubMed]
11. Sobin, L.H.; Gospodarowicz, M.K.; Wittekind, C. *TNM Classification of Malignant Tumours*, 7th ed.; Wiley-Blackwell: Oxford, UK, 2009.
12. Nakata, Y.; Hanai, N.; Nishikawa, D.; Suzuki, H.; Koide, Y.; Fukuda, Y.; Nomura, M.; Kodaira, T.; Shimizu, T.; Hasegawa, Y. Comparison between chemoselection and definitive radiotherapy in patients with cervical esophageal squamous cell carcinoma. *Int. J. Clin. Oncol.* **2017**, *22*, 1034–1041. [CrossRef] [PubMed]
13. Grasl, S.; Janik, S.; Parzefall, T.; Formanek, M.; Grasl, M.C.; Heiduschka, G.; Erovic, B.M. Lymph node ratio as a prognostic marker in advanced laryngeal and hypopharyngeal carcinoma after primary total laryngopharyngectomy. *Clin. Otolaryngol.* **2020**, *45*, 73–82. [CrossRef] [PubMed]
14. Kitamura, Y.; Oshikiri, T.; Takiguchi, G.; Urakawa, N.; Hasegawa, H.; Yamamoto, M.; Kanaji, S.; Yamashita, K.; Matsuda, T.; Fujino, Y.; et al. Impact of lymph node ratio on survival outcome in esophageal squamous cell carcinoma after minimally invasive esophagectomy. *Ann. Surg. Oncol.* **2021**, *28*, 4519–4528. [CrossRef] [PubMed]
15. Wang, J.; Tan, Y.; Shen, Y.; Lv, M.; Li, J.; Sun, J. Oncological safety of submental island flap for reconstruction of pathologically node-negative and node-positive T1-2 oral squamous cell carcinoma-related defects: A retrospective study and comparison of outcomes. *Oral Oncol.* **2020**, *102*, 104507. [CrossRef] [PubMed]
16. Thomas, S.; Varghese, B.T.; Ganesh, S.A.; Desai, K.P.; Iype, E.M.; Balagopal, P.G.; Sebastian, P. Oncological Safety of submental artery island flap in oral reconstruction—analysis of 229 cases. *Indian J. Surg. Oncol.* **2016**, *7*, 420–424. [CrossRef] [PubMed]

 *biomedicines*

Article

T-Cell Infiltration and Immune Checkpoint Expression Increase in Oral Cavity Premalignant and Malignant Disorders

Subin Surendran [1], Usama Aboelkheir [1], Andrew A. Tu [1], William J. Magner [1], S. Lynn Sigurdson [1], Mihai Merzianu [2], Wesley L. Hicks, Jr. [1], Amritha Suresh [1,3], Keith L. Kirkwood [4] and Moni A. Kuriakose [1,3,*]

[1]   Head & Neck Surgery, Roswell Park Comprehensive Cancer Center, Buffalo, NY 14263, USA; subin.thenkunnelsurendran@roswellpark.org (S.S.); usama_aboelkheir@yahoo.com (U.A.); tu.andrew.2@gmail.com (A.A.T.); william.magner@roswellpark.org (W.J.M.); lynn.sigurdson@roswellpark.org (S.L.S.); wesley.hicks@roswellpark.org (W.L.H.J.); amritha.suresh@ms-mf.org (A.S.)
[2]   Pathology, Roswell Park Comprehensive Cancer Center, Buffalo, NY 14263, USA; mihai.merzianu@roswellpark.org
[3]   Integrated Head and Neck Oncology Program, Mazumdar Shaw Medical Foundation Bangalore, Bangalore 560099, India
[4]   Periodontics and Endodontics, University at Buffalo School of Dental Medicine, Buffalo, NY 14214, USA; klkirk@buffalo.edu
*   Correspondence: moni.kuriakose@roswellpark.org; Tel.: +1-716-845-3158

Citation: Surendran, S.; Aboelkheir, U.; Tu, A.A.; Magner, W.J.; Sigurdson, S.L.; Merzianu, M.; Hicks, W.L., Jr.; Suresh, A.; Kirkwood, K.L.; Kuriakose, M.A. T-Cell Infiltration and Immune Checkpoint Expression Increase in Oral Cavity Premalignant and Malignant Disorders. *Biomedicines* **2022**, *10*, 1840. https://doi.org/10.3390/biomedicines10081840

Academic Editor: Randolph C. Elble

Received: 30 June 2022
Accepted: 26 July 2022
Published: 30 July 2022

Publisher's Note: MDPI stays neutral with regard to jurisdictional claims in published maps and institutional affiliations.

Abstract: The immune cell niche associated with oral dysplastic lesion progression to carcinoma is poorly understood. We identified T regulatory cells (Treg), CD8[+] effector T cells (Teff) and immune checkpoint molecules across oral dysplastic stages of oral potentially malignant disorders (OPMD). OPMD and oral squamous cell carcinoma (OSCC) tissue sections (N = 270) were analyzed by immunohistochemistry for Treg (CD4, CD25 and FoxP3), Teff (CD8) and immune checkpoint molecules (PD-1 and PD-L1). The Treg marker staining intensity correlated significantly ($p < 0.01$) with presence of higher dysplasia grade and invasive cancer. These data suggest that Treg infiltration is relatively early in dysplasia and may be associated with disease progression. The presence of CD8[+] effector T cells and the immune checkpoint markers PD-1 and PD-L1 were also associated with oral cancer progression ($p < 0.01$). These observations indicate the induction of an adaptive immune response with similar Treg and Teff recruitment timing and, potentially, the early induction of exhaustion. FoxP3 and PD-L1 levels were closely correlated with CD8 levels ($p < 0.01$). These data indicate the presence of reinforcing mechanisms contributing to the immune suppressive niche in high-risk OPMD and in OSCC. The presence of an adaptive immune response and T-cell exhaustion suggest that an effective immune response may be reactivated with targeted interventions coupled with immune checkpoint inhibition.

Keywords: oral cancer; oral potentially malignant disorders; tumor microenvironment; immunotherapy; immune cell infiltration; immune checkpoint; oncogenesis; tumorigenesis

1. Introduction

Oral cavity potentially malignant disorders (OPMD), such as leukoplakia and erythroplakia, are often the precursors to invasive squamous cell carcinoma of the oral cavity. Oral squamous cell carcinoma (OSCC) accounts for more than 300,000 cases each year worldwide [1] with a static 5-year survival rate of <50% [2,3]. The pathology of these oral cavity disorders may exhibit progression from hyperplasia and dysplasia to frank invasive cancer [4]. Dysplasia is pathologically divided into mild, moderate and severe. It is extremely difficult to clinically distinguish which of these lesions will progress to malignancy. Furthermore, there is a scarcity of literature addressing the underlying mechanisms, particularly those of the tumor microenvironment (TME) in disease progression.

Characterizing tumor niche, especially the immune cells, during early carcinogenesis will enable a heightened understanding of the cell types and their interactions during the early events of tumor immune evasion. We are interested in understanding how the TME contributes to oral carcinogenesis. While the oral carcinoma immune profile is well-characterized [5,6], comparatively few studies have addressed the role of the immune cell niche in premalignant lesions [7,8].

Two publications have described the accumulation of regulatory T cell (Treg) and immune checkpoint expression in the oral cancer TME; however, they only compared OSCC to normal tissue [6,9]. Immune checkpoints dampen adaptive immune responses and contribute to tumor evasion. A single publication demonstrated Treg accumulation in OPMD and OSCC [8]. Our previous work examined CXCL12 and CXCR4, a chemokine axis known to recruit Treg, in OPMD and OSCC [10].

Two additional publications addressed PD-1 and T cells in limited oral leukoplakia cohorts without precise dysplasia classification [11,12]. Our aim was to address gaps in the literature through determination of Treg, CD8 and immune checkpoint levels during progression to oral cancer. In the current study, we postulate that there are specific changes in the OPMD TME that are associated with the progression of these lesions to carcinoma.

We hypothesize that regulatory T cells and immune checkpoint molecules accumulate in the OPMD TME and correlate with oral dysplastic progression to OSCC. To test this hypothesis, we investigated Treg (CD4, CD25 and FoxP3), Teff (CD8) and immune checkpoint molecule (PD-1 and PD-L1) expression across the spectrum of oral dysplasia (mild, moderate and severe) to invasive carcinoma. Our study of OPMD and its TME immune components may be an important step towards identifying pathways of tumor malignancy and immune escape.

2. Materials and Methods

2.1. Patient Cohorts

The study was approved by the Institutional Review Board (RPCI IRB No: I66805) and conducted on de-identified formalin-fixed paraffin-embedded (FFPE) oral lesion samples from patients with histologically confirmed oral potentially malignant disorders (OPMD) or oral cancer (OSCC) who presented to the Department of Head and Neck Surgery, Roswell Park Comprehensive Cancer Center (2006 to 2013). Biopsies with clinically and histologically normal mucosa were used as controls. Samples were histologically classified as normal mucosa; parakeratosis/hyperplasia; mild, moderate or severe dysplasia; and carcinoma based on pathology assessments. An ordinal score between 1 and 6 was assigned to stratify the lesions as normal tissues (1), parakeratosis/hyperplasia (2), mild (3), moderate (4), severe dysplasia (5) and carcinoma (6).

2.2. Immunohistochemical Staining

FFPE blocks were sectioned (5 μm) for slides and then deparaffinized sequentially in xylene and ethanol. Antigen retrieval was performed in TRS (Target Retrieval Solution, High pH, Dako, Denmark) for 30 min steaming followed by 20 min at room temperature. The slides were incubated with 0.3% hydrogen peroxide in methanol to block endogenous peroxidases, rinsed with tris-buffered saline (TBS, Dako, Denmark) and incubated overnight at 4 °C with primary antibodies (Supplementary Table S1). Antibody detection was performed using mouse- and rabbit-specific HRP/DAB (ABC) secondary antibody detection kits (Abcam-ab64264) following the manufacturer's protocol. All slides were counterstained with Mayer's hematoxylin, mounted and scanned using (Aperio ScanScope XT 1509, AT2, Leica Biosystems, IL, USA) at 20(×) magnification.

2.3. Scoring and Analysis of Immunohistochemical Data

The presence of tumor-infiltrating lymphocytes (TIL) was quantitatively scored by counting the cells positively stained for FoxP3, CD4, CD8, CD25 and PD-1 [6,13]. An independent pathologist defined the area of sections to be assessed based on histology, and two independent reviewers blinded to clinicopathological data of the samples assessed the staining of each slide. Each image from the Aperio Scanscope (20(×) magnification) was raised 20(×) in Imagescope analysis software (Leica biosystems, IL, USA) resulting in a high-power field (hpf) with a total magnification of 400(×). The number of positively stained cells was counted in each hpf until the entire section was evaluated. In every section, 3–15 hpfs were scored and averaged to obtain the final count. The values from both reviewers were averaged, and discrepancies in scoring were re-assessed by both reviewers.

The expression of PD-L1 was quantitated using the H-score method. The sum of the intensity (0 = negative staining; 1+ = weak staining; 2+ = moderate staining; and 3+ = strong staining) and percentage of positively stained tumor cells was calculated with the H-score ranging from 0 (negative) to 300: H-score = (1 × % weakly stained cells) + (2 × % moderately stained cells) + (3 × % strongly stained cells) [13].

2.4. Statistical Analysis

All statistical analyses were performed using IBM SPSS Statistics v26 (IBM, Armonk, NY, USA), including standard frequency and descriptive assessments as well as bivariate correlation analysis and multivariate analysis. Pearson (scale) and Spearman (ordinal) correlations were analyzed as applicable. Significance was determined with the Chi-square test (two-tailed sigma <0.05). Survival analysis was performed using the Kaplan–Meier algorithm with significance determined by the Mantel–Cox Log Rank test $p < 0.05$. Power analyses were performed in G*Power 3.1.

3. Results

3.1. Patient Characteristics

A cohort of 270 OPMD and malignant tissue samples obtained from 128 patients was evaluated for CD4, CD8, FoxP3, CD25, PD-1 and PD-L1. Patient demographics and smoking status are described in Table 1. Lesions with mild or no dysplasia were classified as low-risk, while moderate dysplasia, severe dysplasia and carcinoma in situ were considered high-risk lesions (Table 2). The malignant lesions included micro-invasive carcinoma and carcinoma regardless of stage and differentiation.

Table 1. Patient demographics.

Characteristic	Cases	Percentage
Sex		
Male	71	55.5%
Female	57	45.5%
Age Group (years)		
28–50	30	23.4%
51–60	40	31.3%
61–70	35	27.3%
71+	23	18.0%
Race		
White	116	90.6%
Black	6	4.7%
Asian	3	2.3%
American Indian or Alaskan Native	1	0.8%
Hispanic	1	0.8%
Other	1	0.8%
Ever Smoker		
Yes	82	64.1%
No	45	35.9%
Total Patients	128	100%
Mean Age = 59.9, Median 60.0		

Table 2. Pathology distribution of samples.

Pathology	N	Percentage
Normal	87	32.2%
Hyperplasia/Parakeratosis	45	16.7%
Mild Dysplasia	69	25.6%
Moderate Dysplasia	22	8.1%
Severe Dysplasia/Carcinoma in situ	13	4.8%
Carcinoma	34	12.6%
Total Samples	**270**	

3.2. Immune Marker Detection across Oral Cancer Stages

The first goal of our study was to identify the presence of Teff, Treg and PD-1/PD-L1 in OPMD and OSCC microenvironments. To quantify the presence of regulatory T cells, we performed immunohistochemical staining for CD4, CD25 and FoxP3 on the FFPE sections (Figure 1) (Table 3). We also stained for effector T cell marker CD8 and immune checkpoint components PD-1 and PD-L1 in these same samples. Increasing the expression of all these markers was evident throughout disease progression as reflected by the average number of positively stained cells for immune markers and PD-1 and H-score for PD-L1. (Figure 1) (Table 3). Correlation coefficients were determined between the IHC scores and ordinal designations of pathology (Figure 1) (Table 4). The Spearman correlations suggest that the levels of each of these markers in the oral microenvironment increase with disease progression.

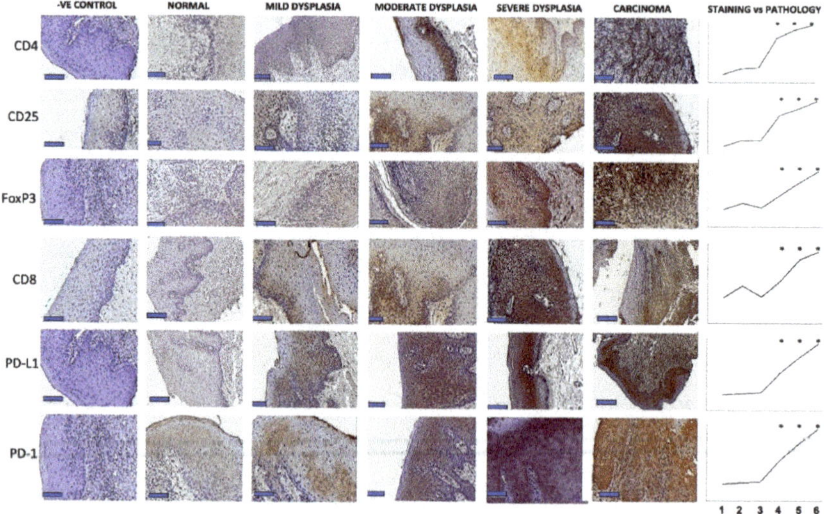

Figure 1. Infiltrating lymphocytes and immune checkpoint expression levels increase in parallel with oral cancer progression. Immunohistochemical analysis of immune markers in patient clinical samples pathologically defined as normal mucosa, pro-gressive grades of dysplasia (mild, moderate, and severe dysplasia) and carcinoma. Representative images are shown at total magnification $400(\times)$ (scale bar, 100 μm). Average IHC staining for each immune marker was calculated and correlated with pa-thology (ordinal scale). The Y-axis in the graph indicates the IHC score and the X-axis indicates the pathological distribution of patient cohorts 1-normal, 2- parakeratosis (images not shown), 3-mild dysplasia, 4-moderate dysplasia, 5-severe dysplasia and 6-carcinoma. Pathology cohorts whose IHC staining value is significantly different from the normal group are indicated (*, $p < 0.05$).

Table 3. T cell, checkpoint molecule and chemokine quantitation in oral potentially malignant and malignant lesions.

Pathology		CD25 (N:247)	CD4 (N:263)	FoxP3 (N:248)	CD8 (N:247)	PD-L1 (N:256)	PD-1 (N:253)	CXCR4 ^ (N:209)	CXCL12 ^ (N:211)
Normal	Mean [#]	15.6	17.0	39.3	25.2	34.3	32.0	2.7	115.0
	SEM	1.7	1.6	3.5	1.9	5.3	5.3	0.3	6.6
Parakeratosis	Mean	31.4	32.0	54.4	37.0	36.4	25.5	2.3	112.8
	SEM	6.7	5.1	6.8	4.0	7.7	6.0	0.4	10.2
Mild dysplasia	Mean	28.5	31.0	43.0	26.0	39.0	38.6	3.2	127.7
	SEM	3.8	2.7	4.0	2.1	4.6	4.5	0.3	6.9
Moderate dysplasia	Mean	83.3	92.3	72.2	41.6	82.1	97.0	3.4	93.6
	SEM	11.0	11.8	12.8	5.5	13.8	14.7	0.7	22.7
Severe dysplasia	Mean	114.1	129.1	101.1	69.5	127.3	112.5	5.0	198.1
	SEM	12.8	13.2	12.8	7.9	17.0	20.6	1.0	20.4
Carcinoma	Mean	103.2	109.4	120.4	65.1	134.3	139.3	5.0	153.5
	SEM	9.6	8.7	6.3	5.8	14.1	16.2	0.7	16.4

^, Published data [10]; [#], average number of positively stained cells/hpf (CD25, CD4, FoxP3, CD8, PD-1, CXCR4), average Histoscore/hpf (PD-L1, CXCL12); and SEM, Standard Error of the Mean.

Table 4. Correlations between pathology and IHC assessments.

	Pathology	CD25	CD4	FoxP3	CD8	PD-L1	PD-1	CXCR4 ^	CXCL12 ^
Pathology	1.00								
CD25	**0.64**	1.00							
CD4	**0.70**	**0.96**	1.00						
FoxP3	**0.52**	**0.71**	**0.69**	1.00					
CD8	**0.49**	**0.79**	**0.78**	**0.54**	1.00				
PD-L1	**0.52**	**0.54**	**0.58**	**0.48**	**0.51**	1.00			
PD-1	**0.54**	**0.53**	**0.55**	**0.45**	**0.46**	**0.74**	1.00		
CXCR4 ^	0.27	0.17 [#]	0.23	0.14	0.22	0.35	0.19 [#]	1.00	
CXCL12 ^	0.20	0.14 [#]	0.17	0.15	0.12	0.32	0.29	0.32	1.00

IHC vs. pathology (scored on an ordinal scale) associations are presented as Spearman correlation coefficients whereas correlations between markers are presented as Pearson correlation coefficients. bold, $p < 0.01$; [#], $p < 0.05$; and ^ published IHC data re-analyzed in the context of new data presented here [10].

The intensity of CD25 staining correlated significantly ($p < 0.01$) with the severity of dysplasia (Figure 1) (Table 4). A similar pattern was observed for CD4 wherein high-risk dysplastic and carcinoma samples showed higher infiltration compared to normal tissues ($p < 0.01$) (Figure 1) (Table 4). The presence of FoxP3[+] cells strongly correlated with pathology ($p < 0.01$) (Figure 1) (Table 4). As Tregs are expected to express CD4, CD25 and FoxP3, we tested the correlations between these markers (Table 4). The strong positive correlations between CD4, CD25 and FoxP3 (correlation coefficients 0.64–0.96, $p < 0.01$) support our hypothesis that Tregs are recruited to OPMD and OSCC. Increased Treg levels with advanced pathology indicate their probable role in immune escape and potential role in immunotherapy resistance.

CD8[+] T cells are the effector T cells critical to immune-mediated tumor control and to immunotherapy responsiveness. To quantify Teff infiltration into OPMD and oral cancer lesions, we performed CD8 immunohistochemistry on our panel of oral tissues. Our results identified cytotoxic T cells within the oral tissues and OSCC TME (Figure 1) (Table 3), and the presence of CD8[+] cells positively correlated with dysplastic progression (Spearman coeff = 0.49, $p < 0.01$) (Table 4).

It has been previously demonstrated that activated effector T cells can be inhibited in the TME by tumor expression of PD-1 ligands [14–16]. To determine the expression of this immune checkpoint in OPMD and OSCC, we stained our oral tissue panel for PD-1 and PD-L1 (Figure 1) (Table 3). PD-L1 IHC scores in each OPMD/OSCC category were higher than the normal samples and reached statistical significance at moderate dysplasia. This increase in PD-L1 staining associated with dysplastic progression was validated by its Spearman correlation coefficient (0.52 $p < 0.01$) (Table 4).

Similarly, PD-1 detection began to increase in the mild dysplasia samples and was significantly elevated in moderate and severe dysplasia as well as carcinoma samples (Spearman correlation coefficient 0.54, $p < 0.01$) (Table 4). Importantly, our PD-1 and CD8 quantitation closely correlated with each other ($p < 0.001$) suggesting the expected colocalization. Similarly, CD4 cells were also tightly correlated with PD-1 expression. PD-1 detection did not correlate with CXCL12, which is expressed by stromal cells, including cancer-associated fibroblasts supporting the specificity of our IHC assessment.

3.3. Correlations between Treg, Teff and Immune Checkpoints

Our analysis demonstrated that Treg and Teff were present at multiple stages of oral mucosal dysplasia and cancer. It is possible that these populations play a mechanistic role in tumor development, as these patterns appear to be strikingly similar. To quantify these patterns, we examined the Pearson correlation coefficients between CD8 and Treg IHC values (CD8:FoxP3 correlation coefficient 0.54, $p < 0.01$) (Table 4). The close alignment of these populations may indicate a distinct mechanism for suppressing effective CD8$^+$ cytotoxic T lymphocyte (CTL)/Teff responses in the oral TME.

The PD-L1 levels were also closely correlated with CD8 levels (Pearson correlation coefficient 0.51, $p < 0.01$) (Table 4) suggesting that, although Teff were present in the lesion and tumor microenvironments, they could be subject to immune checkpoint inhibition and rendered ineffective due to exhaustion.

3.4. CXCR4-CXCL12 Correlation with Immune Landscape

Our previously published work in OPMD that dealt with cancer stem cells and markers of chemokine pathways indicated that the presence of CXCL12 (SDF-1) and CXCR4 in oral premalignant and malignant sections correlated with disease progression [10]. The patient samples described here were included in our previous cohorts. The CXCL12:CXCR4 axis is known to facilitate homing of Treg; thus, we incorporated these published IHC data into our new analysis. CXCR4 cytoplasmic staining was positively correlated with CD25 (Treg) as well as the immune checkpoint marker PD-1 ($p < 0.05$) (Table 4).

3.5. Survival Analysis

In a small subset of OSCC patients, survival analysis was performed. IHC scores for each marker were analyzed based on quartiles and survival analyses were performed for the highest vs. lowest staining levels. Since few patients in this study had overall survival data available, the date of last follow up was considered as a surrogate marker (N = 19). Kaplan–Meier survival analyses revealed intriguing trends suggesting that lower levels of Treg recruitment (CD25 and FoxP3) were associated with improved survival; however, none reached statistical significance (Figure 2).

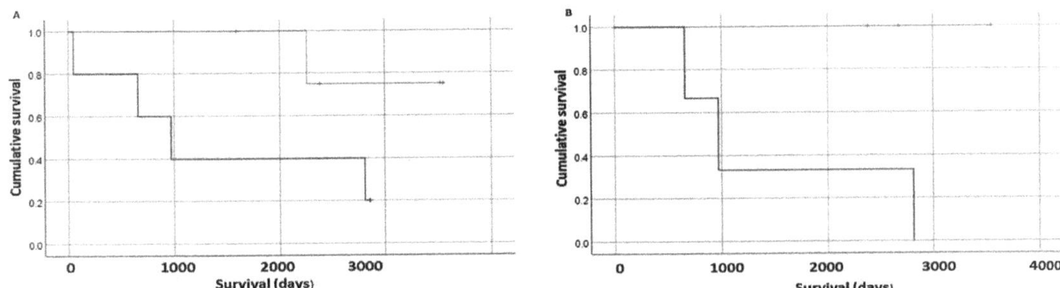

Figure 2. Survival analysis based on Treg surrogate markers. (**A**), Kaplan–Meier analysis of OPMD/OSCC patients stratified by first vs. fourth quartile of CD25 staining. The red line represents the highest quartile CD25 detected, and the blue line represents the lowest quartile of CD25 detection. The survival difference did not reach statistical significance—Mantel–Cox Log Rank test $p = 0.092$. (**B**) Kaplan–Meier analysis of OPMD/OSCC patients stratified by first vs. fourth quartile of FoxP3 staining. The red line represents the highest quartile FoxP3 detected, and the blue line represents the lowest quartile of FoxP3 detection. The survival difference did not reach statistical significance—Mantel–Cox Log Rank test $p = 0.063$.

4. Discussion

Immune dysfunction within the OPMD and OSCC TME may contribute to disease evolution [17–19]. Immune changes during dysplastic progression have not been well defined. Therefore, a better understanding of these immune landscapes would provide insights into underlying mechanisms and potential therapeutic targets. The TME includes both pro-tumorigenic and anti-tumorigenic immune cells. The regulation and balance of these cell populations influence the immune responses within the tumor cell/stromal framework, including mechanisms that support escape from immune surveillance [20,21] and subsequent malignant transformation.

Although the oral carcinoma immune profile is well characterized [5,6], few studies have addressed the role of the immune cell niche during oral carcinogenesis. Kouketsu et al. demonstrated Treg accumulation in OPMD and OSCC [8]. Chen et al. reported CD8 and PD-L1 increased expression in OSCC versus healthy tissues [12]. Their samples included leukoplakia, of which six samples were considered dysplastic.

We confirmed and extended these observations by incorporating chemokine pathway, immune checkpoint and Teff infiltration detection in a cohort of patient samples that included normal, mild, moderate and severe dysplasia as well as carcinoma. The data reported here is the first thorough examination of CD4, CD8 and immune checkpoint changes through pathologically defined stages of oral dysplasia. Where data were available, we analyzed the impact of these immune microenvironment changes on patient survival.

Lymphocyte infiltration into the TME is generally viewed as an indication of host immunity against the tumor, and its presence may alter tumor biology [22,23]. $CD8^+$ Teff (CTL) are immune anti-tumor effector cells whereas regulatory T cells (Treg, $CD25^+CD4^+FoxP3^+$) can suppress this cytotoxic T-cell function. The interplay between Treg and Teff is crucial in immune evasion by cancer cells, anti-tumor response and immune homeostasis. The presence of $CD8^+$ T cells is associated with a more favorable clinical prognosis [24,25]. However, Treg express inhibitory coreceptors and release immune-suppressive cytokines, thereby, hampering cytotoxic T lymphocytes, natural killer (NK) cells, dendritic cells (DC) and B cell function as well as dampening anti-tumor immunity [26–30].

Treg cells express the chemokine receptor CXCR4 and can migrate along a gradient of its ligand CXCL12 [31]. This axis can contribute to Treg recruitment to the TME and the resulting immune-suppressive tumor environment. Our previously published work demonstrated the enhanced expression of CXCL12 and CXCR4 during malignant progression of oral dysplastic lesions [10]. In aggregate, the data presented by this and other studies reflect the importance of both the immune response and epithelial stromal interactions in the regulation/progression of disease.

This study quantified the presence of specific T cell subsets (CD4, CD25, FoxP3 and CD8) in the stromal milieu along with activation of immune modulator pathways during OPMD progression and in OSCC. Our results identified the presence of an increasingly immunosuppressive microenvironment during dysplastic progression as indicated by Treg recruitment and activation of the PD-1 immune checkpoint axis. This immune shift correlated with increases in CXCR4, CXCL12 and CD44 shown in our previous work [10]. $CD8^+$ T cell recruitment in early lesions increased along the spectrum of progression and in OSCC.

The presence of $CD8^+$ T cells in dysplastic lesions and OSCC may be a positive prognostic indication; however, the coordinate induction of immune checkpoint expression suggested an additional layer of immune suppression in oral dysplasia and carcinoma. The joint recruitment of Treg and Teff as well as the induction of immune checkpoint expression establish the balance between a productive immune response and immunotherapy resistance. Understanding this balance will provide novel targets for therapeutic intervention to improve patient responses. The identification of these features in dysplastic lesions may support earlier intervention that might disrupt this reinforced suppressive environment.

Several studies have shown strong tumor infiltration by $CD8^+$ cells in head and neck cancer [32,33]. A tongue cancer study indicated that malignant transformation was accompanied by an increase in $CD4^+$ and B cells, in addition to $CD8^+$ and $CD14^+$ cells [18]. We focused on $CD4^+$ Treg, $CD8^+$ Teff cells and the PD-1 immune checkpoint and their potential association with pre-malignant and malignant oral cavity lesions. The data reported here revealed a strong correlation between CD4, CD25 and FoxP3 markers, indicating a high incidence of Treg cells, which in turn correlated with dysplastic progression to carcinoma. We also found significantly increased numbers of $CD8^+$ cells with increasing severity of dysplasia in our patient cohorts, which is consistent with studies performed on other malignancies [34–37].

Recent studies emphasized the role of $CD8^+$ T cells in the control of tumor growth and the prolongation of patient survival [35,38]. The concurrent presence of Treg and $CD8^+$ T cells in our study, however, suggests a potentially responsive immune environment that has shifted to immunosuppression. Additionally, the significant drop in the CD8/Treg infiltration in the high-risk group as compared to the low-risk cohort suggested that the low-risk cohort (mild to moderate dysplasia) might benefit from immunological modulation.

The correlation between FoxP3 and disease grade indicated an increase in Treg recruitment to oral cavity lesions during disease progression. Since Treg can counteract the effect of immune checkpoint blockade, their presence could impair immune attack and result in decreased survival. The limited patient follow up in our dataset revealed a trend toward Treg recruitment having a negative impact on survival. Future studies with more extensive patient observation will strengthen this association.

Immune checkpoints are critical regulators of immune homeostasis, have been co-opted by tumors to support immune evasion and are now therapeutic targets showing variable disease site efficacy. PD-L1 overexpression in solid tumors, including head and neck cancers [39], can provide direct tumor protection and reduce the activity of PD-1 expressing tumor-infiltrating effector $CD8^+$ T cells [40,41]. PD-L1 expression in oral dysplasia has not been well characterized.

Our study indicated that the PD-1/PD-L1 axis correlated with disease progression. The combined increase in the expression of PD-1/PD-L1 and Treg suggested a highly immune suppressive environment in this cohort. Although ICI immunotherapy has been approved for head and neck squamous cell carcinoma, a clinical response is only observed in <20% of patients [42,43]. The primary challenges in checkpoint inhibitor therapy are the overlapping, parallel and synergistic immune escape pathways that tumor cells utilize to modulate the niche and survive immune surveillance and therapy.

Our data describe Treg, Teff and immune checkpoint changes across pathologically defined stages of dysplasia and provide a longitudinal view of the immune microenvironment during OPMD/OSCC progression. However, our approach cannot address every marker or cell type of interest. Future studies using high-throughput methods will be necessary for complete characterization of the dysplastic immune microenvironments. Interest has recently been expressed in the potential use of immunotherapy to prevent progression of dysplastic lesions [44]. Further characterization of these pathways and their modulators in dysplasia and OSCC is likely to identify novel targets for both chemoprevention and chemotherapy.

5. Conclusions

Our results established a combined presence of Treg and CD8$^+$ T cells and immune checkpoint molecule expression in OPMD/OSCC microenvironments. Our data demonstrated dynamic changes in the numbers and balance among these immune components across the spectrum of dysplastic progression and carcinoma. These data support the feasibility of immunotherapy for OPMD and its potential application in OSCC chemoprevention.

Our results also suggest that T-cell exhaustion may contribute to immune escape and immunotherapy resistance in dysplastic progression to OSCC. These data highlight the need for the parallel targeting of mechanisms contributing to the immune suppressive niche in high-risk OPMD and in OSCC. The accumulation of Treg, as well as CD8$^+$ Teff relatively early in the dysplastic progression suggests that a combinatorial intervention, including immune checkpoint inhibition, may prevent progression by interrupting immune escape or enhancing the immunotherapeutic response.

Supplementary Materials: The following supporting information can be downloaded at: https://www.mdpi.com/article/10.3390/biomedicines10081840/s1.

Author Contributions: Conceptualization: S.S., U.A. and M.A.K.; Methodology: S.S., A.S. and W.J.M.; Software: S.S., U.A., A.A.T. and W.J.M.; Validation: A.S., W.J.M. and M.M.; Formal analysis: S.S., U.A., A.A.T., W.J.M. and A.S.; Investigation: S.S., A.S. and W.J.M.; Resources: M.M.; Data curation: W.J.M.; Writing—original draft preparation: S.S. and U.A.; Writing—review and editing: S.L.S., A.S., K.L.K., W.J.M., M.M., W.L.H.J. and M.A.K.; Visualization: S.S., U.A., A.A.T. and M.M.; Supervision: A.S., W.J.M., S.L.S., M.A.K. and W.L.H.J.; Project administration: M.A.K., W.L.H.J. and S.L.S.; Funding acquisition: M.A.K. and W.L.H.J. All authors have read and agreed to the published version of the manuscript.

Funding: This work was partially supported by Roswell Park Comprehensive Cancer Center recruitment start up grant #71 4069 01 (M.A.K.) and United States Department of Defense (DoD) grant W81XWH-20-PRCRP-IPA-CA200341 (K.L.K.).

Institutional Review Board Statement: The study was conducted in accordance with the Declaration of Helsinki and approved by the Institutional Review Board of Roswell Park Comprehensive Cancer Center, Buffalo, NY, USA (RPCI IRB No: I66805).

Informed Consent Statement: Informed consent was obtained from all subjects involved in the study.

Data Availability Statement: The data presented in this study are available on request from the corresponding author.

Acknowledgments: Erika K VanDette, Pathology Core Facility, Roswell Park Comprehensive Cancer Center. Wiam Bshara, Pathology Core Facility, Roswell Park Comprehensive Cancer Center. Biospecimens for this study were provided by the Pathology Network Shared Resource, which is funded by the National Cancer Institute (NCI P30CA16056) and is a Roswell Park Comprehensive Cancer Center Cancer Center Support Grant shared resource.

Conflicts of Interest: The authors declare no conflict of interest.

References

1. Jiang, Z.; Wu, C.; Hu, S.; Liao, N.; Huang, Y.; Ding, H.; Li, R.; Li, Y. Research on neck dissection for oral squamous-cell carcinoma: A bibliometric analysis. *Int. J. Oral Sci.* **2021**, *13*, 13. [CrossRef] [PubMed]
2. Jie, W.P.; Bai, J.Y.; Li, B.B. Clinicopathologic analysis of oral squamous cell carcinoma after interstitial brachytherapy. *Technol. Cancer Res. Treat.* **2018**, *17*, 1533033818806906. [CrossRef] [PubMed]
3. Xu, H.; Jin, X.; Yuan, Y.; Deng, P.; Jiang, L.; Zeng, X.; Li, X.-S.; Wang, Z.-Y.; Chen, Q.-M. Prognostic value from integrative analysis of transcription factors c-Jun and Fra-1 in oral squamous cell carcinoma: A multicenter cohort study. *Sci. Rep.* **2017**, *7*, 7522. [CrossRef] [PubMed]
4. Condurache, H.; Oana, M.; Botez, A.E.; Olinici, D.T.; Onofrei, P.; Stoica, L.; Grecu, V.B.; Toader, P.M.; Gheucă-Solovăstru, L.; Cotrutz, E.C. Molecular markers associated with potentially malignant oral lesions (Review). *Exp. Ther. Med.* **2021**, *22*, 834. [CrossRef]
5. Diao, P.; Jiang, Y.; Li, Y.; Wu, X.; Li, J.; Zhou, C.; Jiang, L.; Zhang, W.; Yan, E.; Zhang, P.; et al. Immune landscape and subtypes in primary resectable oral squamous cell carcinoma: Prognostic significance and predictive of therapeutic response. *J. ImmunoTherapy Cancer* **2021**, *9*, e002434. [CrossRef]
6. Stasikowska-Kanicka, O.; Wągrowska-Danilewicz, M.; Danilewicz, M. Immunohistochemical analysis of Foxp3(+), CD4(+), CD8(+) cell Infiltrates and PD-L1 in oral squamous cell carcinoma. *Pathol. Oncol. Res.* **2018**, *24*, 497–505. [CrossRef] [PubMed]
7. Foy, J.P.; Bertolus, C.; Ortiz-Cuaran, S.; Albaret, M.A.; Williams, W.N.; Lang, W.; Destandau, S.; Souza, G.; Sohier, E.; Kielbassa, J.; et al. Immunological and classical subtypes of oral premalignant lesions. *Oncoimmunology* **2018**, *7*, e1496880. [CrossRef] [PubMed]
8. Kouketsu, A.; Sato, I.; Oikawa, M.; Shimizu, Y.; Saito, H.; Tashiro, K.; Yamashita, Y.; Takahashi, T.; Kumamoto, H. Regulatory T cells and M2-polarized tumour-associated macrophages are associated with the oncogenesis and progression of oral squamous cell carcinoma. *Int. J. Oral Maxillofac. Surg.* **2019**, *48*, 1279–1288. [CrossRef] [PubMed]
9. Lequerica-Fernández, P.; Suárez-Canto, J.; Rodriguez-Santamarta, T.; Rodrigo, J.P.; Suárez-Sánchez, F.J.; Blanco-Lorenzo, V.; Domínguez-Iglesias, F.; García-Pedrero, J.M.; de Vicente, J.C. Prognostic Relevance of CD4(+), CD8(+) and FOXP3(+) TILs in Oral Squamous Cell Carcinoma and Correlations with PD-L1 and Cancer Stem Cell Markers. *Biomedicines* **2021**, *9*, 653. [CrossRef]
10. Surendran, S.; Siddappa, G.; Mohan, A.; Hicks, W., Jr.; Jayaprakash, V.; Mimikos, C.; Mahri, M.; Almarzouki, F.; Morrell, K.; Ravi, R.; et al. Cancer stem cell and its niche in malignant progression of oral potentially malignant disorders. *Oral. Oncol.* **2017**, *75*, 140–147. [CrossRef] [PubMed]
11. Chaves, A.L.F.; Silva, A.G.; Maia, F.M.; Lopes, G.F.M.; de Paulo, L.F.B.; Muniz, L.V.; Dos Santos, H.B.; Soares, J.M.A.; Souza, A.A.; de Oliveira Barbosa, L.A.; et al. Reduced CD8(+) T cells infiltration can be associated to a malignant transformation in potentially malignant oral epithelial lesions. *Clin. Oral. Investig.* **2019**, *23*, 1913–1919. [CrossRef] [PubMed]
12. Chen, X.J.; Tan, Y.Q.; Zhang, N.; He, M.J.; Zhou, G. Expression of programmed cell death-ligand 1 in oral squamous cell carcinoma and oral leukoplakia is associated with disease progress and CD8+ tumor-infiltrating lymphocytes. *Pathol. Res. Pract.* **2019**, *215*, 152418. [CrossRef] [PubMed]
13. Baş, Y.; Koç, N.; Helvacı, K.; Koçak, C.; Akdeniz, R.; Şahin, H.H.K. Clinical and pathological significance of programmed cell death 1 (PD-1)/programmed cell death ligand 1 (PD-L1) expression in high grade serous ovarian cancer. *Transl. Oncol.* **2021**, *14*, 100994. [CrossRef] [PubMed]
14. Simon, S.; Labarriere, N. PD-1 expression on tumor-specific T cells: Friend or foe for immunotherapy? *Oncoimmunology* **2017**, *7*, e1364828. [CrossRef]
15. He, X.; Xu, C. Immune checkpoint signaling and cancer immunotherapy. *Cell Res.* **2020**, *30*, 660–669. [CrossRef] [PubMed]
16. Dong, Y.; Sun, Q.; Zhang, X. PD-1 and its ligands are important immune checkpoints in cancer. *Oncotarget* **2017**, *8*, 2171–2186. [CrossRef]
17. Quan, H.; Shan, Z.; Liu, Z.; Liu, S.; Yang, L.; Fang, X.; Li, K.; Wang, B.; Deng, Z.; Hu, Y.; et al. The repertoire of tumor-infiltrating lymphocytes within the microenvironment of oral squamous cell carcinoma reveals immune dysfunction. *Cancer Immunol. Immunother.* **2020**, *69*, 465–476. [CrossRef] [PubMed]
18. Gannot, G.; Gannot, I.; Vered, H.; Buchner, A.; Keisari, Y. Increase in immune cell infiltration with progression of oral epithelium from hyperkeratosis to dysplasia and carcinoma. *Br. J. Cancer* **2002**, *86*, 1444–1448. [CrossRef]
19. Sharma, S.H.; Thulasingam, S.; Nagarajan, S. Chemopreventive agents targeting tumor microenvironment. *Life Sci.* **2016**, *145*, 74–84. [CrossRef] [PubMed]
20. Elmusrati, A.; Wang, J.; Wang, C.-Y. Tumor microenvironment and immune evasion in head and neck squamous cell carcinoma. *Int. J. Oral Sci.* **2021**, *13*, 24. [CrossRef]
21. Tanaka, T.; Ishigamori, R. Understanding carcinogenesis for fighting oral cancer. *J. Oncol.* **2011**, *2011*, 603740. [CrossRef] [PubMed]
22. Suárez-Sánchez, F.J.; Lequerica-Fernández, P.; Rodrigo, J.P.; Hermida-Prado, F.; Suárez-Canto, J.; Rodríguez-Santamarta, T.; Domínguez-Iglesias, F.; García-Pedrero, J.M.; de Vicente, J.C. Tumor-infiltrating CD20+ B lymphocytes: Significance and prognostic implications in oral cancer microenvironment. *Cancers* **2021**, *13*, 395. [CrossRef] [PubMed]
23. Van der Leun, A.M.; Thommen, D.S.; Schumacher, T.N. CD8(+) T cell states in human cancer: Insights from single-cell analysis. *Nat. Rev. Cancer* **2020**, *20*, 218–232. [CrossRef] [PubMed]
24. Lalos, A.; Tülek, A.; Tosti, N.; Mechera, R.; Wilhelm, A.; Soysal, S.; Daester, S.; Kancherla, V.; Weixler, B.; Spagnoli, G.C.; et al. Prognostic significance of CD8+ T-cells density in stage III colorectal cancer depends on SDF-1 expression. *Sci. Rep.* **2021**, *11*, 775. [CrossRef] [PubMed]

25. Maibach, F.; Sadozai, H.; Seyed Jafari, S.M.; Hunger, R.E.; Schenk, M. Tumor-Infiltrating Lymphocytes and Their Prognostic Value in Cutaneous Melanoma. *Front. Immunol.* **2020**, *11*, 2105. [CrossRef] [PubMed]
26. Gun, S.Y.; Lee, S.W.L.; Sieow, J.L.; Wong, S.C. Targeting immune cells for cancer therapy. *Redox Biol.* **2019**, *25*, 101174. [CrossRef]
27. Badoual, C.; Sandoval, F.; Pere, H.; Hans, S.; Gey, A.; Merillon, N.; Van Ryswick, C.; Quintin-Colonna, F.; Bruneval, P.; Brasnu, D.; et al. Better understanding tumor-host interaction in head and neck cancer to improve the design and development of immunotherapeutic strategies. *Head Neck* **2010**, *32*, 946–958. [CrossRef] [PubMed]
28. Koontongkaew, S. The tumor microenvironment contribution to development, growth, invasion and metastasis of head and neck squamous cell carcinomas. *J. Cancer* **2013**, *4*, 66–83. [CrossRef] [PubMed]
29. Fontenot, J.D.; Gavin, M.A.; Rudensky, A.Y. Pillars Article: Foxp3 programs the development and function of CD4, CD25, regulatory T cells. *J. Immunol.* **2017**, *198*, 986–992. [PubMed]
30. Khan, M.; Arooj, S.; Wang, H. NK Cell-Based immune checkpoint inhibition. *Front. Immunol.* **2020**, *11*, 167. [CrossRef]
31. Kohli, K.; Pillarisetty, V.G.; Kim, T.S. Key chemokines direct migration of immune cells in solid tumors. *Cancer Gene Ther.* **2022**, *29*, 10–21. [CrossRef] [PubMed]
32. Balermpas, P.; Michel, Y.; Wagenblast, J.; Seitz, O.; Weiss, C.; Rödel, F.; Rödel, C.; Fokas, E. Tumour-infiltrating lymphocytes predict response to definitive chemoradiotherapy in head and neck cancer. *Br. J. Cancer* **2014**, *110*, 501–509. [CrossRef] [PubMed]
33. Nguyen, N.; Bellile, E.; Thomas, D.; McHugh, J.; Rozek, L.; Virani, S.; Peterson, L.; Carey, T.E.; Walline, H.; Moyer, J.; et al. Tumor infiltrating lymphocytes and survival in patients with head and neck squamous cell carcinoma. *Head Neck* **2016**, *38*, 1074–1084. [CrossRef]
34. Öhman, J.; Magnusson, B.; Telemo, E.; Jontell, M.; Hasséus, B. Langerhans cells and T cells sense cell dysplasia in oral leukoplakias and oral squamous cell carcinomas–evidence for immunosurveillance. *Scand. J. Immunol.* **2012**, *76*, 39–48. [CrossRef]
35. Rosenberg, S.A.; Yang, J.C.; Sherry, R.M.; Kammula, U.S.; Hughes, M.S.; Phan, G.Q.; Citrin, D.E.; Restifo, N.P.; Robbins, P.F.; Wunderlich, J.R.; et al. Durable complete responses in heavily pretreated patients with metastatic melanoma using T-cell transfer immunotherapy. *Clin. Cancer Res.* **2011**, *17*, 4550–4557. [CrossRef]
36. Mandal, R.; Şenbabaoğlu, Y.; Desrichard, A.; Havel, J.J.; Dalin, M.G.; Riaz, N.; Lee, K.W.; Ganly, I.; Hakimi, A.A.; Chan, T.A.; et al. The head and neck cancer immune landscape and its immunotherapeutic implications. *JCI Insight* **2016**, *1*, e89829. [CrossRef] [PubMed]
37. Leemans, C.R.; Snijders, P.J.F.; Brakenhoff, R.H. The molecular landscape of head and neck cancer. *Nat. Rev. Cancer* **2018**, *18*, 269–282. [CrossRef] [PubMed]
38. Peske, J.D.; Woods, A.B.; Engelhard, V.H. Control of CD8 T-cell infiltration into tumors by vasculature and microenvironment. *Adv. Cancer Res.* **2015**, *128*, 263–307. [CrossRef]
39. Ohigashi, Y.; Sho, M.; Yamada, Y.; Tsurui, Y.; Hamada, K.; Ikeda, N.; Mizuno, T.; Yoriki, R.; Kashizuka, H.; Yane, K.; et al. Clinical significance of programmed death-1 ligand-1 and programmed death-1 ligand-2 expression in human esophageal cancer. *Clin. Cancer Res.* **2005**, *11*, 2947–2953. [CrossRef]
40. Shen, Y.; Teng, Y.; Lv, Y.; Zhao, Y.; Qiu, Y.; Chen, W.; Wang, L.; Wang, Y.; Mao, F.; Cheng, P.; et al. PD-1 does not mark tumor-infiltrating CD8+ T cell dysfunction in human gastric cancer. *J. ImmunoTherapy Cancer* **2020**, *8*, e000422. [CrossRef]
41. Kim, J.M.; Chen, D.S. Immune escape to PD-L1/PD-1 blockade: Seven steps to success (or failure). *Ann. Oncol.* **2016**, *27*, 1492–1504. [CrossRef] [PubMed]
42. Yearley, J.H.; Gibson, C.; Yu, N.; Moon, C.; Murphy, E.; Juco, J.; Lunceford, J.; Cheng, J.; Chow, L.Q.M.; Seiwert, T.Y.; et al. PD-L2 expression in human tumors: Relevance to Anti-PD-1 therapy in cancer. *Clin. Cancer Res.* **2017**, *23*, 3158–3167. [CrossRef] [PubMed]
43. Müller, T.; Braun, M.; Dietrich, D.; Aktekin, S.; Höft, S.; Kristiansen, G.; Göke, F.; Schröck, A.; Brägelmann, J.; Held, S.A.E.; et al. PD-L1: A novel prognostic biomarker in head and neck squamous cell carcinoma. *Oncotarget* **2017**, *8*, 52889–52900. [CrossRef] [PubMed]
44. Silvio, G.; Jack, B. The next frontier: Head and neck cancer immunoprevention. *Cancer Prev. Res.* **2017**, *10*, 681–683.

 *biomedicines*

Article

Prognostic Role of Systemic Inflammatory Markers in Patients Undergoing Surgical Resection for Oral Squamous Cell Carcinoma

Uiju Cho [1], Yeoun-Eun Sung [2], Min-Sik Kim [3] and Youn-Soo Lee [2,*]

[1] Department of Hospital Pathology, St. Vincent's Hospital, College of Medicine, The Catholic University of Korea, Seoul 06591, Korea; hailtoya@catholic.ac.kr

[2] Department of Hospital Pathology, Seoul St. Mary's Hospital, College of Medicine, The Catholic University of Korea, Seoul 06591, Korea; yesung@catholic.ac.kr

[3] Department of Otorhinolaryngology, Seoul St. Mary's Hospital, College of Medicine, The Catholic University of Korea, Seoul 06591, Korea; entkms@catholic.ac.kr

* Correspondence: lys9908@catholic.ac.kr

Abstract: Background: A high platelet–lymphocyte ratio (PLR) is a marker of systemic inflammation and, together with the neutrophil–lymphocyte ratio (NLR), is associated with poor outcomes in several cancers. We investigated the prognostic value of PLR and other systemic inflammatory markers, such as NLR, systemic immune-inflammation index (SII), and systemic inflammation response index (SIRI), in oral squamous cell carcinoma (OSCC) patients undergoing surgical resection. Methods: We derived PLR, NLR, SII, and SIRI from a retrospective chart review of 269 consecutive OSCC patients. The complete blood count examined in the immediate preoperative period was used to compute PLR, NLR, SII, and SIRI. We analyzed the relationship between these systemic inflammatory markers and the clinicopathologic characteristics, disease-specific survival (DSS), and progression-free survival (PFS) of patients. Results: In the univariate analysis, high PLR and SII were significantly associated with worse DSS and PFS (all $p < 0.05$). In the multivariate analysis, PLR (HR 2.36, 95% CI 1.28–4.36 for DSS; HR 1.80, 95% CI 1.06–3.06 for PFS) was an independent predictor of survival outcomes. When PLR was analyzed as a continuous variable, the relationship between the outcome and preoperative PLR was not monotonically linear. In the subgroup analysis, PLR was more strongly associated with DSS and PFS in patients who were male, had stage III/IV OSCC, or had lymph node metastasis. Conclusion: Our data suggest that in OSCC patients, the pretreatment PLR is an independent predictor of DSS and PFS. The PLR is a readily available biomarker that will improve prognostication and risk stratification in OSCC.

Keywords: oral cancer; inflammation; prognosis; surgical resection; overall survival

Citation: Cho, U.; Sung, Y.-E.; Kim, M.-S.; Lee, Y.-S. Prognostic Role of Systemic Inflammatory Markers in Patients Undergoing Surgical Resection for Oral Squamous Cell Carcinoma. *Biomedicines* **2022**, *10*, 1268. https://doi.org/10.3390/biomedicines10061268

Academic Editor: Vui King Vincent-Chong

Received: 6 May 2022
Accepted: 25 May 2022
Published: 29 May 2022

Publisher's Note: MDPI stays neutral with regard to jurisdictional claims in published maps and institutional affiliations.

1. Introduction

Oral squamous cell carcinoma (OSCC) is a carcinoma with squamous differentiation arising from the mucosal epithelium of the oral cavity and mobile tongue. The global incidence of oral cancer, the majority being squamous cell carcinoma, is approximately 3.5 million new cases per year, and it causes 1.7 million deaths per year. Oral cancer accounts for 2.0% of all cancers [1]. In Korea, oral cancer is the second most common cancer among head and neck cancers, and the incidence of oral cancer has been slightly increasing in recent decades [2,3]. The incidence rate is especially rising more steeply in the third- or fourth-decade age groups [4]. Smoking, drinking alcohol, lifestyle changes, the popularization of early diagnosis, and genetic factors could be the causes of such an increase [4,5]. Oral cancer is more common among men than women, and most common in the fifth and sixth decades [6]. The survival rate of oral cancer is approximately 50% [7]. Advancements in traditional treatment modalities, i.e., surgery, chemotherapy, and radiotherapy, have not been able to noticeably increase the survival rate, yet the side effects of these treatments are significant.

Currently, the prediction of tumor progression or recurrence depends largely on classic histologic parameters, such as tumor size, depth of invasion, pattern of invasion, and nodal status [6]. Many novel biomarkers have been investigated to achieve better risk stratification for adjuvant treatment modalities or more aggressive treatment in patients with distant metastasis [8]. PD-L1 is a recently discovered prognostic biomarker and immune checkpoint inhibitor; therefore, anti-PD-L1 therapy would be a promising treatment for OSCC [9,10]. However, none of the novel biomarkers have been recommended as prognosticators valid for clinical use to date.

The cell-mediated inflammatory response has been shown to play a critical role in cancer development and growth. For example, tumor-associated neutrophils are considered potent stimulators of angiogenesis, and their protumoral cytokines promote tumor growth [11,12]. Additionally, extensive disruption of hematopoiesis occurs as cancer progresses [11,13]. Changes in the systemic inflammatory response to tumor cells, especially white blood cells and platelets, have drawn attention as valuable prognostic biomarkers [11]. The systemic inflammatory markers, i.e., neutrophil–lymphocyte ratio (NLR) and platelet–lymphocyte ratio (PLR), have been used as prognostic biomarkers in various types of cancers [14–18]. These serum inflammatory markers are easily available because they can be retrieved from routine blood tests [19].

Until now, the NLR, which uses differential white cell counts, has been the most extensively investigated marker in operable and inoperable cancers [19,20]. Elevated PLR and NLR were associated with poor survival outcomes in previous studies [21,22]. Furthermore, investigators explored the combination of the scores with acute-phase protein-based scores (Glasgow Prognostic Score) [23] or developed novel inflammatory markers to provide additional prognostic value in different cancers [24]. Among such novel markers, the most recently developed are the systemic immune-inflammation index (SII) [24,25] and systemic inflammation response index (SIRI) [26,27]. They are derived from three types of inflammatory cells (lymphocytes, neutrophils, and platelets or monocytes), and were shown to be independent predictors of overall survival in patients with lung [24,27], breast [26], esophageal [25], and urologic cancers [28]. High preoperative SII and SIRI were also shown to be independent prognostic factors in patients with OSCC, but the data are still very limited [29–32].

In OSCC, data to support the clinical value of different systemic inflammatory markers are still accumulating. However, the results are controversial and need further research [8,33,34]. Furthermore, the data to evaluate the clinical value of systemic inflammatory markers are still insufficient, since studies of inflammatory markers have focused on NLR, and no study has simultaneously compared the prognostic values of NLR, PLR, and the new emerging markers, SII and SIRI, in OSCC.

In this study, we aimed to validate the prognostic value of a panel of systemic inflammatory markers, NLR, PLR, SII and SIRI. In addition, we evaluated the relationships of clinicopathologic parameters and systemic inflammatory markers with survival in OSCC patients.

2. Materials and Methods

2.1. Study Population

We retrospectively identified and enrolled 269 patients with oral cavity and mobile tongue squamous cell carcinoma who had undergone surgical resection at Seoul St. Mary's Hospital between January 2003 and December 2019. We excluded patients with other malignancies, with autoimmune diseases, who had received neoadjuvant therapy, or who had insufficient preoperative blood tests carried out to calculate systemic inflammation markers for the study.

This study was conducted in accordance with the amended Declaration of Helsinki. The study was approved by our Institutional Review Board (Seoul St. Mary's Hospital, IRB

No. 86651124), and the requirement for informed consent was waived by the Institutional Review Board.

2.2. Data Collection

The following data were collected from the patients' medical records: date of the primary cancer diagnosis, age at diagnosis, anatomical sublocation, tumor size, tumor differentiation, depth of invasion, lymphatic invasion, vascular invasion, perineural invasion, presence of lymph node metastasis, distant metastases, and adjuvant therapy. M stage was defined as M0 unless distant metastasis was specified in the medical records. The slides were reviewed by an expert pathologist (S.Y.E.) and restaged according to the American Joint Committee on Cancer (AJCC) staging manual, 8th edition [35]. The differential white blood cell (WBC) count that was measured within one month before the surgery as part of the routine preoperative workup was collected from the medical report. Systemic inflammatory markers were defined as follows: NLR (absolute neutrophil count/absolute lymphocyte count), PLR (absolute platelet count/absolute lymphocyte count), SII (platelet count × neutrophil count/lymphocyte count), and SIRI (neutrophil count × monocyte count/lymphocyte count).

The cutoff values for platelet count, NLR, PLR, SII, and SIRI were determined from receiver operating characteristic (ROC) curves for overall survival considering both sensitivity and specificity. The cutoff values and the area under the curve (AUC) values are shown in Table 1. The patients were divided into two groups (the low group and high group) based on NLR, PLR, SII, and SIRI. The disease-specific survival (DSS) and progression-free survival (PFS) rates of the patients were compared by patient characteristics, including the NLR, PLR, SII, and SIRI.

Table 1. Cutoff values of systemic inflammatory markers determined by receiver operating curves for overall survival.

	Cutoff Value	AUC	Sensitivity	Specificity	Accuracy
Platelet	296.5	0.5667	0.2923	0.8676	0.7286
NLR	1.7584	0.5407	0.6769	0.4363	0.4944
PLR	159.4521	0.5983	0.4	0.8088	0.71
SII, 10^9/L	548.9451	0.5615	0.5077	0.6667	0.6283
SIRI, 10^9/L	0.8938	0.5422	0.5385	0.5833	0.5725

AUC, area under the curve; NLR, neutrophil–lymphocyte ratio; PLR, platelet–lymphocyte ratio; SII, systemic inflammation index; SIRI, systemic inflammation response index.

2.3. Statistics

The characteristics of the systemic inflammatory markers are shown as both medians and means. Student's t-test was used to compare continuous characteristics between the two groups. Pearson's test was used to analyze the correlation between two continuous variables. DSS was considered the period between surgery and the date of the last follow-up, or cancer-specific death. PFS was considered the period between surgery and the date of recurrence, locoregional progression, metastasis, or death. The DSS and PFS rates were analyzed using Kaplan–Meier survival curves and compared statistically with the log-rank test. Age, sex, and all variables with significant prognostic values in the univariate analysis were subjected to multivariate analyses using the Cox proportional hazards model. A two-sided $p < 0.05$ was considered statistically significant. The potential nonlinear relationships between the continuous PLR and the survival outcomes were flexibly analyzed using a restricted cubic spline (RCS) with four knots [36]. The median value was used as a continuous predictor. Statistical analyses were conducted using SPSS 21.0 for Windows (IBM Corporation, Armonk, NY, USA) and R version 4.1.2.

3. Results

3.1. Patient Characteristics

The mean age of the patients was 55.1 ± 15.2 years, ranging from 18 to 90 years. The majority were male (64.3%) and had tongue cancer (74.3%). The distribution of the pathologic stage among the patients was as follows: stage I, 29.4%; stage II, 18.2%; stage III, 20.5%; and stage IV, 32.0% (Table 2). We calculated a mean platelet count of 241.45 ± 69.75 × 10^9/L, a mean NLR of 2.58 ± 2.02, a mean PLR of 140.70 ± 61.26, a mean SII of 619.53 ± 494.38 × 10^9/L, and a mean SIRI of 1.33 ± 1.64 × 10^9/L (Table 3). An increase in the lymphocyte count was correlated with the platelet count ($r = 0.33$, $p < 0.001$) and the monocyte count ($r = 0.22$, $p < 0.001$) but not with the neutrophil count ($r = 0.0022$, $p = 0.97$) (Supplementary Figure S1). Additionally, an increase in PLR was correlated with NLR ($r = 0.48$, $p < 0.001$) (Supplementary Figure S2).

Table 2. Clinicopathologic characteristics of patients with oral squamous cell carcinoma.

Characteristic	Number (%)
Total	269
Age (mean, years)	55.1 ± 15.2
Sex	
Male	173 (64.3%)
Female	96 (35.7%)
Location	
Mobile tongue	200 (74.3%)
Other (palate, lip, retromolar area, etc.)	69 (25.7%)
Tumor size (cm)	2.7 ± 1.7
Depth of invasion (cm)	1.0 ± 0.9
Differentiation	
Well	133 (49.4%)
Moderate	120 (44.6%)
Poor	16 (6.0%)
T stage	
T1	82 (30.5%)
T2	73 (27.1%)
T3	87 (32.3%)
T4	27 (10.0%)
N stage	
N0	146 (61.1%)
N1	22 (9.2%)
N2	23 (9.6%)
N3	46 (19.3%)
N4	2 (0.8%)
Stage	
I	79 (29.4%)
II	49 (18.2%)
III	55 (20.5%)
IV	86 (32.0%)
Adverse pathologic features	
Lymphatic invasion	73 (27.1%)
Vascular invasion	8 (3.0%)
Perineural invasion	77 (28.6%)
Adjuvant therapy	
Radiation therapy alone	56 (20.8%)
Chemotherapy and radiation therapy	60 (22.3%)
None	153 (56.9%)

Table 3. Summary statistics of inflammatory markers in patients with oral squamous cell carcinoma.

Parameter	Mean ± SD	Median (Range)	Cutoff Value	Population (n = 269) with Given Cutoff, Number (%)
Differential white blood cell count				
Neutrophil count, 10^9/L	4.23 ± 2.14	3.64 (0.87–12.90)	NA	NA
Lymphocyte count, 10^9/L	1.89 ± 0.67	1.82 (0.52–0.44)	NA	NA
Monocyte count, 10^9/L	0.47 ± 0.19	0.42 (0.00–1.42)	NA	NA
Platelet count, 10^9/L	241.45 ± 69.75	234.0. (37.20–652.00)	>296.5	46 (17.10%)
Calculated ratio and index				
NLR	2.58 ± 2.02	1.94 (0.37–16.00)	>1.76	159 (59.11%)
PLR	140.70 ± 61.26	130.99 (22.17–551.81)	>159.45	65 (24.16%)
SII, 10^9/L	619.53 ± 494.38	452.42 (66.78–3515.33)	>548.95	101 (37.55%)
SIRI, 10^9/L	1.33 ± 1.64	0.83 (0–16.04)	>0.89	120 (44.61%)

3.2. Correlation between Inflammatory Markers and Clinical Factors

We examined the correlation between systemic inflammatory markers and clinicopathologic parameters. The mean WBC count was higher in patients with a depth of invasion >1 cm and advanced T and AJCC stages (1 and 2 vs. 3 and 4) (all $p < 0.05$) (Supplementary Table S1). Platelet levels were higher in the younger patient group ($p = 0.0284$), but other markers showed no difference between the two age groups (Supplementary Table S1). There were no significant differences between the high and low NLR groups in the clinicopathologic parameters. High PLR was correlated with >1 cm depth of invasion and advanced T and AJCC stages (all $p < 0.05$). Likewise, advanced stage was correlated with high SII and SIRI ($p = 0.0015$ and 0.0131, respectively) (Table 4).

Table 4. Correlation between clinicopathologic parameters and systemic inflammation markers.

Parameter	No.	NLR High (>1.76)		PLR High (>159.45)		SII High (>548.95 × 10^9/L)		SIRI High (>0.89 × 10^9/L)	
		(n = 159)	p	(n = 65)	p	(n = 101)	p	(n = 120)	p
Age									
≤55	134	84 (52.8%)	0.2970	33 (50.8%)	0.9140	53 (52.5%)	0.5604	67 (55.8%)	0.0964
>55	135	75 (47.2%)		32 (49.2%)		48 (47.5%)		53 (44.2%)	
Sex									
Male	173	52 (32.7%)	0.2195	22 (33.9%)	0.7219	32 (31.7%)	0.2878	34 (28.3%)	0.0239
Female	96	107 (67.3%)		43 (66.2%)		69 (68.3%)		86 (71.7%)	
Location									
Mobile Tongue	200	116 (73.0%)	0.5292	45 (69.2%)	0.2779	73 (72.3%)	0.5462	84 (70.0%)	0.1427
Other	69	43 (27.0%)		20 (30.8%)		28 (27.7%)		36 (30.0%)	
Depth of invasion									
≤1 cm	165	93 (58.5%)	0.2489	30 (46.2%)	0.0039	55 (54.5%)	0.0723	66 (55.0%)	0.0554
>1 cm	104	66 (41.5%)		35 (53.9%)		46 (45.5%)		54 (45.0%)	
Lymphatic invasion									
Absent	196	112 (70.4%)	0.2828	45 (69.2%)	0.4496	65 (64.4%)	0.015	80 (66.7%)	0.0403
Present	73	47 (29.6%)		20 (30.8%)		36 (35.6%)		40 (33.3%)	

Table 4. *Cont.*

Parameter	No.	NLR High (>1.76)		PLR High (>159.45)		SII High (>548.95 × 10^9/L)		SIRI High (>0.89 × 10^9/L)	
		(*n* = 159)	*p*	(*n* = 65)	*p*	(*n* = 101)	*p*	(*n* = 120)	*p*
Vascular invasion									
Absent	261	105 (95.5%)	0.2783	198 (97.1%)	0.9553	99 (98.0%)	0.7141	119 (99.2%)	0.0789
Present	8	5 (4.6%)		6 (2.9%)		2 (2.0%)		1 (0.8%)	
Perineural invasion									
Absent	192	85 (77.3%)	0.0751	150 (73.5%)	0.1662	66 (65.3%)	0.0904	74 (63.3%)	0.0090
Present	77	25 (22.7%)		54 (26.5%)		35 (34.7%)		44 (36.7%)	
T stage									
T1 and T2	155	88 (55.4%)	0.3640	27 (41.5%)	0.0026	52 (51.5%)	0.1143	62 (51.7%)	0.0761
T3 and T4	114	71 (44.7%)		38 (58.5%)		49 (48.5%)		58 (48.3%)	
Lymph node metastasis									
Absent	176	75 (68.2%)	0.4295	40 (61.5%)	0.4490	60 (59.4%)	0.1074	72 (60.0%)	0.0930
Present	93	35 (31.8%)		25 (38.5%)		41 (40.6%)		47 (40.0%)	
Stage									
I, II	128	69 (43.4%)	0.0983	22 (33.9%)	0.0109	37 (36.6%)	0.0053	47 (39.2%)	0.0131
III, IV	141	90 (56.0%)		43 (66.2%)		64 (63.4%)		73 (60.8%)	
Distant metastasis									
Absent	254	152 (95.6%)	0.3132	60 (92.3%)	0.3932	95 (94.1%)	0.8400	115 (95.8%)	0.3659
Present	15	7 (4.4%)		5 (7.7%)		6 (5.9%)		5 (4.2%)	

NLR, neutrophil–lymphocyte ratio; PLR, platelet–lymphocyte ratio; SII, systemic inflammation index; SIRI, systemic inflammation response index.

The survival analysis results are shown in Table 5. The median follow-up period was 36 months (range 0~185 months). Of the 269 patients, 65 patients died during the follow-up period, and 93 patients experienced disease progression. The 2-year and 5-year DSS rates of the OSCC patients were 78.8% and 75.6%, respectively.

Table 5. Survival analysis of patients with oral squamous cell carcinoma according to clinicopathologic parameters and systemic inflammatory markers.

Variables	Disease-Specific Survival			Progression-Free Survival		
	Univariate Analysis	Multivariate Analysis		Univariate Analysis	Multivariate Analysis	
	p	*p*	Hazard Ratio (95% CI)	*p*	*p*	Hazard Ratio (95% CI)
Age (>55 years)	0.1293	0.2252	1.38 (0.82–2.34)	0.1329	0.3500	1.23 (0.79–1.92)
Gender (Male)	0.2874	0.9221	0.97 (0.54–1.73)	0.3462	0.7593	0.93 (0.58–1.48)
T stage (reference 1)	<0.0001	0.8286		<0.0001	0.6317	
2		0.6694	0.62 (0.07–5.63)		0.9388	0.92 (0.11–7.81)
3		0.4558	0.39 (0.03–4.70)		0.8899	1.18 (0.12–11.61)
4		0.6402	0.54 (0.04–7.15)		0.5508	2.04 (0.20–21.25)

Table 5. *Cont.*

Variables	Disease-Specific Survival			Progression-Free Survival		
	Univariate Analysis	Multivariate Analysis		Univariate Analysis	Multivariate Analysis	
	p	p	Hazard Ratio (95% CI)	p	p	Hazard Ratio (95% CI)
N stage (reference 0)	<0.0001	0.6277		<0.0001	0.9787	
1		0.1091	0.32 (0.08–1.29)		0.6247	0.77 (0.27–2.17)
2		0.6264	0.69 (0.15–3.13)		0.6875	0.77 (0.21–2.78)
3		0.6979	0.76 (0.18–3.12)		0.8974	0.92 (0.27–3.10)
4		0.9726	0.00 (0–1000)		0.9609	0.00 (0–1000)
DOI (>1 cm)	<0.0001	0.4120	1.75 (0.46–6.60)	<0.0001	0.7644	0.86 (0.32–2.33)
Stage (reference I)	<0.0001	0.3739		<0.0001	0.7194	
II		0.4611	2.53 (0.22–29.72)		0.7179	1.52 (0.16–14.96)
III		0.2311	4.22 (0.40–44.57)		0.5945	1.83 (0.20–16.92)
IV		0.1079	8.59 (0.62–118.32)		0.3427	3.19 (0.29–34.96)
Lymphatic invasion	0.00086	0.1813	1.56 (0.81–2.98)	0.0010	0.0757	1.64 (0.95–2.84)
Vascular invasion	0.7050	-	-	0.6535	-	-
Perineural invasion	<0.0001	0.1651	1.51 (0.84–2.71)	0.0030	0.9179	1.03 (0.62–1.70)
Differentiation (reference well)	0.0022	0.0663		0.0025	0.1059	
Moderate		0.4403	0.80 (0.46–1.41)		0.8445	1.05 (0.66–1.65)
Poor		0.0622	2.33 (0.96–5.67)		0.0380	2.28 (1.05–4.97)
Platelet high	0.0013	0.1474	1.60 (0.84–3.00)	0.0093	0.5314	1.20 (0.68–2.13)
NLR high	0.1221	-	-	0.0485	0.4553	1.25 (0.69–2.27)
PLR high	0.0004	0.0064	2.33 (1.27–4.28)	0.0020	0.0300	1.80 (1.06–3.06)
SII high	0.0119	0.6822	0.88 (0.46–1.65)	0.0174	0.5836	0.83 (0.44–1.59)
SIRI high	0.0949	-	-	0.1326	-	-

CI, confidence interval; NLR, neutrophil–lymphocyte ratio; PLR, platelet–lymphocyte ratio; SII, systemic inflammation index; SIRI, systemic inflammation response index. In univariate analysis, T stage, N stage, depth of invasion, AJCC stage, lymphatic invasion, perineural invasion, tumor differentiation, and platelet count were associated with DSS and PFS (all $p < 0.05$).

Among the systemic inflammatory markers, the survival analysis revealed poorer DSS and PFS for patients with high PLR and SII (all $p < 0.05$) (Figures 1 and 2). The 5-year DSS rates of the low vs. high PLR groups were 81.2% vs. 58.4% ($p = 0.0004$), and the 5-year PFS rates of the low vs. high PLR groups were 70.2% vs. 45.9% ($p = 0.0002$). NLR was associated with PFS ($p = 0.0485$), and SIRI showed no association with DSS or PFS ($p = 0.012$) (Figures 1 and 2).

We entered factors that were significant in univariate analysis into the multivariate model, and the high PLR remained significant for DSS and PFS (DSS: hazard ratio (HR) = 2.36, 95% CI 1.28–4.36, $p = 0.0059$; PFS: HR = 1.80, 95% CI 1.06–3.06, $p = 0.0300$). Platelet count over 296.5×10^9/L was not an independent prognostic factor for DSS and PFS ($p = 0.141$ and $p = 0.531$, respectively).

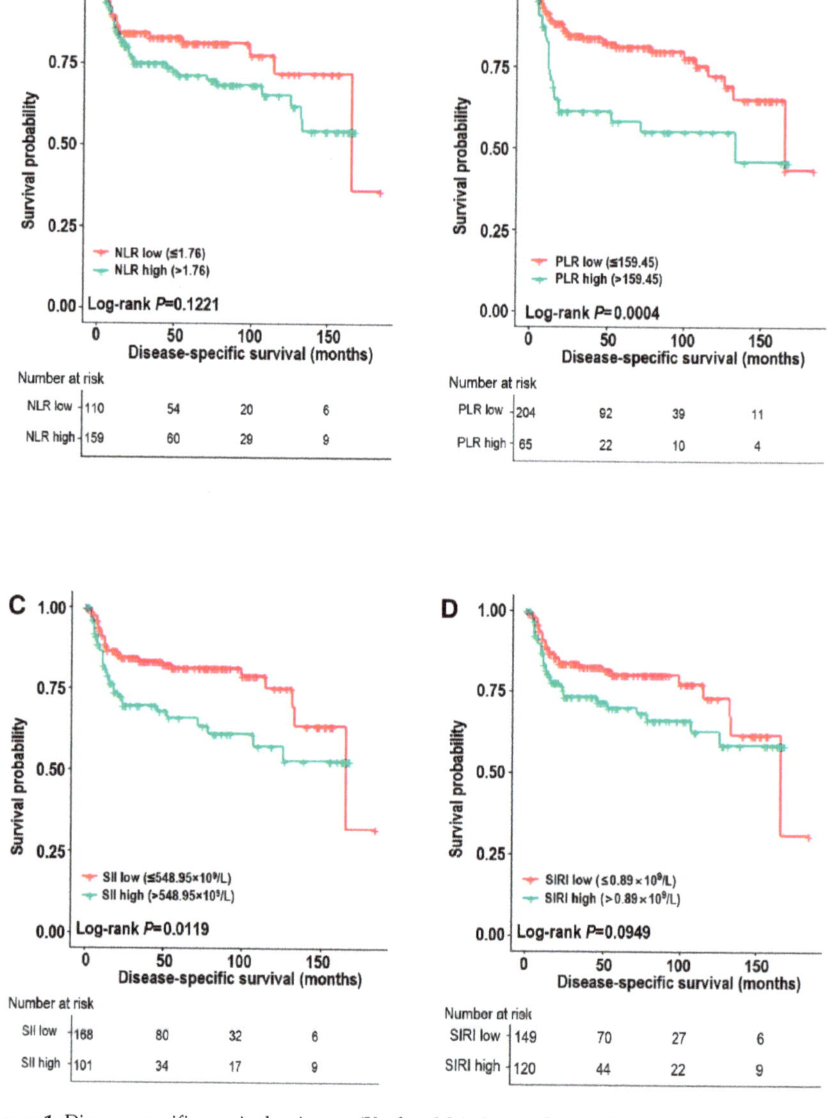

Figure 1. Disease-specific survival estimates (Kaplan–Meier) according to the neutrophil–lymphocyte ratio (NLR) (**A**), platelet–lymphocyte ratio (PLR) (**B**), systemic inflammation index (SII) (**C**), and systemic inflammation response index (SIRI) (**D**).

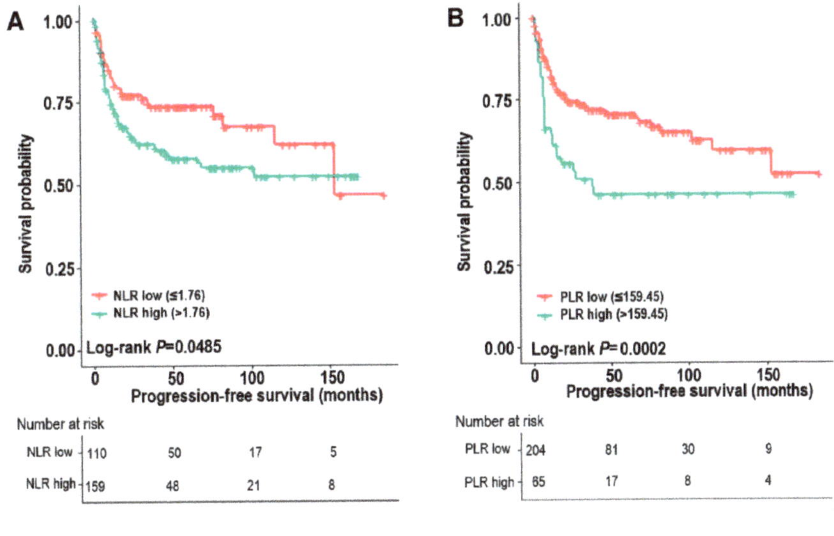

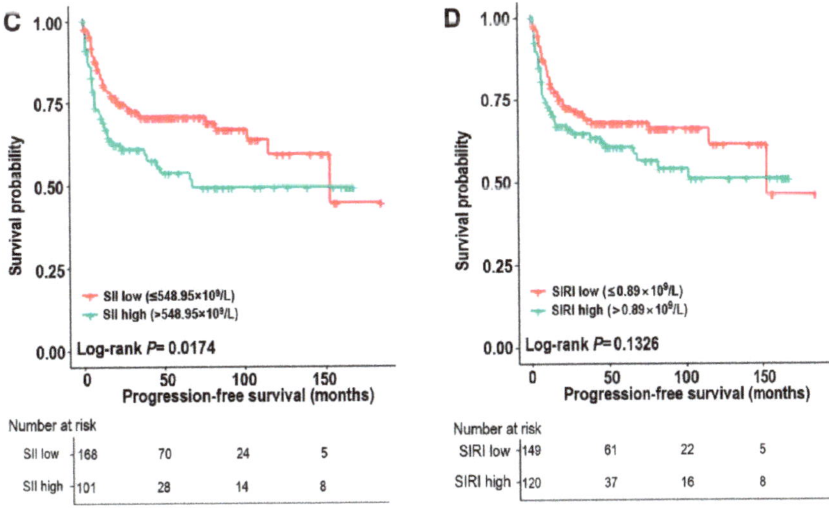

Figure 2. Progression-free survival estimates (Kaplan–Meier) according to the neutrophil–lymphocyte ratio (NLR) (**A**), platelet–lymphocyte ratio (PLR) (**B**), systemic inflammation index (SII) (**C**), and systemic inflammation response index (SIRI) (**D**).

3.3. Analysis of the Relationship between PLR and Survival According to Clinical Factors

We conducted subset analyses of the impact of PLR on survival according to selected clinicopathologic factors using forest plots. Notably, PLR was more strongly associated with DSS and PFS in patients who were male, had stage III/IV OSCC, or had lymph node metastasis (all $p < 0.05$) (Figure 3).

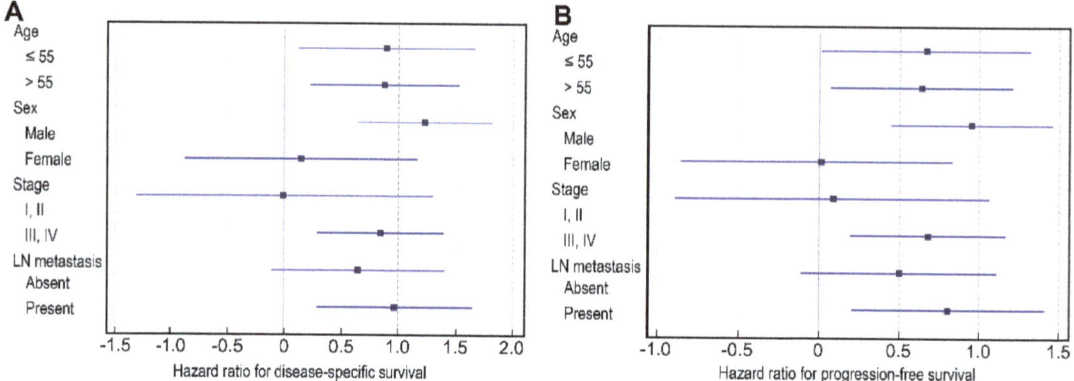

Figure 3. Forest plot of disease-specific survival hazard ratios (**A**) and progression-free survival hazard ratios (**B**) by subgroup. PLR, platelet–lymphocyte ratio; LN, lymph node.

3.4. Nonlinear Association between PLR and Survival

Furthermore, PLR was studied as a continuous variable in univariate analysis. The RCS analysis showed a curvilinear, J-shaped association between the PLR and survival outcomes rather than a straight line (Figure 4). This result suggested a possible nonlinear association between the PLR and the risk of DSS (p-value for nonlinearity = 0.0424). We did not find statistical evidence for nonlinearity with the progression-free survival outcome. We estimated the mortality risk to reach a nadir PLR in the range of 100–120, with inverse associations below that range and positive associations above that range, although the magnitude of associations varied.

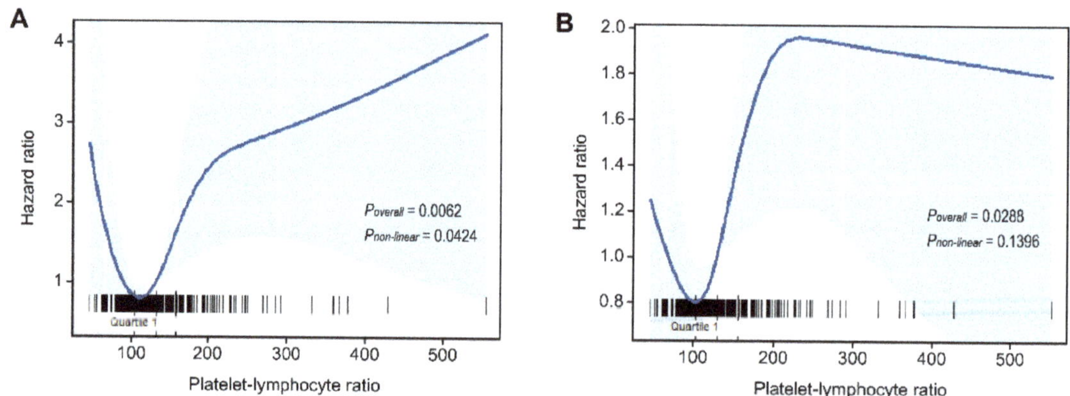

Figure 4. Hazard ratio for the risk of disease-specific death (**A**) and progression (**B**) in oral squamous cell carcinoma patients, evaluated by restricted cubic splines from unadjusted Cox proportional hazard models. The solid line represents the central risk estimate, and the shaded area represents the 95% confidence interval. The rug plot is of cases.

4. Discussion

The inflammatory response is greatly influenced by tumor manifestation [37,38]. Many cancers cause extensive disruption of hematopoiesis [11]. In particular, cell mediation is closely associated with tumor development, growth, and metastasis [39]. There are methods to measure the inflammatory response, including C-reactive protein, erythrocyte sedimentation rate, and peripheral blood cell count [40,41]. NLR, PLR, and monocyte–lymphocyte ratio (MLR) are measured from peripheral blood cell counts, and illustrate how much neutrophils, platelets, and monocytes are increased compared to lymphocytes. These changes represent the cell-mediated systemic inflammatory response [42]. In studies conducted over the last decade, the presence of an elevated NLR and PLR has been associated with poorer outcomes in different types of malignancies [14,42,43]. Some researchers termed systemic inflammatory responses "the tip of the cancer iceberg" [44], but this approach is underutilized in clinical practice.

Recently, new prognostic scores or indexes were created through a combination of nutrition index, performance index, or three or more peripheral blood cell counts [23,45]. SII and SIRI are recently suggested novel prognostic biomarkers, derived from a combination of the absolute neutrophil count, lymphocyte count, and monocyte count. The prognostic value of each has been suggested in patients with lung cancer [24]. In our study, high SII and SIRI were associated with poor DSS in patients with OSCC in the univariate analysis. They were promising prognostic biomarkers in OSCC, but they failed to be identified as independent prognostic factors in the multivariate analysis.

In OSCC, several studies have investigated the prognostic value of NLR and PLR [33,46]. In a meta-analysis of 10 studies, a high NLR was associated with a poor prognosis in patients with OSCC [33]. PLR was also an independent prognostic factor in previous studies [46–48], and PLR was superior in the studies by Tazeen et al. [46] and Rosculet et al. [48]. In concordance with previous studies, our data showed that PLR was more strongly associated with overall survival and PFS than NLR, SII, and SIRI in patients with OSCC. NLR, SII, and SIRI were associated with either DSS or PFS in the univariate analysis, but PLR remained the only significant prognostic indicator after adjustments in the multivariate analysis. The subset analysis revealed that PLR had a greater prognostic impact in patients with advanced disease (stage III and IV) than in those with localized disease (stage I and II) (HR = −0.01, 95% CI −1.32 to 1.32, p = 0.9932 vs. HR = 0.84, 95% CI 0.29 to 1.40, p = 0.0028). This may be explained by the fact that cancer cells interact little with the microenvironment and local inflammatory cells in the low stages, and subsequently elicit little systematic inflammation [11,39,49]. Our data suggest that more active surveillance or treatment may be required for patients with high PLR, especially for those with stage III or IV OSCC. We need to actively seek alternative therapies such as immunotherapy or molecular targeted therapy in this patient group because their prognosis is predicted to be poorer after surgery.

Furthermore, we analyzed PLR as a continuous variable as well as a dichotomized variable. Interestingly, the prognostic correlation of PLR did not show monophasic linearity. The J-shaped relationship between DSS and PLR indicates an unfavorable prognosis for not only the high PLR group but also the extremely low PLR group. A J-shaped association is often observed in epidemiologic relationships, such as the association of body mass index and mortality [50]. A few researchers demonstrated a similar nonlinear association between NLR and prognosis in patients with gastric, breast, and oral cancers in the United States general population [51–55]. The nonlinear relationship between the survival outcome and pretreatment PLR has never been demonstrated previously. Our data showed that PLR has a nonlinear prognostic pattern similar to NLR. The increased risk of DSS at extremely low PLRs might indicate the strong cancer-related disturbance of normal hematopoiesis in these patients and its effect on survival, or the pervasive effect of unknown underlying noncancer-related health conditions. Nevertheless, an explicit rationale for such a pattern has yet to be established. It is necessary to further research the nonlinear prognostic pattern of PLR and to consider how to employ the best PLR thresholds in clinical practice.

The underlying mechanisms of PLR as a prognostic marker have not been fully elucidated, but there are several explanations. Bodies of evidence have shown that platelet activation is a key biological process for cancer occurrence, progression, and metastasis [56,57]. The phenomena related to platelet activation through interaction with cancer cells are thrombocytosis and thromboembolism [58]. Thrombocytosis is often observed in patients with advanced malignancy, and thromboembolism occurs frequently in cancer patients, with lethal consequences [57]. Aggregated platelets could enhance tumorigenesis by releasing pro-angiogenic mediators within the tumor microvasculature [59]. Furthermore, platelets could influence the metastatic potential of cancer cells through several biological pathways, i.e., secretion of cellular growth factors, helping stable tumor cell adhesion to endothelial cells, and impeding cell-mediated immunity against tumor cells [59–64]. A newly discovered link between platelets and cancer cells is tumor-educated platelets [65]. Tumor-educated platelets were most recently found through liquid biopsy research. Tumor-educated platelets are functional cells with a distinct tumor-driven phenotype that are thought to acquire tumor-derived factors and undergo signal-dependent changes in RNA processing within blood circulation [65]. Tumor-educated platelets have been shown to participate in multiple steps of metastasis, leading to lethal consequences [66,67]. There is a hypothesis that microvesicles containing RNA and proteins taken up by platelets promote tumor growth and immune evasion [68,69]. The relationship between tumor-educated platelets and PLR elevation is unclear. Future research on this matter would be interesting and would help elucidate the biological mechanisms underlying PLR as a biomarker.

The strength of our study is that various, not a single, systemic inflammatory markers were analyzed and that this was performed in a homogenous, rich sample group. Exploration of the nonlinear association between PLR and survival is another strength of this study.

However, our study had several limitations. First, its single-center, retrospective design may have caused potential bias. The systemic inflammatory marker cutoffs were derived from the AUC of these parameters against overall survival. Moreover, these cutoffs were not assessed in another data set for validation. Therefore, to use PLR in clinical practice, further prospective study or consensus on the optimal cutoff would be desirable. Secondly, our study could not set appropriate criteria for application, considering the major limitation of the systemic inflammatory markers, i.e., the influence of infection. Accumulation of scientific evidence and an understanding of the role of systemic markers through prospective multicenter trials would be needed to address this limitation.

In conclusion, this study evaluated the prognostic value of a panel of systemic inflammatory markers, including NLR, PLR, SII, and SIRI, in OSCC. Our data demonstrated that PLR was a valuable independent prognostic biomarker in patients with OSCC, especially in those with advanced disease. High PLR was associated with worse survival in these patients. We also demonstrated a nonlinear correlation between PLR and survival. A J-shaped association implied that an extremely low PLR was also a poor prognostic factor in patients with OSCC. Preoperative assessments of cellular biomarkers from peripheral blood could provide high-quality prognostic information, and they represent another promising approach for improving patient stratification.

Supplementary Materials: The following supporting information can be downloaded at: https://www.mdpi.com/article/10.3390/biomedicines10061268/s1, Figure S1: Pearson's correlation between lymphocyte count and platelet count, neutrophil count, and monocyte count; Figure S2: Pearson's correlation between platelet-lymphocyte ratio and neutrophil-lymphocyte ratio; Table S1: Correlation between clinicopathologic parameters and white blood cell and platelet count.

Author Contributions: Conceptualization, U.C. and Y.-S.L.; methodology, U.C.; software, U.C.; validation, U.C. and Y.-S.L.; formal analysis, U.C.; investigation, U.C.; resources, Y.-E.S. and M.-S.K.; data curation, U.C. and Y.-E.S.; writing—original draft preparation, U.C.; writing—review and editing, Y.-S.L.; visualization, U.C.; supervision, Y.-S.L.; project administration, Y.-S.L.; funding acquisition, U.C. All authors have read and agreed to the published version of the manuscript.

Funding: The authors wish to acknowledge the financial support of the St. Vincent's Hospital Research Institute of Medical Science (SVHR-2019-12).

Institutional Review Board Statement: The study was approved by our Institutional Review Board (Seoul St. Mary's Hospital, IRB No. KC22RASI0082, 11 March 2022).

Informed Consent Statement: The requirement for informed consent was waived by the Institutional Review Board.

Data Availability Statement: The data presented in this study are available on request from the corresponding author.

Conflicts of Interest: The authors declare no conflict of interest.

References

1. Bray, F.; Ferlay, J.; Soerjomataram, I.; Siegel, R.L.; Torre, L.A.; Jemal, A. Global cancer statistics 2018: GLOBOCAN estimates of incidence and mortality worldwide for 36 cancers in 185 countries. *CA Cancer J. Clin.* **2018**, *68*, 394–424. [CrossRef] [PubMed]
2. Suh, J.D.; Cho, J.H. Trends in head and neck cancer in South Korea between 1999 and 2012. *Clin. Exp. Otorhinolaryngol.* **2016**, *9*, 263–269. [CrossRef] [PubMed]
3. Hong, S.; Won, Y.J.; Park, Y.R.; Jung, K.W.; Kong, H.J.; Lee, E.S. Cancer statistics in korea: Incidence, mortality, survival, and prevalence in 2017. *Cancer Res. Treat.* **2020**, *52*, 335–350. [CrossRef] [PubMed]
4. Jung, Y.S.; Seok, J.; Hong, S.; Ryu, C.H.; Ryu, J.; Jung, K.W. The emergence of oral cavity cancer and the stabilization of oropharyngeal cancer: Recent contrasting epidemics in the South Korean population. *Cancer* **2021**, *127*, 1638–1647. [CrossRef]
5. Sun, J.R.; Kim, S.M.; Seo, M.H.; Kim, M.J.; Lee, J.H.; Myoung, H. Oral cancer incidence based on annual cancer statistics in Korea. *J. Korean Assoc. Oral Maxillofac. Surg.* **2012**, *38*, 20. [CrossRef]
6. El-Naggar, A.K.; Chan, J.K.C.; Grandis, J.R.; Takata, T.; Slootweg, P.J. *WHO Classification of Head and Neck Tumours*; IARC: Lyon, France, 2017.
7. Botta, L.; Gatta, G.; Trama, A.; Bernasconi, A.; Sharon, E.; Capocaccia, R.; Mariotto, A.B.; RARECAREnet Working Group. Incidence and survival of rare cancers in the US and Europe. *Cancer Med.* **2020**, *9*, 5632–5642. [CrossRef]
8. Almangush, A.; Heikkinen, I.; Mäkitie, A.A.; Coletta, R.D.; Läärä, E.; Leivo, I.; Salo, T. Prognostic biomarkers for oral tongue squamous cell carcinoma: A systematic review and meta-analysis. *Br. J. Cancer* **2017**, *117*, 856–866. [CrossRef]
9. Lenouvel, D.; Gonzalez-Moles, M.A.; Ruiz-Avila, I.; Gonzalez-Ruiz, L.; Gonzalez-Ruiz, I.; Ramos-Garcia, P. Prognostic and clinicopathological significance of PD-L1 overexpression in oral squamous cell carcinoma: A systematic review and comprehensive meta-analysis. *Oral Oncol.* **2020**, *106*, 104722. [CrossRef]
10. Kujan, O.; van Schaijik, B.; Farah, C.S. Immune Checkpoint Inhibitors in Oral Cavity Squamous Cell Carcinoma and Oral Potentially Malignant Disorders: A Systematic Review. *Cancers* **2020**, *12*, 1937. [CrossRef]
11. Hiam-Galvez, K.J.; Allen, B.M.; Spitzer, M.H. Systemic immunity in cancer. *Nat. Rev. Cancer* **2021**, *21*, 345–359. [CrossRef]
12. Selders, G.S.; Fetz, A.E.; Radic, M.Z.; Bowlin, G.L. An overview of the role of neutrophils in innate immunity, inflammation and host-biomaterial integration. *Regen. Biomater.* **2017**, *4*, 55–68. [CrossRef] [PubMed]
13. Kersten, K.; Coffelt, S.B.; Hoogstraat, M.; Verstegen, N.J.; Vrijland, K.; Ciampricotti, M.; Doornebal, C.W.; Hau, C.S.; Wellenstein, M.D.; Salvagno, C. Mammary tumor-derived CCL2 enhances pro-metastatic systemic inflammation through upregulation of IL1β in tumor-associated macrophages. *Oncoimmunology* **2017**, *6*, e1334744. [CrossRef] [PubMed]
14. Cupp, M.A.; Cariolou, M.; Tzoulaki, I.; Aune, D.; Evangelou, E.; Berlanga-Taylor, A.J. Neutrophil to lymphocyte ratio and cancer prognosis: An umbrella review of systematic reviews and meta-analyses of observational studies. *BMC Med.* **2020**, *18*, 360. [CrossRef] [PubMed]
15. Patel, A.; Ravaud, A.; Motzer, R.J.; Pantuck, A.J.; Staehler, M.; Escudier, B.; Martini, J.F.; Lechuga, M.; Lin, X.; George, D.J. Neutrophil-to-lymphocyte ratio as a prognostic factor of disease-free survival in postnephrectomy high-risk locoregional renal cell carcinoma: Analysis of the S-TRAC trial. *Clin. Cancer Res.* **2020**, *26*, 4863–4868. [CrossRef]
16. Chrom, P.; Stec, R.; Bodnar, L.; Szczylik, C. Incorporating neutrophil-to-lymphocyte ratio and platelet-to-lymphocyte ratio in place of neutrophil count and platelet count improves prognostic accuracy of the international metastatic renal cell carcinoma database consortium model. *Cancer Res. Treat.* **2018**, *50*, 103–110. [CrossRef]
17. Mleko, M.; Pitynski, K.; Pluta, E.; Czerw, A.; Sygit, K.; Karakiewicz, B.; Banas, T. Role of systemic inflammatory reaction in female genital organ malignancies–state of the art. *Cancer Manag. Res.* **2021**, *13*, 5491–5508. [CrossRef]
18. Schneider, M.; Schäfer, N.; Bode, C.; Borger, V.; Eichhorn, L.; Giordano, F.A.; Güresir, E.; Heimann, M.; Ko, Y.D.; Lehmann, F.; et al. Prognostic value of preoperative inflammatory markers in melanoma patients with brain metastases. *J. Clin. Med.* **2021**, *10*, 634. [CrossRef]
19. Dolan, R.D.; McSorley, S.T.; Horgan, P.G.; Laird, B.; McMillan, D.C. The role of the systemic inflammatory response in predicting outcomes in patients with advanced inoperable cancer: Systematic review and meta-analysis. *Crit. Rev. Oncol./Hematol.* **2017**, *116*, 134–146. [CrossRef]
20. Guthrie, G.J.K.; Charles, K.A.; Roxburgh, C.S.D.; Horgan, P.G.; McMillan, D.C.; Clarke, S.J. The systemic inflammation-based neutrophil–lymphocyte ratio: Experience in patients with cancer. *Crit. Rev. Oncol./Hematol.* **2013**, *88*, 218–230. [CrossRef]
21. Cho, U.; Park, H.S.; Im, S.Y.; Yoo, C.Y.; Jung, J.H.; Suh, Y.J.; Choi, H.J. Prognostic value of systemic inflammatory markers and development of a nomogram in breast cancer. *PLoS ONE* **2018**, *13*, e0200936. [CrossRef]
22. Chon, S.; Lee, S.; Jeong, D.; Lim, S.; Lee, K.; Shin, J. Elevated platelet lymphocyte ratio is a poor prognostic factor in advanced epithelial ovarian cancer. *J. Gynecol. Obstet. Hum. Reprod.* **2021**, *50*, 101849. [CrossRef] [PubMed]
23. Golder, A.M.; McMillan, D.C.; Park, J.H.; Mansouri, D.; Horgan, P.G.; Roxburgh, C.S. The prognostic value of combined measures of the systemic inflammatory response in patients with colon cancer: An analysis of 1700 patients. *Br. J. Cancer* **2021**, *124*, 1828–1835. [CrossRef] [PubMed]

24. Zhang, Y.; Chen, B.; Wang, L.; Wang, R.; Yang, X. Systemic immune-inflammation index is a promising noninvasive marker to predict survival of lung cancer: A meta-analysis. *Medicine* **2019**, *98*, e13788. [CrossRef] [PubMed]
25. Wang, L.; Wang, C.; Wang, J.; Huang, X.; Cheng, Y. A novel systemic immune-inflammation index predicts survival and quality of life of patients after curative resection for esophageal squamous cell carcinoma. *J. Cancer Res. Clin. Oncol.* **2017**, *143*, 2077–2086. [CrossRef] [PubMed]
26. Wang, L.; Zhou, Y.; Xia, S.; Lu, L.; Dai, T.; Li, A.; Chen, Y.; Gao, E. Prognostic value of the systemic inflammation response index (SIRI) before and after surgery in operable breast cancer patients. *Cancer Biomark.* **2020**, *28*, 537–547. [CrossRef] [PubMed]
27. Li, S.; Yang, Z.; Du, H.; Zhang, W.; Che, G.; Liu, L. Novel systemic inflammation response index to predict prognosis after thoracoscopic lung cancer surgery: A propensity score-matching study. *ANZ J. Surg.* **2019**, *89*, E507–E513. [CrossRef] [PubMed]
28. Huang, Y.; Gao, Y.; Wu, Y.; Lin, H. Prognostic value of systemic immune-inflammation index in patients with urologic cancers: A meta-analysis. *Cancer Cell Int.* **2020**, *20*, 499. [CrossRef]
29. Diao, P.; Wu, Y.; Li, J.; Zhang, W.; Huang, R.; Zhou, C.; Wang, Y.; Cheng, J. Preoperative systemic immune-inflammation index predicts prognosis of patients with oral squamous cell carcinoma after curative resection. *J. Transl. Med.* **2018**, *16*, 1–11. [CrossRef]
30. Lu, Z.; Yan, W.; Liang, J.; Yu, M.; Liu, J.; Hao, J.; Wan, Q.; Liu, J.; Luo, C.; Chen, Y. Nomogram based on systemic immune-inflammation index to predict survival of tongue cancer patients who underwent cervical dissection. *Front. Oncol.* **2020**, *10*, 341. [CrossRef]
31. Valero, C.; Zanoni, D.K.; McGill, M.R.; Ganly, I.; Morris, L.G.T.; Quer, M.; Shah, J.P.; Wong, R.J.; León, X.; Patel, S.G. Pretreatment peripheral blood leukocytes are independent predictors of survival in oral cavity cancer. *Cancer* **2020**, *126*, 994–1003. [CrossRef]
32. Nie, Z.; Zhao, P.; Shang, Y.; Sun, B. Nomograms to predict the prognosis in locally advanced oral squamous cell carcinoma after curative resection. *BMC Cancer* **2021**, *21*, 372. [CrossRef] [PubMed]
33. Wang, Y.; Wang, P.; Andrukhov, O.; Wang, T.; Song, S.; Yan, C.; Zhang, F. Meta-analysis of the prognostic value of the neutrophil-to-lymphocyte ratio in oral squamous cell carcinoma. *J. Oral Pathol. Med.* **2018**, *47*, 353–358. [CrossRef] [PubMed]
34. Hasegawa, T.; Iga, T.; Takeda, D.; Amano, R.; Saito, I.; Kakei, Y.; Kusumoto, J.; Kimoto, A.; Sakakibara, A.; Akashi, M. Neutrophil-lymphocyte ratio associated with poor prognosis in oral cancer: A retrospective study. *BMC Cancer* **2020**, *20*, 568. [CrossRef] [PubMed]
35. Edition, S.; Edge, S.B.; Byrd, D.R. *AJCC Cancer Staging Manual*; Springer: Cham, Switzerland, 2017.
36. Durrleman, S.; Simon, R. Flexible regression models with cubic splines. *Stat. Med.* **1989**, *8*, 551–561. [CrossRef]
37. Diakos, C.I.; Charles, K.A.; McMillan, D.C.; Clarke, S.J. Cancer-related inflammation and treatment effectiveness. *Lancet Oncol.* **2014**, *15*, e493–e503. [CrossRef]
38. Mantovani, A.; Allavena, P.; Sica, A.; Balkwill, F. Cancer-related inflammation. *Nature* **2008**, *454*, 436–444. [CrossRef]
39. Janssen, L.M.E.; Ramsay, E.E.; Logsdon, C.D.; Overwijk, W.W. The immune system in cancer metastasis: Friend or foe? *J. Immunother. Cancer* **2017**, *5*, 79. [CrossRef]
40. Park, G.; Song, S.Y.; Ahn, J.H.; Kim, W.L.; Lee, J.S.; Jeong, S.Y.; Park, J.W.; Choi, E.K.; Choi, W.; Jung, I.H. The pretreatment erythrocyte sedimentation rate predicts survival outcomes after surgery and adjuvant radiotherapy for extremity soft tissue sarcoma. *Radiat. Oncol.* **2019**, *14*, 116. [CrossRef]
41. Alexandrakis, M.G.; Passam, F.H.; Ganotakis, E.S.; Sfiridaki, K.; Xilouri, I.; Perisinakis, K.; Kyriakou, D.S. The clinical and prognostic significance of erythrocyte sedimentation rate (ESR), serum interleukin-6 (IL-6) and acute phase protein levels in multiple myeloma. *Clin. Lab. Haematol.* **2003**, *25*, 41–46. [CrossRef]
42. Dupré, A.; Malik, H.Z. Inflammation and cancer: What a surgical oncologist should know. *Eur. J. Surg. Oncol.* **2018**, *44*, 566–570. [CrossRef]
43. Templeton, A.J.; Ace, O.; McNamara, M.G.; Al-Mubarak, M.; Vera-Badillo, F.E.; Hermanns, T.; Šeruga, B.; Ocana, A.; Tannock, I.F.; Amir, E. Prognostic role of platelet to lymphocyte ratio in solid tumors: A systematic review and meta-analysis. *Cancer Epidemiol. Prev. Biomark.* **2014**, *23*, 1204–1212. [CrossRef] [PubMed]
44. McAllister, S.S.; Weinberg, R.A. The tumour-induced systemic environment as a critical regulator of cancer progression and metastasis. *Nat. Cell Biol.* **2014**, *16*, 717–727. [CrossRef] [PubMed]
45. McSorley, S.T.; Lau, H.Y.N.; McIntosh, D.; Forshaw, M.J.; McMillan, D.C.; Crumley, A.B. Staging the tumor and staging the host: Pretreatment combined neutrophil lymphocyte ratio and modified glasgow prognostic score is associated with overall survival in patients with esophagogastric cancers undergoing treatment with curative intent. *Ann. Surg. Oncol.* **2021**, *28*, 722–731. [CrossRef] [PubMed]
46. Tazeen, S.; Prasad, K.; Harish, K.; Sagar, P.; Kapali, A.S.; Chandramouli, S. Assessment of pretreatment neutrophil/lymphocyte ratio and platelet/lymphocyte ratio in prognosis of oral squamous cell carcinoma. *J. Oral Maxillofac. Surg.* **2020**, *78*, 949–960. [CrossRef]
47. Ong, H.S.; Gokavarapu, S.; Wang, L.Z.; Tian, Z.; Zhang, C.P. Low pretreatment lymphocyte-monocyte ratio and high platelet-lymphocyte ratio indicate poor cancer outcome in early tongue cancer. *J. Oral Maxillofac. Surg.* **2017**, *75*, 1762–1774. [CrossRef]
48. Rosculet, N.; Zhou, X.C.; Ha, P.; Tang, M.; Levine, M.A.; Neuner, G.; Califano, J. Neutrophil-to-lymphocyte ratio: Prognostic indicator for head and neck squamous cell carcinoma. *Head Neck* **2017**, *39*, 662–667. [CrossRef]
49. Gonzalez, H.; Hagerling, C.; Werb, Z. Roles of the immune system in cancer: From tumor initiation to metastatic progression. *Genes Dev.* **2018**, *32*, 1267–1284. [CrossRef]
50. Chokshi, D.A.; El-Sayed, A.M.; Stine, N.W. J-shaped curves and public health. *JAMA* **2015**, *314*, 1339–1340. [CrossRef]

51. Urabe, M.; Yamashita, H.; Uemura, Y.; Tanabe, A.; Yagi, K.; Aikou, S.; Seto, Y. Non-linear association between long-term outcome and preoperative neutrophil-to-lymphocyte ratio in patients undergoing curative resection for gastric cancer: A retrospective analysis of 1335 cases in a tetrachotomous manner. *Jpn. J. Clin. Oncol.* **2018**, *48*, 343–349. [CrossRef]
52. Shimada, H.; Takiguchi, N.; Kainuma, O.; Soda, H.; Ikeda, A.; Cho, A.; Miyazaki, A.; Gunji, H.; Yamamoto, H.; Nagata, M. High preoperative neutrophil-lymphocyte ratio predicts poor survival in patients with gastric cancer. *Gastric Cancer* **2010**, *13*, 170–176. [CrossRef]
53. Koh, C.H.; Bhoo-Pathy, N.; Ng, K.L.; Jabir, R.S.; Tan, G.H.; See, M.H.; Jamaris, S.; Taib, N.A. Utility of pre-treatment neutrophil-lymphocyte ratio and platelet-lymphocyte ratio as prognostic factors in breast cancer. *Br. J. Cancer* **2015**, *113*, 150–158. [CrossRef] [PubMed]
54. Mattavelli, D.; Lombardi, D.; Missale, F.; Calza, S.; Battocchio, S.; Paderno, A.; Bozzola, A.; Bossi, P.; Vermi, W.; Piazza, C.; et al. Prognostic nomograms in oral squamous cell carcinoma: The negative impact of Low Neutrophil to Lymphocyte Ratio. *Front. Oncol.* **2019**, *9*, 339. [CrossRef]
55. Song, M.; Graubard, B.I.; Rabkin, C.S.; Engels, E.A. Neutrophil-to-lymphocyte ratio and mortality in the United States general population. *Sci. Rep.* **2021**, *11*, 464. [CrossRef] [PubMed]
56. Nash, G.F.; Turner, L.F.; Scully, M.F.; Kakkar, A.K. Platelets and cancer. *Lancet Oncol.* **2002**, *3*, 425–430. [CrossRef]
57. Gay, L.J.; Felding-Habermann, B. Contribution of platelets to tumour metastasis. *Nat. Rev. Cancer* **2011**, *11*, 123–134. [CrossRef]
58. Yu, L.; Guo, Y.; Chang, Z.; Zhang, D.; Zhang, S.; Pei, H.; Pang, J.; Zhao, Z.J.; Chen, Y. Bidirectional interaction between cancer cells and platelets provides potential strategies for cancer therapies. *Front. Oncol.* **2021**, *11*, 764119. [CrossRef]
59. Sierko, E.; Wojtukiewicz, M.Z. Platelets and angiogenesis in malignancy. *Semin Thromb Hemost* **2004**, *30*, 95–108. [CrossRef]
60. Egan, K.; Crowley, D.; Smyth, P.; O'Toole, S.; Spillane, C.; Martin, C.; Gallagher, M.; Canney, A.; Norris, L.; Conlon, N.; et al. Platelet adhesion and degranulation induce pro-survival and pro-angiogenic signalling in ovarian cancer cells. *PLoS ONE* **2011**, *6*, e26125. [CrossRef]
61. Palumbo, J.S.; Talmage, K.E.; Massari, J.V.; La Jeunesse, C.M.; Flick, M.J.; Kombrinck, K.W.; Jirouskova, M.; Degen, J.L. Platelets and fibrin(ogen) increase metastatic potential by impeding natural killer cell-mediated elimination of tumor cells. *Blood* **2005**, *105*, 178–185. [CrossRef]
62. Suzuki, K.; Aiura, K.; Ueda, M.; Kitajima, M. The influence of platelets on the promotion of invasion by tumor cells and inhibition by antiplatelet agents. *Pancreas* **2004**, *29*, 132–140. [CrossRef]
63. Klinger, M.H.; Jelkmann, W. Role of blood platelets in infection and inflammation. *J. Interferon Cytokine Res.* **2002**, *22*, 913–922. [CrossRef] [PubMed]
64. Nieswandt, B.; Hafner, M.; Echtenacher, B.; Mannel, D.N. Lysis of tumor cells by natural killer cells in mice is impeded by platelets. *Cancer Res.* **1999**, *59*, 1295–1300. [PubMed]
65. Roweth, H.G.; Battinelli, E.M. Lessons to learn from tumor-educated platelets. *Blood* **2021**, *137*, 3174–3180. [CrossRef] [PubMed]
66. Li, N. Platelets in cancer metastasis: To help the "villain" to do evil. *Int. J. Cancer* **2016**, *138*, 2078–2087. [CrossRef] [PubMed]
67. Saito, R.; Shoda, K.; Maruyama, S.; Yamamoto, A.; Takiguchi, K.; Furuya, S.; Hosomura, N.; Akaike, H.; Kawaguchi, Y.; Amemiya, H. Platelets enhance malignant behaviours of gastric cancer cells via direct contacts. *Br. J. Cancer* **2021**, *124*, 570–573. [CrossRef] [PubMed]
68. Skog, J.; Würdinger, T.; Van Rijn, S.; Meijer, D.H.; Gainche, L.; Curry, W.T.; Carter, B.S.; Krichevsky, A.M.; Breakefield, X.O. Glioblastoma microvesicles transport RNA and proteins that promote tumour growth and provide diagnostic biomarkers. *Nat. Cell Biol.* **2008**, *10*, 1470–1476. [CrossRef]
69. Liu, C.; Yu, S.; Zinn, K.; Wang, J.; Zhang, L.; Jia, Y.; Kappes, J.C.; Barnes, S.; Kimberly, R.P.; Grizzle, W.E. Murine mammary carcinoma exosomes promote tumor growth by suppression of NK cell function. *J. Immunol.* **2006**, *176*, 1375–1385. [CrossRef]

MDPI

St. Alban-Anlage 66

4052 Basel

Switzerland

www.mdpi.com

Biomedicines Editorial Office

E-mail: biomedicines@mdpi.com

www.mdpi.com/journal/biomedicines